Annual International Conference on Emerging Research Areas on Computing & Communication Systems for a Fourth Industrial Revolution (AICERA 2020)

Held online

IOP Conference Series: Materials Science and Engineering
Volume 1085

Kanjirapally, India
14 - 16 December 2020

ISBN: 978-1-7138-3348-2
ISSN: 1757-8981

Printed from e-media with permission by:

Curran Associates, Inc.
57 Morehouse Lane
Red Hook, NY 12571

Some format issues inherent in the e-media version may also appear in this print version.

Institute of Physics
Dirac House, Temple Back
Bristol BS1 6BE UK

Phone: 44 1 17 929 7481
Fax: 44 1 17 920 0979

techtracking@iop.org

Additional copies of this publication are available from:

Curran Associates, Inc.
57 Morehouse Lane
Red Hook, NY 12571 USA
Phone: 845-758-0400
Fax: 845-758-2633
Email: curran@proceedings.com
Web: www.proceedings.com

TABLE OF CONTENTS

ADVANCED ENGINEERING APPLICATIONS

DATA MINING & COMPUTATIONAL INTELLIGENCE TECHNIQUES

GREEN COMPUTING AND NEURAL NETWORKS

Preface

AICTE Sponsored International E-Conference on Computing & Communication Systems for a Fourth Industrial Revolution, the 9th edition of Annual International Conference on Emerging Research Areas (AICERA 2020), held virtually on 14-16 December 2020 in Amal Jyothi College of Engineering Kanjirappally, Kerala, India. The AICERA 2020 was organized by Amal Jyothi College of Engineering in association with All India Council for technical Education (AICTE)New Delhi and technically sponsored by IOP Publishing.

The Conference aims at bringing together the researchers, scientists, engineers, and scholar students in all areas of Engineering and Technology, and provides an international forum for the dissemination of original research results, new ideas and practical development experiences which concentrate on both theory and practices. AICERA-2020 aims to create a platform for open communication and idea exchange on the recent Technological and Engineering developments. This conference provides an international forum for sharing knowledge and innovations in the field of Computer Science and Information Technology. The three-day event has provided an exposure to erudite work by researchers in the field of engineering. It also facilitates interaction among academicians, engineers and research scholars. Sessions on the side-line discussed recent advances, challenges, breakthroughs and research innovations on various engineering aspects of green technologies. The conference consists of keynote lectures, tutorials, workshops and oral presentations on all aspects of Engineering and Science. Researchers were invited to speak and present the latest developments and technical solutions in their recent multidisciplinary area.

Proceedings contain 39 papers presented on the conference both offline and online. All papers have been peer reviewed and passed several stages of edition procedure. We express our sincere gratitude to all people who contributed to this process and did all their best to enhance the scientific merit and quality of the Proceedings. We are also grateful to the authors for their contributions, to the speakers and participants of the AICERA 2020.

ORGANIZING COMMITTEE

Chief Patron

H.E Mar Jose Pulickal	Bishop, Diocese of Kanjirappally
Very Rev. Fr. Boby Mannamplackal	Managing Trustee

Patron

Fr. Dr. Mathew Paikatt	Manager
Fr. Jacob Kodimarathummoottil	Bursar

Conference Chair

Dr. Z. V. Lakaparampil	Principal
Prof. Tomy Joseph	Registrar

IOP Conf. Series: Materials Science and Engineering 1085 (2021) 011001 doi:10.1088/1757-899X/1085/1/011001

Dr. Jacob Philip	Dean (Academic)
Dr. Sony C. George	Dean (Research)
Fr. Dr. James Philip	Dean (Basic Skills & Welfare)
Fr. Rubin Thottupuram	HOD (Computer Applications)
Prof. Manoj T. Joy	HOD (Computer Science)
Prof. Shiju George	HOD (Information Technology)

Organising Chair

Dr. Juby Mathew	Dept. of Computer Applications

Organizing Co-Chairs

Prof. Syam Gopi	Dept. of Computer Science
Dr. Arun K. S.	Dept. of Computer Science
Prof. Lisha Varghese	Dept. of Computer Applications
Prof. Joms Antony	Dept. of Information Technology

Programme Committee

Prof. Sandhya Ramakrishnan	Dept. of Information Technology
Prof. Jayakrishna V.	Dept. of Computer Science
Prof. Sona Maria Sebastian	Dept. of Computer Applications

Proceedings and Publication

Prof. Shiney Thomas	Dept. of Computer Science
Prof. Krishnalal G.	Dept. of Computer Science
Prof. Jetty Benjamin	Dept. of Computer Applications
Prof. Neethu C Sekhar	Dept. of Computer Science
Prof. Jis Joe Mathew	Dept. of Computer Science
Prof. Tom Kurian	Dept. of Computer Science
Prof. Midhuna Jyothi R.	Dept. of Information Technology

Publicity & Communication

Prof. Santhoshkumar G S	Dept. of Computer Science
Prof. Resmipriya M G	Dept. of Computer Science
Prof. Bini M Issac	Dept. of Computer Science
Prof. Saumya Sadanandan	Dept. of Information Technology
Prof. Jobin T. J	Dept. of Computer Applications

Website Administration

Mr. Sebin Sebastian	Dept. of Computer Applications

Registration Chair

Prof. Fabeela Ali Rawther	Dept. of Computer Science
Prof. Elisabeth Thomas	Dept. of Computer Science
Prof. Jerin Thomas	Dept. of Computer Science
Prof. Jincy Lukose	Dept. of Information Technology
Prof. Dilu Mariya Joseph	Dept. of Computer Applications
Prof. Grace Joseph	Dept. of Computer Applications
Prof. Rini Kurian	Dept. of Computer Applications

Presentation and Online Management

Fr. Siju John	Dept. of Computer Science
Prof. Merin Manoj	Dept. of Computer Applications
Prof. Aleena Xavier T	Dept. of Computer Science
Sr. Elsin Chakkalackal S.H	Dept. of Computer Applications
Prof. Meera Rose Mathew	Dept. of Computer Applications
Prof. Gloriya Mathew	Dept. of Computer Applications

Brochure, Certificate, Design and Printing

Prof. Jetty Benjamin	Dept. of Computer Applications
Prof. Nimmy Francis	Dept. of Computer Applications
Mr. Rajeshkumar N.	Dept. of Computer Science

TECHNICAL REVIEW COMMITTEE

EXTERNAL REVIEWERS

1. Dr. Sabu M.K, Associate Professor, Department of Computer Applications, Cochin University of Science and Technology, Kochi
2. Dr. Sajimon Abraham, Associate Professor, SMBS, Mahatma Gandhi University, Kottayam
3. Dr. R. Vijayakumar, Rtd Professor, School of Computer Sciences, Mahatma Gandhi University, Kottayam
4. Dr. Gobi Mohan, Professor, Bharathiar University, Coimbatore

5. Dr. P SojanLal, Professor of Computer Science, MBITS, Kothamangalam, Kerala
6. Dr. Madhu S. Nair, Associate Professor, Department of Computer Applications, CUSAT
7. Dr. Binu Thomas, Associate Professor, Marian College, Kuttikanam
8. Dr. Terry Jacob Mathew, Associate Professor, MACFAST College, Tiruvalla

INTERNAL REVIEWERS

1. Dr. Lisa Mathew, Professor, Department of Basic Science
2. Dr. Jeba Sonia J, Associate Professor, Department of Computer Science and Engineering
3. Dr. Arun K S, Assistant Professor, Department of Computer Science and Engineering
4. Dr. Jina Varghese, Assistant Professor, Department of Computer Science and Engineering
5. Dr. Aju Mathew George, Assistant Professor, Department of Civil Engineering
6. Prof. Shiney Thomas, Associate Professor, Department of Computer Science and Engineering
7. Prof. Shiju George, Associate Professor Department of Information Technology
8. Prof. Santhoshkumar G S, Associate Professor, Department of Computer Science and Engineering
9. Prof. Syam Gopi, Associate Professor, Department of Computer Science and Engineering
10. Prof. Arun S, Associate Professor, Department of Electrical and Electronics Engineering
11. Prof. Fabeela Ali Rawther, Assistant Professor, Department of Computer Science and Engineering
12. Prof. Jerin Thomas, Assistant Professor, Department of Computer Science and Engineering
13. Prof. Anishamol Abraham, Assistant Professor, Department of Computer Science and Engineering
14. Prof. Jayakrishna V, Assistant Professor, Department of Computer Science and Engineering
15. Prof. Krishnalal G, Assistant Professor, Department of Computer Science and Engineering
16. Prof. Aleena Xavier T, Assistant Professor, Department of Computer Science and Engineering
17. Prof. Shinto Sebastian, Assistant Professor, Department of Electronics and Communication Engineering
18. Prof. Victor Jose, Assistant Professor, Department of Electrical and Electronics Engineering
19. Prof. Bobin K Mathew, Assistant Professor, Department of Electrical and Electronics Engineering
20. Prof. Rani Chacko, Assistant Professor, Department of Electrical and Electronics Engineering
21. Prof. Vinu Sankar, Assistant Professor, Department of Electrical and Electronics Engineering
22. Prof. Sreeram H, Assistant Professor, Department of Automobile Engineering
23. Prof. Lisha Varghese, Assistant Professor, Department of Computer Applications
24. Prof. Shelly Shiju George, Assistant Professor, Department of Computer Applications

25. Prof. Jetty Benjamin, Assistant Professor, Department of Computer Applications
26. Prof. Dilu Mariya Joseph, Assistant Professor, Department of Computer Applications
27. Prof.Gloriya Mathew, Assistant Professor, Department of Computer Applications
28. Prof.Jobin T J, Assistant Professor, Department of Computer Applications

Peer Review Policy

All papers published in this volume of *IOP Conference Series: Materials Science and Engineering* have been peer reviewed through processes administered by the Editors. Reviews were conducted by expert referees to the professional and scientific standards expected of a proceedings journal published by IOP Publishing.

• **Type of peer review:** Single-blind

• **Describe criteria used by Reviewers when accepting/declining papers. Was there the opportunity to resubmit articles after revisions?**
The paper is relevant to the topic of the conference and contains new research result(s), is clearly written and structured; figures and tables are informative and correctly organized; references are correct and contain new literature sources. After revision the papers were submitted to the editor; in some cases, two to three revisions were done.

• **Conference submission management system:**
Easychair system was used. The papers were submitted to the Easy chair and then distributed to reviewers by the editor. The reviewers sent their reviews to the editor who then finalised the editing process.

• **Number of submissions received:** 91

• **Number of submissions sent for review:** 64

• **Number of submissions accepted:** 39

• **Acceptance Rate (Number of Submissions Accepted / Number of Submissions Received X 100):** 42.85%

• **Average number of reviews per paper:** 2

• **Total number of reviewers involved:** 36

• **Any additional info on review process (ie plagiarism check system):** Turnitin tool used for plagiarism check

• **Contact person for queries:** Dr Juby Mathew, *jubymathew@amaljyothi.ac.in*

Dynamic Multi-agent Real Time Scheduling Framework for Production Management

Girish Talmale [1*] **, Urmila Shrawankar** [1]

Department of Computer Science and Engineering, G H Raisoni College of Engineering, Nagpur, India

* Email : girishtalmale@gmail.com

Abstract. Production management is a complex problem. The problem becomes more complex in open, uncertain, dynamic environments and distributed productions. The existing static scheduling methods are not efficient to deal with this problem efficiently. This paper proposed multi-agent approach for real time scheduling in production management to efficiently manage production. The proposed scheduling optimizes the performance of production management in dynamic uncertain environment. Dynamic multi-agent approach in real time scheduling consist of the agent collect and analysis the production data in real time and ensure real time response to emergency events. Real time scheduling agents increase the resource utilization and success rate in production management by using real time feedback.

1. Introduction

In production management, different manufacturing resources are scheduled to manufacturing processes with time slots in plan. The production environment is open, uncertain and dynamic in nature. To cope with such an uncertain production environment dynamic scheduling of manufacturing resources is required. The production scheduling is challenging problem particularly in open, dynamic and uncertain environment where the resources may change in time, some resources can be added into the manufacturing system run time or removed from the system. The tasks also change its parameter run time. The existing centralized production management is mainly based on material requirement planning and enterprise resource planning. Theses existing production management scheduling techniques having many limitations as follows.

1. The current approach not efficient to handle uncertain events
2. Not efficiently deals with production disturbances such as failure of machine, late supply of materials and unavailability of employee
3. Existing schedule not efficient to handle run time changes in the manufacturing processes
4. Optimal utilization of resources not possible in the conventional methods.
5. Current production management methods are incorrect, inconsistent, and incomplete and produce time delay.
6. In production management methods not based on dynamic real time response and adaptive control.

To address theses problem this paper proposed multi-agent approach in real time scheduling for efficient production management. Multi-agent approach in real time scheduling increase the resource utilization and the reliability of resources. The real time scheduling monitors the performance of shop floor and react with the changes in real time.

1.1. Background Work

Multi agent technology recently used in production management for optimization. Different multi agent based production management techniques are described in the past [1]. Dynamic scheduling approach is used to manage the production demands in real time[2]. The accuracy of the thesis system is based not only on correct output but also the time when the output produced [3]. The dynamic scheduling problem is of two type one is dynamic planning and dynamic best effort approach [4]. In a dynamic planning approach the job is scheduled only when the deadline can be achieved and in dynamic best effort scheduling job can schedule without checking for deadline guarantee[5]. The dynamic scheduling also categories into two types based on hardness of deadline [6]. In hard dynamic scheduling the job deadline is crucial and hazardous if the deadline of the job is not achieved. Soft dynamic scheduling systems and deadlines are not crucial. The production management is dynamic hard scheduling with dynamic planning approach. Recently multi agent based technology used intelligent dynamic scheduling in production management in factory automation. Multi agent scheduling with contract negotiation protocol which is used to solve many scheduling problems with failure detection. Decentralized manufacturing using multi agent is used to address dynamic resource sharing. Real time scheduling using a multi agent system is presented for transportation problems. Distributed agent based scheduling based approach is presented for a flexible manufacturing system [7]. Radio frequency based identification system used with multi agent system for tracking and tracing manufacturing problems. Some authors also presented a simulation based approach to address internal disturbances in the manufacturing process [8] [9]. Memory aware resource scheduling algorithm is proposed for energy efficient cloud data centers [10].

2. Dynamic Real Time Multi-agent Scheduling Framework

In a dynamic multi-agent approach in real time scheduling framework, the agents perceive the environment and the status of the production system [11]. Environment in production management consists of job arrival, machine computing capacity, resource allocation etc. The physical system in production management is represented as machine and system state. The system state represents the time required to process the unit processing, batches to be processed, length of queue and the critical information.

The agent action is based on events and falls into operation such as preventive, active, protective and informative as it senses the system as well as machine state and takes dynamic real time decision and adaptive control. The real time multi-agent scheduling framework for production management is as shown in Figure 1. The proposed real time multi-agent scheduling framework consists of four agents such as Machine State Agents, Machine Status Evaluation Agent, Production Resource Monitoring Agent and Dynamic Real Time Scheduling Agent.

2.1. Machine Status Agent

This agent is responsible for collection of real time status information of machine different health parameters. The machine status agents collect the different parameters of machines such as machine age, machine working parameter using real time sensing mechanism. The machine status agent also processes raw data of machine into meaningful information which is used for real time application services. The machine status agent consists of two main modules

Data Sensing Module: It senses the real time status information from machines through different sensors. The real time operating system enabled sensing mechanism is used to sense the real time status information of the machine. The different sensors used to get the status of a machine are pressure sensor, vibration sensor, RFID, motion sensors etc.

Real Time Analytic Module: This module responsible for analysis of real time information captured by a data sensing module.

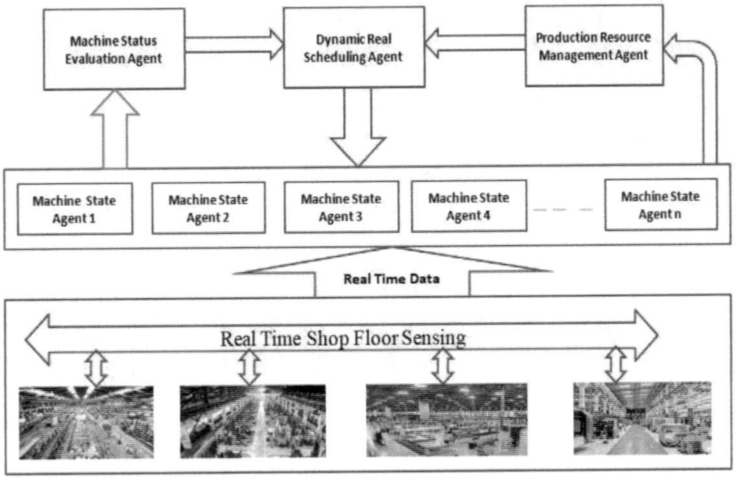

Figure 1. Dynamic Multi agent Real Time Scheduling Framework

Figure.2. show the machine status agents with data sensing and real time analytics module.

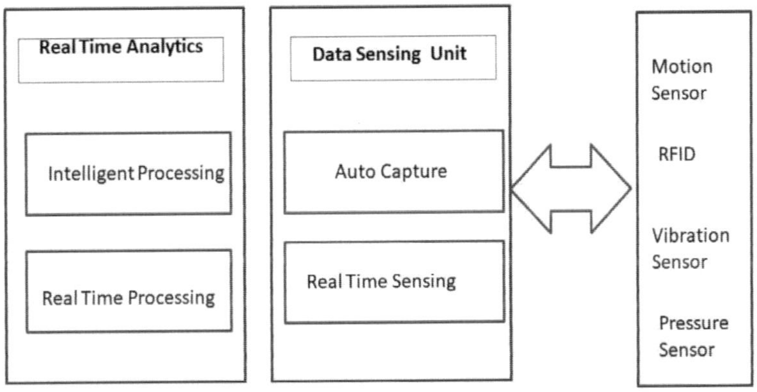

Figure 2. Machine Status Agent

2.2. Machine Status Evaluation Agent (MSEA)

It is responsible for evaluation of machine status information given of machine status agents. This agent evaluates the status information and assigns tasks to the machine for optimal scheduling.

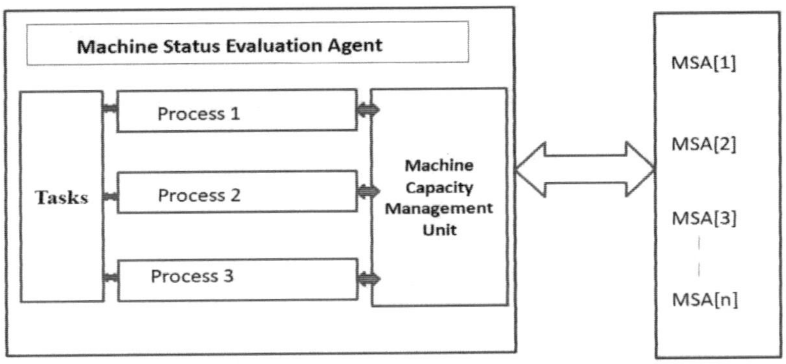

Figure 3. Machine Status Evaluation Agent

IOP Conf. Series: Materials Science and Engineering 1085 (2021) 012001 doi:10.1088/1757-899X/1085/1/012001

It assigns tasks of production management processes to machines as shown in Figure 3. The set of tasks of production processes assigned to the machines as per the machine capacity management unit information and select the optimal machine to which the tasks are assigned. The machine capacity management unit is used to balance load of machines. Machine capacity management unit calculate the machine load by interacting with machine status agents to get the information related current load of the machines and process assigned to machine having less load.

2.3. Production Resource Monitoring Agent

Resource management and monitoring is done by production resource management agents. It also tracked production changes, disturbances in manufacturing processes. Production resource monitoring agents act as an interface between the machine status agent and dynamic real time scheduling agent. It collects the real time information from the machine status agent and sends this to dynamic real time scheduling agents. The production resource monitoring agent gets the details about the production order, its schedule etc. The collection of all the machine status agents is collected and sent to dynamic real time scheduling agents.

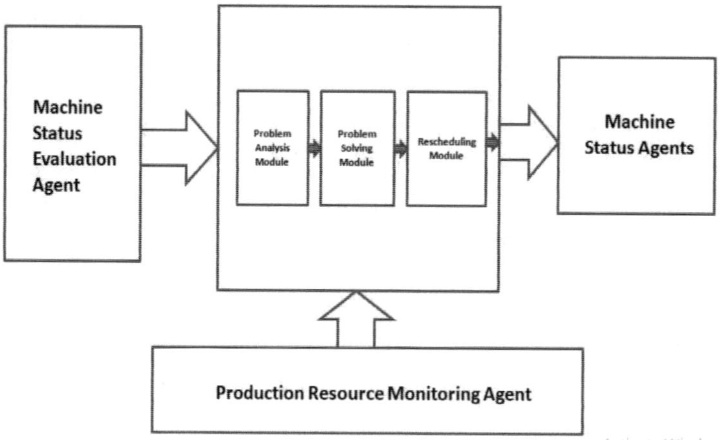

Figure 4. Production Resource Monitoring Agent

Real time information tracking and integration unit of production resource monitoring agent collect real time information from the machine status tracking agents and integrate it. Date source unit collect production order information and send the information to dynamic real time scheduling agents for taking the optimal scheduling information

2.4. Dynamic Real Time Scheduling Agent

It runtime decides the order of execution of tasks of processes in production management. The dynamic real time scheduling agent takes input from machine status evaluation agents and processes monitoring information to schedule to the given tasks. The output of the dynamic real time scheduling agent is that the given process of tasks assigned to the machine within the deadline. Fig.5 shows the dynamic real time scheduling agent which consist of three module problem analysis module, problem solving module and rescheduling module.

3. Results and Discussion

System model of dynamic real time scheduling is presented in Table 1. The different production management consists of a set of processes of tasks and machines. The deadline of each task is used to ensure that all the tasks must finish its execution before its deadline for real time operation. The objective of dynamic real time scheduling is to complete execution of process of tasks within the deadline. The scheduler dynamically checks the load of the machine and assigns the process of task to the machine with minimum load.

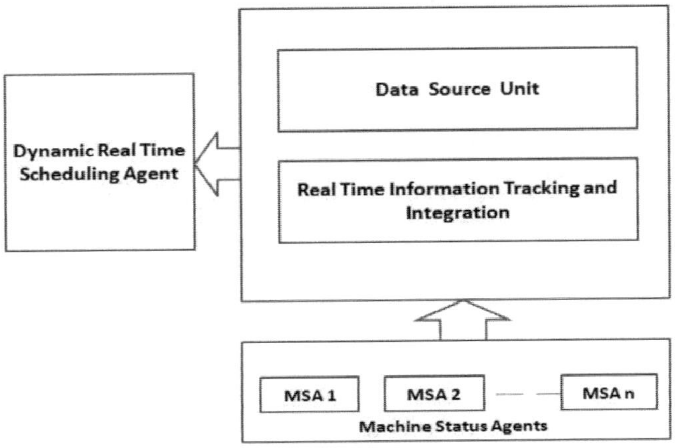

Figure.5. Dynamic Real Time Scheduling Agent

System Parameters	Description
T={T1,T2,-----Tn}	Set of production tasks T
Pi={P1i,P2i,-----Pni}	Set of processes of task Ti
M={M1,M2-----Mk}	Set of Machines
d={d1,d2,-------dn}	Deadline of tasks
A(Ti,Pji,Mk)	Arrival time of Ti,Pj and Mk
F(Ti,Pji,Mk)	Finished Time of Ti,Pj,Mk
E(Ti,Pj,Mk)	Execution Time of Process pj of Tasks Ti on Machine Mk

Algorithm: Dynamic Multi Agent Real Time Scheduler

```
Input: The Tasks Set T = {T1, T2 − − − −Tn}
       Machines Set M = {M1, M2, − − − − Mn}
       Set of Processes of Each Tasks Ti i. e Pni = {P1i, P2i, − − − − −Pni}
Output: Processes Pji of Tasks schedule on Machine Mk
for all tasks ti from t1 to tn do
    for all processes pji of tasks ti do
        Dynamically calculate the load of Machines M1, M2 − − − −Mk
        ML = {ML1, ML2 − − − − − −MLk}   where ML as machine load
        Assigned Process Pji to Machine with Min{ML1, ML2 − − − −MLk}
        if((di − A(Ti, Pji, Mk) + E(Pj, Ti, Mk) ≥ 0)
                Pji of Ti is Scheduled on Mk
        else
                Pji of Ti miss dealine
        end if
end for
```

The proposed algorithm calculated the machine load dynamically for each individual process of tasks and assigned process to machines with minimum load so that the execution of process must get finished its operation. The proposed dynamic multi agent real time scheduling framework is used to improve resource utilization as compare to existing methodology. The proposed framework simulation is applied to set of production tasks T={500, 1000,1500,2000,2500,3000,3500,4000,4500,5000}.
The resource utilization is measured with the proposed framework and the proposed scheduler increases the resource utilization and success rate.

	Resource Utilization (%)	Success Rate (%)
Conventional Multi agent Real Scheduling	83	86
Proposed Dynamic Multi agent Real Time Scheduling	95	97

4. Conclusions

Production management is challenging problem in uncertain, dynamic, open system and distributed environment. The existing static scheduling approach in production management is not efficient in terms of success rate and resource utilization, The proposed dynamic multi agent real time scheduling framework increase resource utilization and success rate where the overall production management operations is manage and coordinated by different agents such as machine status agent gives status information related with health of machine which is further evaluated by machine status evaluation agent and the information used by production and resource monitoring agent for efficient scheduling of production tasks by dynamic real time scheduling agents.

References

[1] Liu Y, Zhang L, Tao F and Wang L 2019 *Frontiers of Information Technology & Electronic Engineering* **20** 1465-92
[2] Lui Y, Wang L, Wang X, Xu and Zhang L 2018 *Int J Production Research*, **57**(15-16) 4854-4879
[3] Chekired D, Khoukhi L and Mouftah H 2018 *IEEE Transactions Industrial Informatics* **14**(10): 4590-4602
[4] Fu Y, Ding J, Wang H 2018 *Applied Soft Computing* **68** 847-855
[5] Ivanov D, Dolgui A and Sokolov B 2016 *Int J Production Research* **54**(2) 386-402
[6] Li K, Zhang H and Cheng B 2018 *Optimization Letter* **12**(1)73-86
[7] Bastos R, Oliveira F and Oliveira J 2005 *Expert Systems with Applications* **28**(1) 9–19
[8] Cho S and Prabhu V 2007 *Journal of Manufacturing Systems* **26**(2) 65–74
[9] Shukla O, Soni G,Kumar R and Ahluwalia R 2020 *Recent Patents on Engineering* **14** 1-18
[10] Liang B, Dong X, Wang Yand Zhang X 2020 *Future Generation Computer Systems* **113** 329-342
[11] Dominguez R, Cannella S 2020 *Sustainability* **12**(5) 1935

One Hop Neighbor based Broadcast Method for Reactive Routing in Mobile Ad-hoc Network

Mr. Nimit Mittal[1], Dr. Pallavi Pahadiya[2] and Prof. Amit Agrawal[3]

[1] Research Scholar, SIRT, Indore mittal.nimit@ymail.com
[2] Prof. & Head, EC, SIRT, Indore hodec.sirt@sageuniversity.in
[3] Prof. CSE, Medi-Caps University, Indore amitn0187@gmail.com

Abstract. To overcome redundant issues of routing messages for the reactive routing protocol is become research challenge. In general, to discover route or path reactive routing protocols broadcast routing messages which may leads congestions and degrades battery power unnecessary. Though, to resolve the challenges several efficient broadcast methods have been suggested. Each method has its own procedure and results sort of advantage and disadvantages. The motive of the paper is to presents a method which is proposed to minimize redundant issue of routing messages resulted by broadcasting. Proposed method uses one hop neighbor's knowledge to broadcast routing messages for the purpose of route discovery. Proposed method experimented through simulation software that named as NS-2. An experimental accomplish certain factors like number of nodes, routing protocol, MAC protocol etc. Proposed method evaluated on the basis of different criteria such as throughput, control overhead, routing efficiency and node remaining battery power.

1. Introduction

According to medium, network is categorized in wired and wireless network. Further, wireless network divides in fixed infrastructure and self-organizing network. Self-organizing network also called infrastructure less that named as wireless ad-hoc network. The complete classification of network depicted in figure 1.

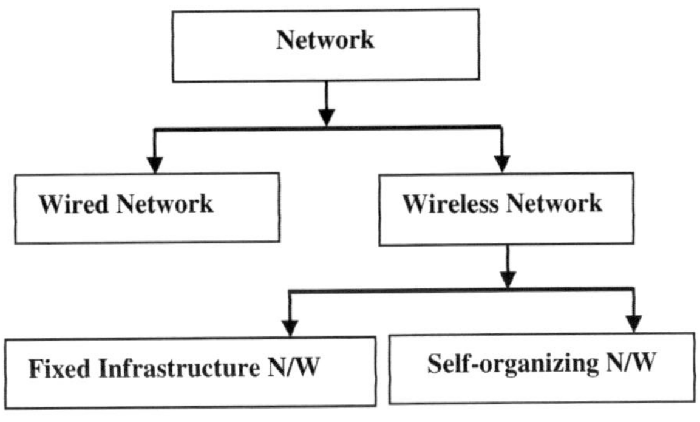

Figure 1. Classification of network according to channel.

When devices are free to move anywhere in ad-hoc networks then it is referred as mobile ad-hoc network. In figure 2 the network scenario of mobile ad-hoc network with four nodes labelled A to D presented. Here node A may communicate with node B or C or D and vice versa because they all are in range of node A.

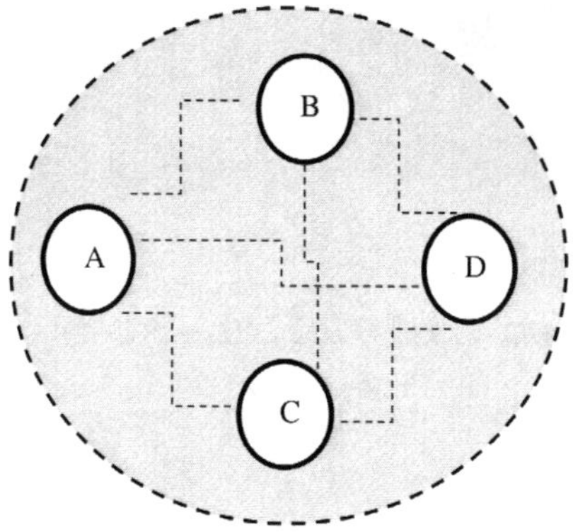

Figure 2. Node 'A' Coverage in mobile ad-hoc network

Due to self-organizing nature, devices in the network are enough sufficient to discover route and initiate transmission. If any change occurred then devices are enough confident to manage the changes. On the other hand, the mobility property of mobile ad-hoc network pays heavy penalty in term of routing overhead, link breaks and wastage of resources. For the route discovery, nodes of network employ numerous routing strategies. One among them is reactive routing that enable nodes to broadcast routing information to all its covering nodes. This phenomenon of routing is repeated until routing information reach to the desired destination node. When destination node receives routing information them reply to reverse manner. Though to complete the procedure nodes get route to its destination. The broadcast procedure for reactive routing protocol results duplicity of routing information as well as unnecessary consumption of resource. The broadcast procedure in routing, leads flooding of routing information to the whole network that occurred many issues.
Flooding employ transmissions of route information message to the whole network via neighbor's knowledge which lead congestion, lifetime of network as well as routing overhead issues. Flooding assist routing protocols to discover route through broadcast of route information message to all nodes neighbors and rebroadcast further [1]. A simple analogy of flooding operation illustrated in figure. 3.

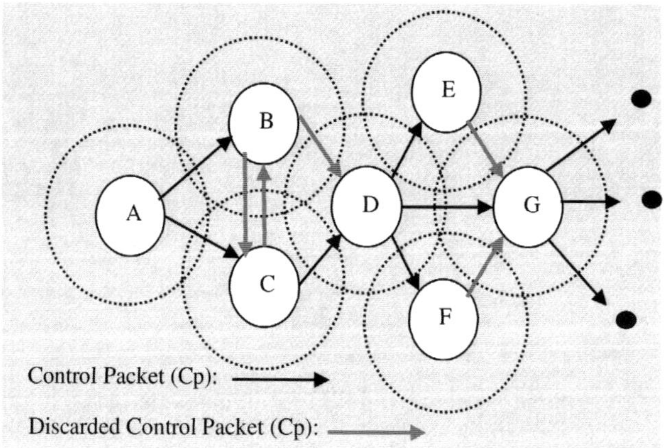

Figure 3. A scenario of flooding Operation

Figure 3 presents a portion of network scenario where 7 nodes which labelled from A to G. Here, node A wish to transmits data to G. For this, Node A initiates broadcast of route information message named control packet to all its neighbors which are B and C. Further, neighbor node B and C receive control message from node A and find itself destination or not. If not then broadcast to further neighbor's and precede same procedure up to node G. When node G found itself is destination then it unicast reply of routing messages to the sender node A. During the route discovery, numerous unwanted retransmissions of routing messages done by intermediate nodes which may lead flooding of routing messages. When intermediate nodes found replica copy of routing messages originated by same sender and for the same destination then messages have discarded by these nodes. To do this, effort applied by nodes in term of battery and transmission power.

Therefore, the nature of broadcast procedure may be restricting up to some level to minimize flooding. Many approaches came to control broadcast issues. Further, broadcasting based on one hop neighbor knowledge, related work, proposed broadcast method, experiment of proposed method and result evaluation presented in subsequent sections.

2. Background

As it is known, node of mobile network is not directly reachable to all other nodes. Means, mobile ad-hoc network formed with limited coverage of node depends on its hardware. Each node has specific range; if any other node belongs in its range then it is direct neighbor or one hop neighbor. Though, this formation makes complete mobile ad-hoc network. "*Each node collects the information of its 1-hop neighbor through transmitting receiving HELLO packet at the MAC layer using MAC layer protocols [2]*". Here one hop neighbor concepts used for broadcasting routing message to discover route. One hop neighbor concept may opt transmitter as well as receiver for the message forwarding function. Both are discussed below.

- Transmitter-based- Here, transmitter take set of neighbors to forward routing messages according some criteria. Methods that comes in [3] [4] [5] are transmitter-based.
- Receiver-based- Inverse of earlier, receiver of routing message decides their neighbors to further broadcasting. Method that defines in [6] is receiver-based.

Forwarding nodes choosing strategies shown in figure 4.

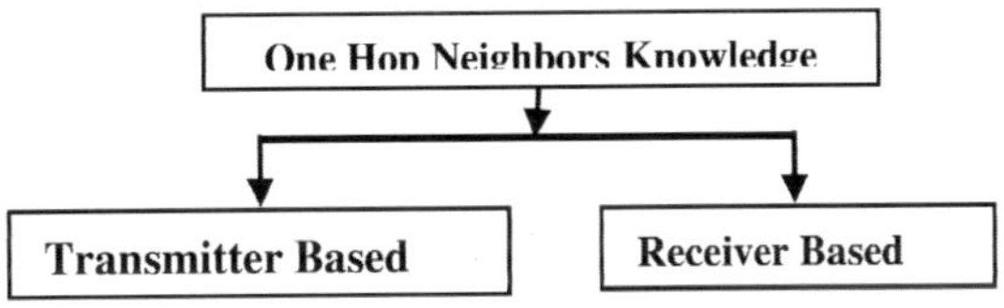

Figure 4. Types of One-Hop Neighbors Knowledge Method

3. Related Work

Several methods that relevant of one hop neighbor and transmitter based have been advised by researchers which discussed here.

Flooding with self-pruning (FSP) [7] was the first receiver-based method. A receiving node match its own 1-hop neighbor with node list in the message, if its entire 1-hop neighbor included in list then it was not forward the message otherwise it forward the message as a sender. Another one is transmitter-based method named vertex forwarding [3] that uses one hop neighbor's knowledge for forwarding routing message. Similarly, another one was an efficient flooding (EF1) method based on 1-hop neighbor's knowledge and transmitter based [8] evolved. It worked in three phase named as forwarding node selection, forwarding node optimization and mobility handling.

Conventional method named selective rebroadcasting extends by Mistral [1] method. The extension ensures balance between control overhead and reliability. Another is, ASTRAL [9] method was advised that ensure efficiency and reliability. It uses forward and backward packet recovery mode.

An Efficient Reliable One-Hop Broadcasting (EROB) [10] was also presented that guarantees the completion of one-hop broadcasting, i.e., all nodes in the source's transmission range will receive the broadcast message. Information broadcasting [11] based a method advised that enable network nodes to take decision about rebroadcast or discard received routing messages. Further, a method named global queue pruning method [12] was come into picture. It restricts transmissions overhead and assures delivery of messages to every node.

Recently, selective epidemic broadcast algorithm [13] was proposed that minimize impacts of broadcast storm problem and enhance delivery rate of data. Latest one is a simple flooding scheme [14] based on residual power of nodes. It broadcast RREQ packets to its entire one hop neighbor with respect of node remaining power. It implemented in both static as well as mobile nature of network.

4. Proposed Method

For restricting impacts of flooding and make efficient broadcasting of routing messages, a method is proposed here. A proposed method uses one hop neighbor perimeter of network and apply decision making at the neighbor's which receive broadcasting messages for further forwarding. Earlier a lot of works have done for efficient flooding. Each one has considered different criteria and procedures. Proposed method is well suitable for mobile nature of network because it considers mobility parameter of nodes for taking decision to forward messages. To accomplish task, initially neighbor discovery performed through HELLO messages at data link layer. In neighbor discovery, HELLO Message broadcast by node to all one hop neighbor. Then each neighbor update neighbor table with entry of Hello message sender id. Similar process done by all the nodes and neighbor to completes network topology. Though proposed method, add two more fields in HELLO message named rem_pw which indicate remaining battery power and vel which indicate velocity of node. Means, when nodes, initiates Hello message then it put its remaining power and velocity in the designated fields to inform about these to all the neighbors. Here the purpose of broadcasting of HELLO Message is only for topology greation or neighbours list preparation not for measure congestion and battery fail. The construction of neighbours table of node is used in routing strategy to select set of neighbours for flooding of routing messages on the basis of mentioned parameters. Though here, each node in the network have neighbor table with remaining power and velocity of each neighbor. These two fields help in broadcasting route messages to selective neighbors according to high remaining power and less velocity. Receiving node of route message decides to further forwarding messages to selective set of neighbors on the basis of these fields. In the similar fashion, broadcasting of route message accomplished in efficient way. After the experiment of proposed method might be results are as expected in terms of routing efficiency, throughput etc.

The format of HELLO message amended as per requirement of proposed method, shown in figure 5.

Source Node ID	Hello_Int	Ref_Int	VLT	Seq No	Neighbor1 ID	REM_PW	Vel	Options

Figure 5. Modified HELLO Message

The complete flow of proposed method represented in flow chart that depicted in figure 6.

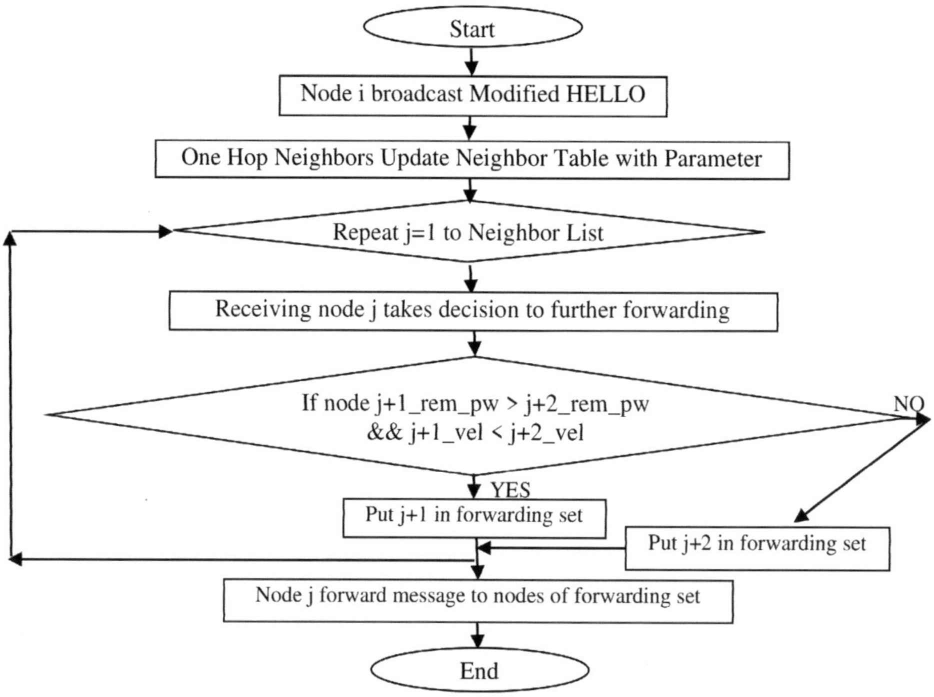

Figure 6. Flow Chart of Proposed Method

IOP Conf. Series: Materials Science and Engineering 1085 (2021) 012002 doi:10.1088/1757-899X/1085/1/012002

5. Experiment & Evaluation

Proposed method is experimented using simulation software named NS-2. NS-2.35 version was used in simulation that available in ns-2.35-allinone.tar or zip. NS-2.35 have built-in energy model named energy.cc file. Battery power is computed using energy model and used in routing protocol with the help of patch implementation. Here simulation done considering mobile network scenario with number of nodes, routing protocol, simulation time, traffic type etc.

5.1. Simulation Metrics

The Table 1 presents, simulation metrics for proposed method in NS-2 environment.

Table 1. Simulation metrics in NS-2.

Parameters Name	Value
Number of mobile nodes	100
Topography Area	1200×1000
Simulation time (seconds)	90
Transmission area (meters)	200
Propagation Model	Two Ray Ground
Traffic type	CBR
Packet size (bytes)	512
Routing Protocol	AODV
Connection Type	TCP

5.2. Simulation Metrics

Simulation of proposed method has done with 100 mobile nodes and simulation time in seconds. In this, a scenario constructed where nodes are scattered in specified topography area with some sender and receivers. CBR traffic is generated using TCP connection. AODV routing protocol modified to accomplish proposed method. Figure7 show scenario of network simulation.

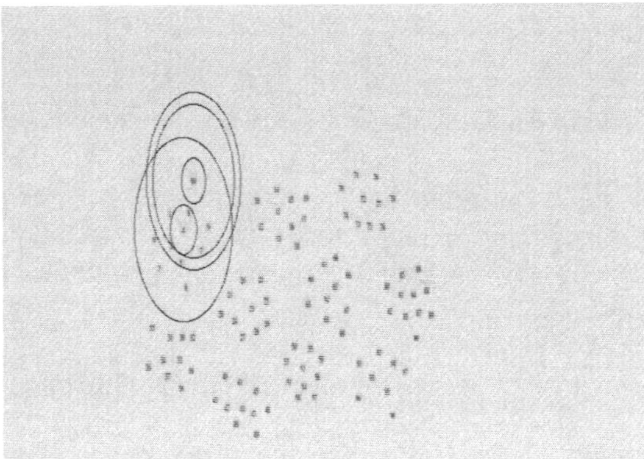

Figure 7. Simulation scenario with 100 mobile nodes

5.3. Evaluation

The results of proposed and experimented method are evaluated with base method by considering few network parameters such as through put, control overhead, routing efficiency, and remaining battery power of nodes.

5.3.1. Throughput

Throughput defines, useful time over the total time multiplied by bandwidth to transmit data packet. In other words, throughput is number or fractions of data packet received per unit time. Figure 8 shows the throughput analysis between proposed method and base method.

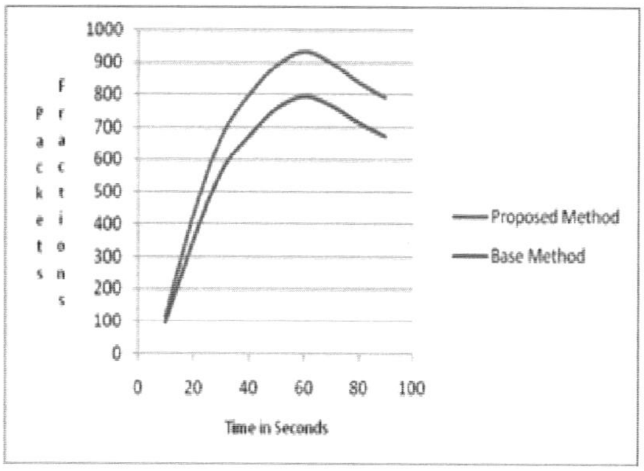

Figure 8. Throughput analysis

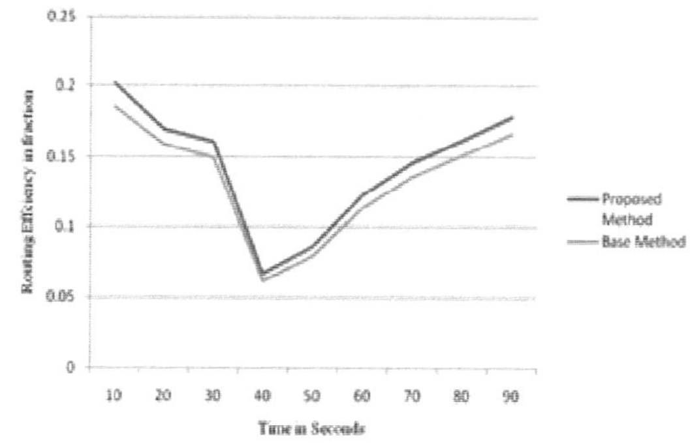

Figure 9. Routing efficiency analysis

5.3.2. Routing Efficiency

It is determining as number of data fractions out of whole number of packets transmitted during the communications. Here routing efficiency was measure via AODV routing protocol with modification of needed things and compare with conventional AODV protocol. Figure 9 shows the routing efficiency analysis.

5.3.3. Control Overhead

Basically, throughput and other parameter affects from control overhead. If control overhead is high then performance is low and if it is less then performance is high. Control overhead defined as number of routing and other control packets used in one communication. It is depicted as number of control packets over the time. Figure 10 shows control overhead analysis.

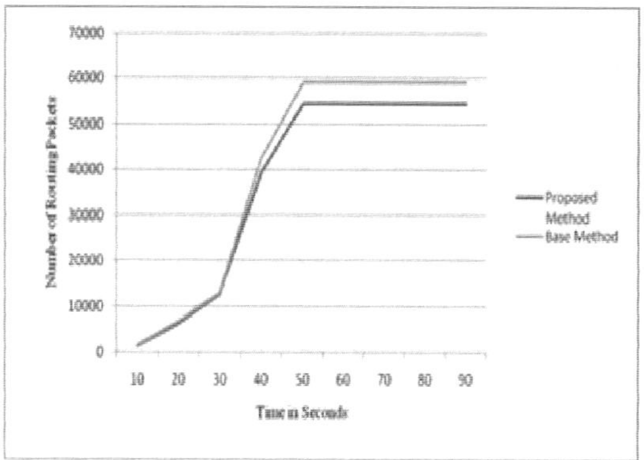

Figure 10. Control overhead analysis

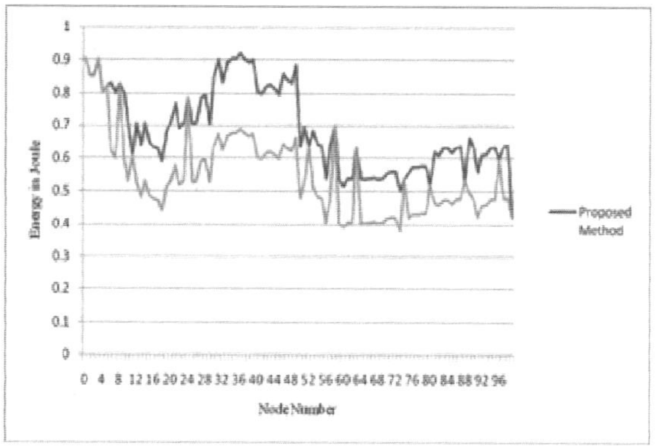

Figure 11. Residual battery power Analysis

5.3.4. Residual Battery Power

Particularly for mobile ad-hoc network the life of network is depends on consumption and residual battery power of nodes. If residual battery power is large then network lifetime is more otherwise it is less. Figure 11 shows residual battery power analysis.

6. Conclusion

Due to self-organizing nature, devices in the network are enough sufficient to discover route and initiate transmission. If any change occurred then devices are enough confident to manage the changes. On the other hand, the mobility property of mobile ad-hoc network pays heavy penalty in term of routing overhead, link breaks and wastage of resources. For the route discovery, nodes of network employ numerous routing strategies. One among them is reactive routing that enable nodes to broadcast routing information to all its covering nodes. The broadcast procedure for reactive routing protocol results duplicity of routing information as well as unnecessary consumption of resource. Though proposed method has minimized impacts of broadcast procedure of reactive routing strategy. Proposed method evaluated with base method over throughput, routing efficiency and so on. Further proposed method has scope to extend in term of parameter, routing protocols and others.

7. References

[1] Stefan Pleisc, Mahesh Balakrishnan, Ken Birman, Robbert van Renesse, May 22–25, 2006, MISTRAL: Efficient Flooding in Mobile Adhoc Networks, *MobiHoc'06, ACM, Florence, Italy.*

[2] Perkins CE, Bhagwat P 1994 Highly Dynamic Destination-Sequenced Distance-Vector Routing (DSDV) for Mobile Computers, *Proceedings of ACM SIGCOMM* pp.234–244.

[3] C. Ho, K. Obraczka, G. Tsudik, and K. Viswanath 1999 Flooding for Reliable Multicast in Multi-hop Ad Hoc Networks, in *Proc. of the Int'l Workshop on Discrete Algorithms and Methods for Mobile Computing and Communication*, pp. 64-71.

[4] S. Even, 1979 Graph algorithms, *Computer Science Press*, 204-209.

[5] D. Johnson and D. A. Maltz 1996 Dynamic Source Routing in Ad Hoc Wireless Networks, in Mobile Computing, T. Imielinski and H. F. Korth, Eds. *Kluwer Academic Publishers, Dordrecht, The Netherlands*. pp. 153-181.

[6] J. Jetcheva, Y. Hu, D. Maltz, and D. Johnson 17 November 2000, A Simple Protocol for Multicast and Broadcast in Mobile Ad Hoc Networks, *Internet Draft: draft-ietf-manet-simple-mbcast-01.txt.*

[7] S. Ni, Y. Tseng, Y. Chen, and J. Sheu, The broadcast storm problem in a mobile ad hoc network, *Proc. of ACM/IEEE MOBICOM'99*, pp. 151-162.

[8] S. Guha and S. Khuller 1996 Approximation algorithms for connected dominating sets, in *Proceedings of European Symposium on Algorithms (ESA).*

[9] Jinman Jung, Yookun Cho, Yeongkwun Kim, Injoo Kim March 22-26, 2010 ASTRAL: An Adaptive, Efficient, and Reliable Flooding Mechanism for MANET, *SAC'10, Sierre, Switzerland.*

[10] Seungjin Park, Seong-Moo Yoo 2013 An efficient reliable one-hop broadcast in mobile ad hoc networks, *Ad Hoc Networks Elsevier* pp. 19–28.

[11] M. Chekhara, K. Zine-Dine, M. Bakhouya, A. Aaroud, D. El Ouadghiri 2016 An Efficient Broadcasting Scheme in Mobile Ad-Hoc Networks, *The 7th International Conference on Emerging Ubiquitous Systems and Pervasive Networks (EUSPN)*, Published by Elsevier B.V.

[12] Slawomir Nowak, Mateusz Nowak, Krzysztof Grochla and Piotr Pecka 2016 Global Queue Pruning Method for Efficient Broadcast in Multihop Wireless Networks, *ISCIS CCIS* **659**, pp. 214–224

[13] M. Chitra, S. Siva Sathya 2017 Selective epidemic broadcast algorithm to suppress broadcast storm in vehicular ad hoc networks, Egyptian Informatics J, hosted by *Elsevier B.V.*

[14] Satoshi Yamazaki, Yu Abiko and Hideki Mizuno 2020 A Simple and Energy-Efficient Flooding Scheme for Wireless Routing, *Hindawi Wireless Communications and Mobile Computing*.

IOP Conf. Series: Materials Science and Engineering 1085 (2021) 012003 doi:10.1088/1757-899X/1085/1/012003

Diagnosis of Age Related Macular Degeneration by Curve Fitting RPE Layer

D Pavithra[1], R Vanithamani[2] and E Karolinekersin[3]

[1,2,3]Avinashilingam Institute for Home Science and Higher Education for Women, Coimbatore, Tamil Nadu, India

E-mail: pavithradharmaraj06@gmail.com, vanithamani_bmie@avinuty.ac.in, Karolinekersin_bmie@avinuty.ac.in

Abstract. Optical Coherence Tomography (OCT) of the retina allow high resolution and non-invasive imaging for diagnosis of macular diseases such as Age Related Macular Degeneration (ARMD). The aim of this work is to suppress the speckle noise, segment the retinal pigment epithelium layer and identify the possibilities for diagnosis of ARMD by curve fitting method. Speckle pattern, which is multiplicative in nature, degrades the quality of OCT images. Bilateral and homomorphic-wavelet filters are adopted to minimize the speckle noise in OCT images. The performance of these filters is tested on a set of OCT images collected from an open source database. The experimental results prove that homomorphic-wavelet filter is better in reducing the speckle noise. Structural changes of retinal pigment epithelium layer, a key factor of ARMD, is identified by Graph-based method, which is utilized further in diagnosis of ARMD by curve fitting method. The sum of squares due to error (SSE) values of ARMD images are very higher compared to normal images, which can be used as a feature for diagnosis.

1. Introduction

Globally, ARMD is one of the prime causes of severe vision impairment in people over the age of sixty. In India, the prevalence of ARMD is around forty percent and it is expected to increase with the rise of population. It is also observed that the prevalence of ARMD is lower in subjects with diabetes and very much negligible in subjects with diabetic retinopathy. In ARMD, a specific part of the eye called macula undergoes progressive degradation due to ageing. There are two main categories namely neovascular and non-neovascular ARMD. Early stage of ARMD is usually non symptomatic, but leads to severe vision loss if left untreated. The main characteristic of non-neovascular ARMD is accumulation of extracellular aggregates called macular drusen which leads to Retinal Pigment Epithelium (RPE) abnormality. RPE is a pigmented retinal layer that nourishes the underlying cells of the retina and its most obvious function is light absorption. It improves the quality of image produced by optical system. RPE is liable to undergo several structural changes due to accumulation of age pigment lipofuscin, accumulation of basal deposits or basal infoldings. These useless residues disturbs the normal functioning of the optical system. Any structural changes to RPE layer degrades the image quality. Study of ARMD pathology using in vitro models have established that autophagy might play a central role in causing death of RPE cells and affects the retinal glial cells to some extent. It is also found that genetic factors has significant effect, as the descendants are more prone to ARMD if some of their ancestors had an experience of this disease. Anti-vascular endothelial growth factor therapy is a commercially available treatment for ARMD that improves or stabilizes vision in majority of the cases. Its key factors include early detection and better understanding of size and location of abnormality. OCT has become a critical tool for baseline evaluation of retinal layers. As healthcare systems are vastly improving, automated screening tools can help ophthalmologists in diagnosis of ARMD at an early stage.

2. Related work

In the last decade, a large number of noise removal and segmentation methods have been developed for OCT images. Different researchers had diverse ideas about speckle noise reduction and segmentation of retinal layers which is discussed in this section. Padmasini N et al. (2014) utilized maximum fuzzy entropy principle for image enhancement followed by anisotropic diffusion filtering to suppress speckle noise in OCT images. Whereas, Naz S et al. (2016) utilized bilateral filter for de-noising and Ilyasova N et al. (2019) used Gaussian kernel with an optimum value of 3.5 for standard deviation as pre-processing step. Murakami T and Ogawa K (2018) proposed a transition invariant method to avoid artifacts at the corners of the block. Initially, the image was circularly shifted at four positions followed by wavelet decomposition using Daubechies kernel. From the de-noised image, pixel positions were relocated and averaged at each pixel position. Sui X et al. (2018) proposed a combination of wavelet-domain sparsity and total variation regularization that alleviates artifacts introduced by pure wavelet thresholding and preserves the edges of retinal layers while removing speckle noise. A common problem addressed in all these works was reduction of speckle noise. As the degree of diagnosis depends on better visualization of retinal layers, the second point discussed in this section is the segmentation techniques. Yang Q et al. (2010) gathered dual-scale gradient information using canny edge detection operator and then performed shortest path search using dynamic programming. This algorithm successfully segmented low intensity and low contrast OCT images in a very short time with good accuracy. Kafieh R et al. (2013) proposed diffusion mapping and k-means clustering with coarse graining for segmentation. This method relied on regional image texture through space-frequency analysis rather than edge information. Naz S et al. (2016) utilized thresholding for segmentation, polynomial curve fitting model to plot curve on RPE and their difference values are used as a feature to classifier for further diagnosis. Su L et al., (2017) compared different edge detection algorithms in segmenting retinal OCT images and suggested two pass method over canny and edge flow technique. Ilyasova et al. (2019) developed a graph based segmentation algorithm by constructing a minimum spanning tree of a connected weighted non-oriented graph and the retina vitreous-body boundary was identified with canny algorithm. De Moura J et al. (2020) identified four layers of the retina where it was subdivided into outer and inner retina. Adapting graph cuts and active contours are utilised for identification and the texture, intensity and clinical features are extracted using learning strategy. Regardless of the existence of general or partial proposals, this work aims to pre-process OCT images in order to reduce speckle noise using different filters and then segment RPE layer for further diagnosis.

3. Proposed Method

The proposed work aims to diagnose ARMD by curve fitting the RPE layer. A flow chart of proposed method is represented in figure 1. Noise removal techniques are discussed in Section 3.1, localization of RPE layer in section 3.2, followed by diagnosis of ARMD using curve fitting method in section 3.3. The images used in this work are taken from the dataset used by Srinivasan P P et al., (2014). This dataset consists of scans acquired from 45 subjects which includes normal, ARMD and diabetic macular edema using 'Spectralis' spectral domain OCT machine.

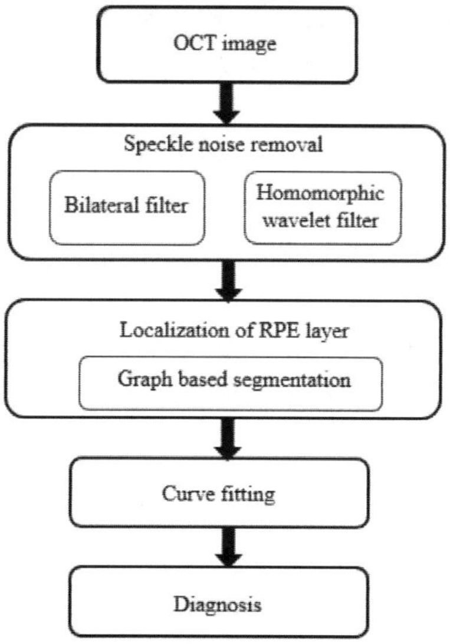

Figure 1. Proposed method

3.1. Speckle noise removal

An OCT image usually consists of speckle pattern which makes it convoluted to extort pathological features that can be utilized for further diagnosis. Hence, bilateral and homomorphic-wavelet filters are adopted as a pre-processing step to suppress the speckle noise associated with OCT images. The bilateral filter takes the weighted average of pixels and preserves edges by considering the variation of intensities alone. The bilateral filter replaces each pixel with its weighted value of neighbourhood depending on the geometric and photometric distance.

Homomorphic wavelet filtering involves the following steps. Initially, the natural logarithm of the input image is taken. Then it computes the wavelet transform, processes the wavelet coefficients followed by reconstructing the image using inverse wavelet transform. Since speckle pattern corresponds to the high frequency component of the image, which can be removed successfully. The exponential of the reconstructed image is the output of homomorphic wavelet filter. The resultant images of bilateral and homomorphic- wavelet filter are shown in figure 2. The performances of these filters are compared with the help of ten OCT images, whose Peak Signal to Noise Ratio (PSNR), Signal to Noise Ratio (SNR) and Structural Similarity Index (SSIM) are calculated. From Table. 1, it is clear that homomorphic wavelet filter removes speckle noise better than bilateral filter and hence, its resultant image is used in the forthcoming steps.

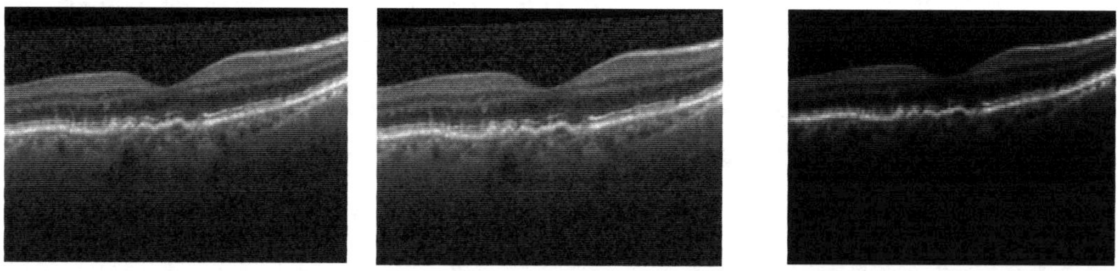

(a) OCT image (b) Bilateral filter (c) Homomorphic-wavelet filter

Figure 2. Despeckled Images

Table 1. Evaluation metrics for Homomorphic-wavelet filter and Bilateral filter

Image	Homomorphic-wavelet filter			Bilateral filter		
	PSNR(dB)	SNR(dB)	SSIM	PSNR(dB)	SNR(dB)	SSIM
1	38.6177	27.6429	0.7035	29.0104	14.8425	0.6701
2	36.6364	26.8430	0.6707	28.2466	13.0292	0.5978
3	37.0047	26.5350	0.6940	29.0250	14.0405	0.5245
4	35.5692	24.0109	0.6095	25.9715	13.0096	0.6631
5	34.0034	23.9171	0.6130	25.0159	12.5961	0.4924
6	36.4364	26.3430	0.6607	28.2066	13.1252	0.5578
7	35.2672	24.1132	0.6435	26.1738	13.1226	0.5934
8	37.7147	27.9428	0.7014	28.9121	14.2407	0.6164
9	31.0034	22.2171	0.5930	24.4119	12.0947	0.4422
10	37.2347	26.7301	0.6847	28.3503	13.8445	0.5528

3.2. Localization of RPE layer

In order to localize the RPE layer, graph based method is adopted. First step is to calculate the graph weights and adjacency matrix. Then, gradient image is created as shown in figure 3(b). In OCT images, retinal layers are generally horizontal structures that are discernable by large variations in pixel intensity in the vertical direction. RPE layer is localized by getting the layer going from light to dark in the vertical direction using shortest path algorithm. As RPE layer is highly reflective, light to dark transition from gradient image can be clearly defined. First and last few points are discarded by assuming them as image borders. The resultant image is shown in figure 3(c), wherein the red line indicates RPE layer.

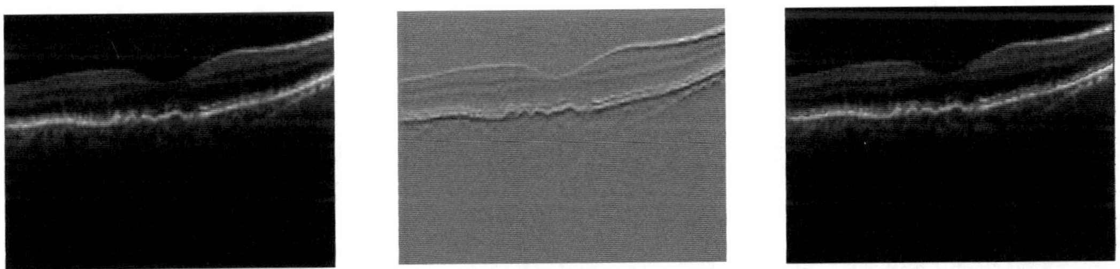

(a) Image with ARMD (b) Gradient Image (c) Resultant Image

Figure 3. Localization of RPE layer

3.3. Curve fitting

Curve fitting is a process of constructing a curve over a series of data points. Its aim is to provide the best possible fit within the available data points and degree of the model. Here, the coordinates of localized RPE layer is considered and a very close curve is drawn over it using a second order Gaussian model. This curve fits correctly as a smooth curve on RPE layer of normal OCT, but doesn't fit correctly on ARMD affected image due to uneven RPE layer (figure 4 and 5).

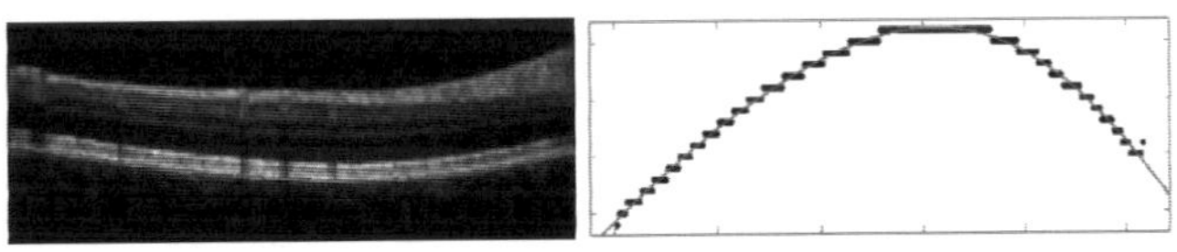

Figure 4. Curve fit for normal

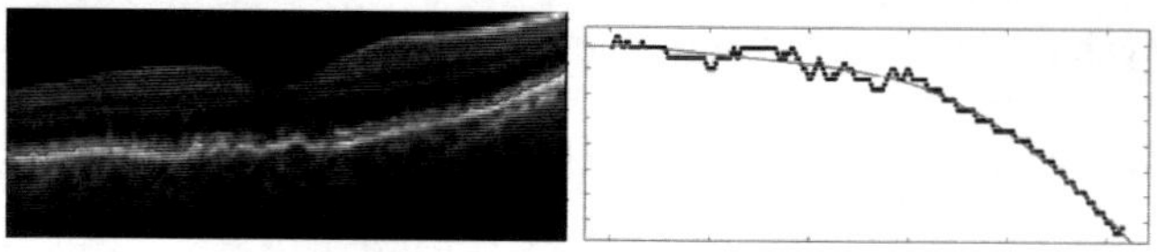

Figure 5. Curve fit for ARMD

The SSE index which specifies the total deviation of response values from the input values are calculated as shown in eqn. 1, where w_i indicates weights, X_i indicates input values and Y_i indicates response values.

$$SSE = \sum_{i=1}^{n} w_i (X_i - Y_i)^2 \quad (1)$$

The plot of SSE values for 15 ARMD and 15 normal images are shown in figure 6. Since the curve cannot fit correctly, SSE values will be higher for ARMD images as compared to normal images. Therefore, curve fit model with SSE index can be used as a feature to diagnose ARMD.

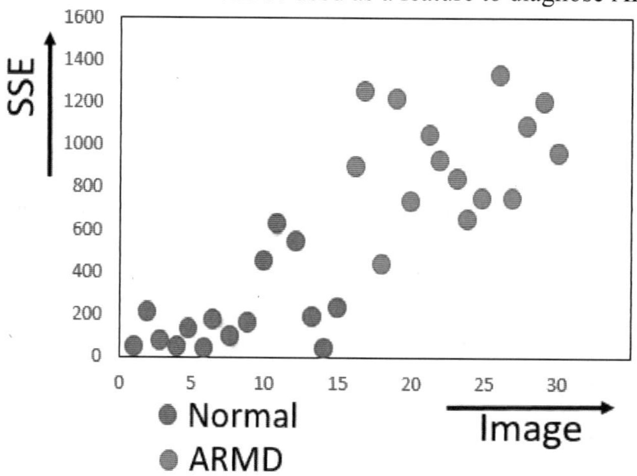

Figure 6. Plot of SSE values of ARMD and normal images

4. Conclusion

In this work, an effective method for de-noising speckle pattern, localizing RPE layer and curve fitting method for diagnosis of ARMD in OCT retinal images is proposed. The graph based method successfully identified the RPE layer and curve fitting approach with discernable SSE index can be used to differentiate images with ARMD from normal images. This work will be a clinically useful tool to assist ophthalmologists in OCT- based ocular pathology identification and helps to improve the efficiency of retinal disease diagnosis. Future work will focus on segmenting additional layers and explore supportive feature extraction methods.

5. References

[1] Yang Q, Reisman C A, Wang Z and Fukuma Y 2010 Automated layer segmentation of macular OCT images using dual-scale gradient information *J. Optics Express* **18**(20), p 21293-21307.
[2] Farsiu S, Izatt J and Toth C 2012 Advances and Emerging Applications of Spectral-domain Optical Coherence Tomography Imaging in Ophthalmology *J. US Ophthalmic Review* **03** p 14.
[3] Kafieh R, Rabbani H, Abramoff M D, Sonka M 2013 Intra-retinal layer segmentation of 3D optical coherence tomography using coarse grained diffusion map *J Medical Image Analysis* **17**(8) p 907-928.

[4] Murakami T and Ogawa K 2018 IEEE *14th Int. Colloquium on Signal Processing & Its Applications* (Batu Feringghi/IEEE) p 31-34.

[5] Padmasini N, Abbirame K S, Yacin S M and Umamaheswari R 2014 *Int. Conf. on Science Engineering and Management Research* (Chennai:IEEE) p 1-5.

[6] Sui X, Ishikawa H, Selesnick I W, Wollstein G and Schuman J S 2018 *IEEE Signal Processing in Medicine and Biology Symp.* (Philadelphia, PA, USA: IEEE) p 1-6.

[7] Ilyasova N, Shirokanev A, Demin N and Paringer R 2019 *5th Int. Conf. on Frontiers of Signal Processing* (Marseille, France:IEEE) p 77-81.

[8] de Moura J, Samagaio G, Novo J, Almuina P, Fernández M I and Ortega M 2020 Joint Diabetic Macular Edema Segmentation and Characterization in OCT Images *J. Digital Imaging* ***33*** *p 1335-1351.*

[9] Naz S, Ahmed A Akram M and Khan S *2016 Sixth Int. Conf. on Image Processing Theory, Tools and Applications (IPTA),* pp. 1-4, IEEE, Oulu, Finland (2016).

[10] Dodo B, Li Y, Eltayef K and Liu X 2018 *Proc. 11th Int. Joint Conf. on Biomedical Engineering Systems and Technologies* - Vol **2**: BIOIMAGING (Portugal), p. 35-42.

[11] Srinivasan P P, Kim L A, Mettu P S, Cousins S W, Comer G M, Izatt J A and Farsiu S 2014 Fully automated detection of diabetic macular edema and dry age-related macular degeneration from optical coherence tomography images. *J Biomedical optics express*, **5**(10), p. 3568–3577.

[12] Rapantzikos K and Zervakis M 2001 *Proc. of Int. Conf. on Image Processing (Cat. No.01CH37205),* vol.**3**, pp. 1055 – 1058, Thessaloniki, Greece.

[13] Amini Z, Kafieh R, Rabbani H 2019 Speckle Noise Reduction and Enhancement for OCT images. In: Chen X., Shi F., Chen H. (eds) Retinal Optical Coherence Tomography Image Analysis. Biological and Medical Physics, *Biomedical Engineering*, Springer, Singapore.

[14] pangyuteng (2020). Graph-based segmentation of retinal layers in oct images (https://www.mathworks.com/matlabcentral/fileexchange/43518-graph-based- segmentation-of-retinal-layers-in-oct-images), MATLAB Central File Exchange. Retrieved September 8, 2020.

[15] Chiu S J et al. 2010 Automatic segmentation of seven retinal layers in SD-OCT images congruent with expert manual segmentation *J Optics express* **18.**18, p 19413-19428.

[16] Kaarniranta K, Tokarz P, Koskela A et al 2017 Autophagy regulates death of retinal pigment epithelium cells in age-related macular degeneration. *J Cell Biol Toxicol* **33** p 113– 128.

[17] Telegina D V, Kozhevnikova O S and Kolosova N G 2018 Changes in Retinal Glial Cells with Age and during Development of Age-Related Macular Degeneration. *Biochemistry* Moscow **83** p 1009–1017.

[18] Su L, Jing Y, Qian G, Sheng Z and Chang'an Z 2017 The Edge Detectors Suitable for Retinal OCT Image Segmentation *J Healthcare Engineering* p 1-13.

IOP Conf. Series: Materials Science and Engineering 1085 (2021) 012004 doi:10.1088/1757-899X/1085/1/012004

Assimilating region of interest for fused biometric modalities through cie l*a*b* color space

P. Bhargavi Devi [1], Dr.K. Sharmila [2]

[1] Research Scholar, Department of Computer Science, VISTAS, bhargavimowly@gmail.com
[2] Research Advisor and Associate Professor, VISTAS, Chennai, sharmila.scs@velsuniv.ac.in

Abstract. Idiosyncrasy in recent times has been a challenge that is not quite easily effectuated in many domains and work structures. Security compromises have led to immeasurable loss and altered the deportment of individuals. In order to counter such circumstances, biometric modality analysis and utilization stimulate an inimitable framework to enhancement of security mitigated impostures. Nonetheless, the existing indagation has always instituted the single biometric modality processing and evaluation. Although the previous work of pre-processing and post-processing for biometric modalities such as fingerprint, iris, gait, and other physiological and behavioral modalities have been considered through explicit algorithmic and computational processing, the rate of precision fallacy and contriving unique identification has been a lingering question. This paper focuses on augmenting the efficacy of identification through the fused biometric modalities of fingerprint and iris, along with unsheathing the region of interest (ROI) by designating color masks using the CIE L*a*b* color space and delta disparity computation. The histogram analysis is implemented to analyze the pixel intensity, and the pixel matched extraction is surface plotted using MATLAB.

1. Introduction

The praxis of effectuating authenticated identity protocols and framework using corporeal data is implemented through the domain called as Biometrics. This field of computation is intriguing for a variety of purposes ranging from its attribute of uniqueness to mitigated use of an added key or security mechanism. Since it is embedded in the human body, it works as an automated system of recognition relying on the comportment attributes. The biometric system institutes a minimal percentage of fallacy and inaccuracy due to the sui generis features that vouchsafes for the ineradicable facets of computation. With such advantageous proclivity, various modalities are explored for diverse applications and working methods. Nonetheless, the most common amongst the different modalities which are studied is the fingerprint and iris. Both of these modalities hold the tendency to develop patterns at the embryonic stages of development and remain to exist through the lifecycle of a human being. As these modalities hold characteristics that are not volatile, they are used as an unswerving recognition system for various applications. The viability of a system is trifurcated into aspects such as – authentication, authorization, and accountability. These components play a vital role in corroborating the communicating devices or the identity of individuals. The below figure delineates the different biometric modalities that are frequently used for various explorations [1].

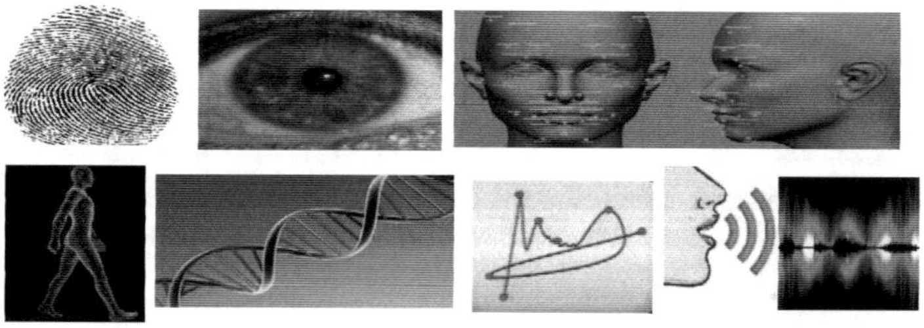

Fig 1. Biometric Modalities

Contrary to the utilization of security cards, keys, passwords, and pins, biometric modalities are innately grafted into the human body. These biometric modalities may vary based on their behavior and can be stratified into physiological and behavioral. The complexity challenge of hacking biometric systems is multifold, as they cannot be cloned, duplicated, and has minimal chances of pilferage. However, these systems also can have a few cons such as ambiguous interpretations for facial recognition when there are identical twins, injuries in the modality may fail the system to stimulate the right identity, and vulnerability exposure when a single modality entity is worked on. To encounter the disadvantages, a multi-modal biometric system has taken over in many applications and working architectures.[2]
Previous work relating to this area of study entailed the fusion of the same modalities along with their processing mechanisms. Nevertheless, this paper instruments the novel approach of explicating the fusion of biometric modalities such as the fingerprint and eye, and aids in extracting the region of interest from the coalesced identity which is created. The International Commission de l'Eclairage (CIE) is a global body that institutes and consigns recommendations about the standards of picture and color intensities [9]. The CIE model of systems effectuates the color space using three coordinate parameters such as the XYZ, L*a*b*, and the L*C*h [10]. This study pivots on the L*a*b* color stack in highlighting the ROI space and the masking of the outer region. The histogram analysis is graphed to analyze the differences, and the pixel intensity disparity is explored by the delta computation. This paper is structured as follows: section II explicates the empirical study of various explorations regarding the ROI implementation by various authors. The third section accentuates the current implementation pertaining to the method of ROI extraction along with incorporating the color space in the masked region. The last two sections illustrate the results and the conclusion for the performed simulations respectively.

2. Empirical review

The region of interest and biometric image processing has been analyzed by diverse researchers. Few of the literature surveys closely relating to this domain of interest is presented below:

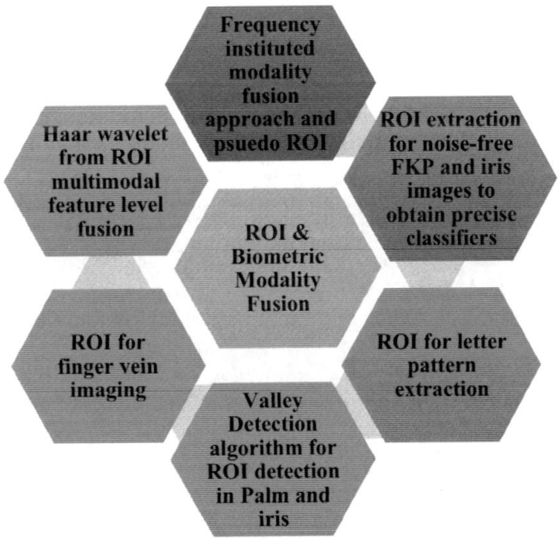

Fig 2. Existing Implementational review

Vincenzo Conti et al [4] expounded on the multimodal identification approach through frequency relied methodologies, the ROI extraction in this study detailed the emphasis on unique delineation of the fingerprint and iris features. The exploration of FKP and iris images to be preprocessed to mitigate noise was scrupulously executed using the Gabor filter [5]. This study was authored by Singh.S et al, and explicated the multi-biometric modality pre-processing approach through ROI and noise removal to obtain better accuracy. The extraction and pre-processing were further juxtaposed with the Neuro-fuzzy neural network classifier to gain better insights. The fusion of palm and iris was researched by Alsubari, A et al,[6] who implemented the feature level fusion using the NVPA method. Various other implementations such as Local binary variance pattern (LBPV) histograms for biometric modality pre-processing in a vector level, Haar wavelet and histogram equalization from ROI Normalization for fingerprint images [7], [8] have been some of the referred articulations for this study.

3. Proposed indagation

The proposed indagation entails the method of extracting the ROI, analyzing the histogram to match the intensity of color pixels in the delta images, and thereby using the intensity to surface plot the graph. The below framework explains the sequential step of processes that the study implements.

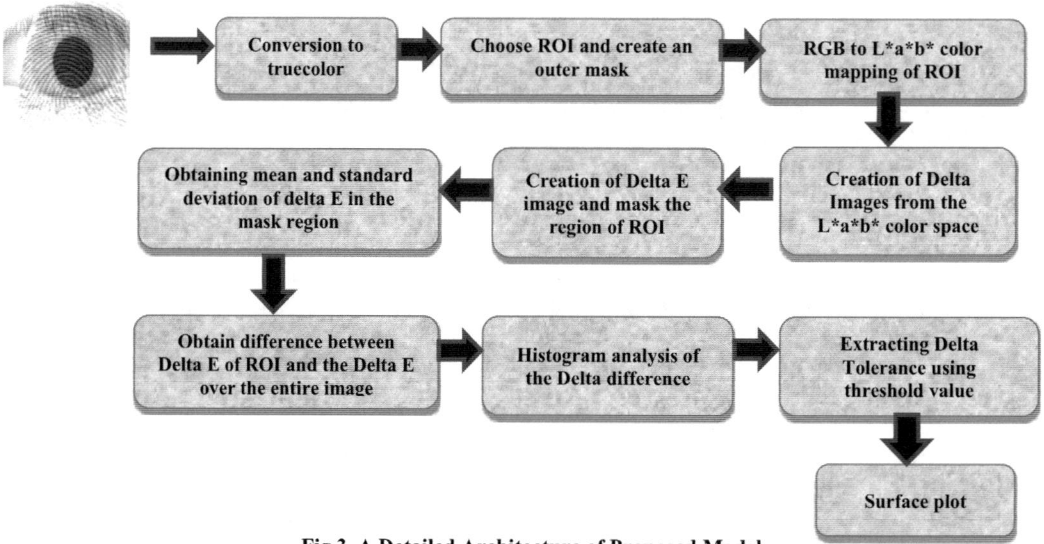

Fig 3. A Detailed Architecture of Proposed Model

The biometric fusion of fingerprint and iris is mapped in the L*a*b* color space [9] to analyze the variations in the pixel intensities of the image. The L*a*b* color space of CIE represents the lightness, red/green value, and the yellow/blue value respectively. The L* is scaled between the darkest color of black to the brightest pixel value for white, with the darkest black at L* = 0, and the brightest white at L* = 100. The color channels, a* and b*, can have neutral gray components with values swinging between the positive red and green at the negative axis, and yellow present in the positive axis and blue lying on the negative pole respectively [10]. The color bands are then extracted into a 2D array comprising the color component in each structure. This evaluation of the color value is usually represented quantitatively with the Delta standards incorporated in the L*a*b* color image. Each of the components are generated as a delta image, and are calculated as follows:

$$\Delta\ \text{Image} = \text{Channel value}(\Delta\text{CV}) - \text{Color space standard}\ (\Delta\text{CS}) \quad (1)$$

The overall Delta E image is computed by

$$\Delta\text{E} = \sqrt{(\Delta\ \text{L} - \Delta\text{LS})^2 + (\Delta\ \text{a} - \Delta\text{aS})^2 + (\Delta\ \text{b} - \Delta\text{bS})^2} \quad (2)$$

where ΔL, Δa and Δb are the channels which are generated and ΔLS, ΔaS and ΔbS are the standards from which they are subtracted. The delta image thus contrived after the computation of mean and standard deviation, is analyzed with the delta image of the ROI extracted region. The disparity thus generated is analogized in a histogram graphing method. The difference between the masked creation delta image and the ROI delta image is termed as the Delta Tolerance level. This delta tolerance level is proffered as the response by the user by analyzing the histogram. The chosen level given as input by the user is then checked for its scale verity, and the pixels within the selected range are extracted. The threshold matched output is thus generated, and the surface plotting of the ROI masked region forms the outcome which is graphically illustrated for better understanding on a 3D axial platform.

4. RESULTS

The various results pertaining to the analysis of the ROI, histogram, and delta masked computation obtained in MATLAB are depicted in the figure below.

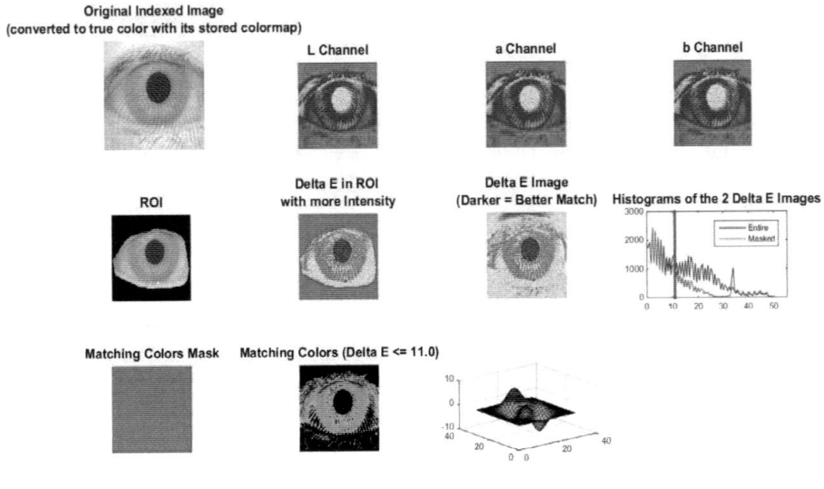

Fig 4. Simulation result

Each of the outputs thus received is relevant to the step-wise architecture delineated in section III. It is to be noted that each of the images taken can have varying pixel ranges, and therefore the delta tolerance would also vary accordingly. The histogram analysis plays a cardinal part in the process of thresholding due to the user opting to move the vertical bar to the value up to which the pixels are to be unsheathed. The images of fingerprint and iris have been procured from web databases and amalgamated before the initial process of ROI segmentation and color-mapping.

5. Conclusion

The study has helped in cognizing the factors pertaining to the extraction of ROI along with emphasizing on the assimilation of intensity variations in pixels. The prominence of this indagation is the creation of an outer mask which helps to circumvent the unwarranted regions of the image and focus entirely on the area which holds the necessary patterns of the modalities. The delta mask creation and delta tolerance values have depicted that although the original images are converted to true color images, the gray component triggers more of the a* channel, rather than the L* and b* channels. The future work of exploration is targeted towards incorporating more real-time images and deriving features from the modalities through the other channels.

References

[1] Arun A. Ross, Karthik Nandakumar, and Anil. K. Jain, Handbook of Multibiometrics", pp 37-58, Springer.

[2] Meghana A Divakar, "Multimodal Biometric System using Index-Based Algorithm for Fast Search", 2nd International Conference on Applied and Theoretical Computing and Communication Technology (iCATccT)

[3] M.Gomez-Barrero, C. Rathgeb, J. Galbally, C. Busch, and J. Fierrez, "Unlinkable and Irreversible biometric template protection based on bloom filters," Information Sciences, vol.370/371, pp. 18–32, 2016.

[4] Vincenzo Conti, Carmelo Militello, Filippo Sorbello, A Frequency-based Approach for Features Fusion in Fingerprint and Iris Multimodal Biometric Identification Systems, IEEE Transactions on Systems, Man, And Cybernetics—Part C: Applications and Reviews, Vol. 40, No. 4, July 2010.

[5] Singh, S., & Kant, C. (2019). FKP and Iris Based Multimodal Biometric System Using PCA with NFNN. SSRN Electronic Journal. doi:10.2139/ssrn.3358136

[6] Alsubari, A., Lonkhande, P., & Ramteke, R. J. (2019). Fuzzy-Based Classification for Fusion of Palmprint and Iris Biometric Traits. Advances in Intelligent Systems and Computing, 113–123. doi:10.1007/978-981-13-6783-0_11

[7] Shams, Mahmoud & Tolba, Ahmad & Sarhan, Shahenda. (2016). Face, Iris, And Fingerprint Multimodal Identification System Based on Local Binary Pattern with Variance Histogram and Combined Learning Vector Quantization. Journal of Theoretical and Applied Information Technology. 89. pp:53-70.

[8] Ujwalla Gawande, Mukesh Zaveri, Avichal Kapur, A Novel Algorithm for Feature Level Fusion

Using SVM Classifier for Multibiometric-Based Person Identification, Article in Applied Computational Intelligence and Soft Computing, published January 2013.

[9] B. Hemery, J.J. Schwartzmann, C. Rosenberger, Study on Color Spaces for Single Image Enrolment Face Authentication, 2010 International Conference on pattern recognition.

[10] Krishnan Nallaperumal, Muthukumar Subramanyam, Ravi Subban, Pasupathi Perumalsamy, Shashikala Durairaj, Gayathri Devi, Selva Kumar, An Analysis of Suitable Color Space for Visually Plausible Shadow-free Scene Reconstruction from Single Image, 2013 IEEE International Conference on Computational Intelligence and Computing Research, 978-1-4799-1597-2/13.

IOP Conf. Series: Materials Science and Engineering 1085 (2021) 012005 doi:10.1088/1757-899X/1085/1/012005

Linear Regression based Power Optimization of Wireless Sensor Network in Smart City

Manish Kumar Sahu , Sunil Patil

Department of Computer Science & Engineering, RKDF University, Bhopal, India

E-mail: manishsahu777@gmail.com

Abstract. The modern smart cities are highly dependent on the performance of the Internet of Things (IoT) based energy-efficient sensor networks. Energy efficiency is a critical and indispensable issue for wireless sensor networks (WSNs). In the collection of sensor nodes, one node is selected to collect data and forward it to the base station. The modern base stations in smart cities are unmanned aerial vehicle (UAV) based. This paper presents a linear regression based model, where the initial residual energy and its corresponding transmission power are submitted to the proposed system, then it generates a prediction model since transmission power depends on the residual energy of the sensor node. Based on this model the transmission power of the sensor node can be calculated for data transmission as a higher residual node is best suitable for data transmission to the base station. As presented in the simulation result, the regression based model gives better performance for energy efficiency in WSN.

1. Introduction

With the population growth and development in surrounding urban regions, the cities should be smart. The services beneficial to society, environment, and economy, are provided by the smart city [1]. These smart city services are significantly interactive, energy and cost-efficient, and more aware in security monitoring. The modern smart city deployments have concentrated on wireless sensor networks (WSNs) based Internet of Things (IoT) to improve the infrastructure efficiency [2]. The complete infrastructure of smart city contains various smart components as presented in Figure 1, it facilitate smart homes [3], smart agriculture [4], smart transportation [5], smart industry [6], smart energy and environment [7], smart health care [8], and many more smart services and applications [9]. So all these are the indispensable constituents to built and develop a smart city. Smart homes provide efficient electricity management for appliances, security cameras, and waste management. Smart agriculture facilitates weather monitoring systems, technology based farming, control of climate changes, water level control, etc. Smart parking, traffic monitoring, logistic, and accident alert systems are provided by a smart transportation system. The modern industry 4.0 [10] highly depends on smart industry and industrial internet of things which uses machine monitoring, smart warehouse, stock, and retail management. Smart power control, earthquake detection, fire control and safety, air quality monitoring, etc. are envisioned by smart energy control and environmental monitoring system. Smart health care system provides patient monitoring, Internet of medical things (IoMT), emergency services and wearable body sensor network (BSN) [11, 12, 13, 14].

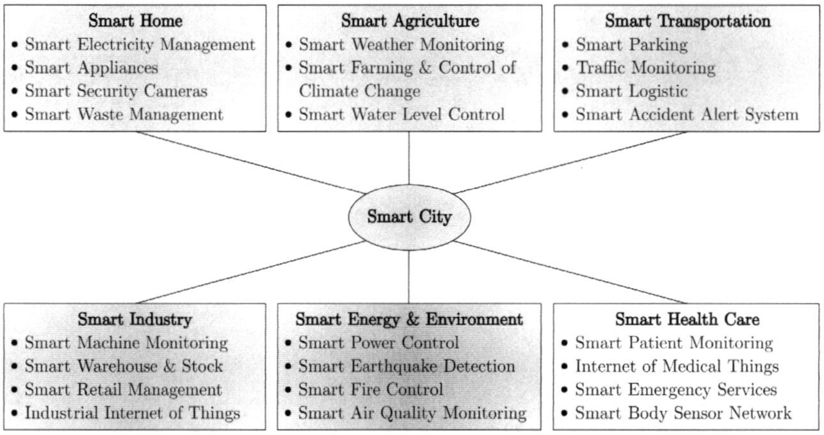

Figure 1. Various Components of Smart City

Concerning wireless sensor network, a smart city contains huge number of heterogeneous sensing and IoT devices. Due to high quantity of sensors connected to IoT network, a vast amount of data is accumulated to the server. The sensor network of smart city is high expandable so new type of several heterogeneous devices can be connected and old unused devices can be removed or repaired in the network [15]. A wide rage of applications are possible in WSN, so the arrangement on different kinds of geographical topologies of sensing nodes and IoT devices are challenging task [16]. The concept of smart city can not be envisaged without energy efficiency, so center point of this paper is the energy efficiency of smart city based WSN. However, the network of IoT devices with collaboration of UAVs create magnificent architecture for future smart city [17].

The section 2 presents surveys and works on energy efficiency issues in WSN for smart city, section 3 explains the proposed system model where linear regression based prediction shows how sensor nodes are selected to transmit data based on its residual energy for efficiency of WSN, section 4 presents the comparative simulation result of proposed system and finally section 5 concludes this research work with future aspect and enhancement.

2. Related Work

The limitation and constraint of energy resource in various sensor networks for smart city applications, many researchers have presented variety of techniques for raising energy efficiency in wireless sensor network (WSN).

Marrero *et al.* [18] presented a novel energy optimization method in the wireless backbone network for smart city. Their mechanism for energy saving also permit to develop a portable type of wireless backbone network which can be applied in various applications and places. For the mesh access point, at the time period of inactive or idle condition of wireless interface, they revealed an critical reduction in power consumption. They formulated an algorithm to manage the state transitions of radio interfaces in the mesh access point that saves and minimizes the overall power consumption of sensor nodes by interoperability of wireless mesh network.

Shafieirad *et al.* [19] analyzed the maximization problem of data quality in "Energy Harvesting Wireless Sensor Network (EH-WSN)". The progressive research and development in smart cities acquire analysis of these networks crucial and important. For smart city, a practical oriented framework for wireless sensor network based application is presented in this research. They proposed a realistic large scale efficient energy-harvesting aware routing scheme for for the maximization problem of data quality in wireless sensor network.

Sirsikar and Chandak [20] proposed a framework called "self-organized wireless sensor network (SOWSN)" and developed two-hop clustering based protocol for residual energy of sensor nodes. To analyze the performance, they compared their proposed protocol for average energy consumption calculation using various sensor nodes at distinct ranges of transmission. The outcomes represent that proposed protocol is capable to minimize the energy consumption and enhances lifetime of the network. When the quantity of sensor node is raised, then it minimizes the energy consumption at specific range, that presents the network scalability.

Fialho and Fortes [21] presented a IoT based smart city application using network of low power sensor nodes. Based on long range (LoRa) protocol, in this communication system, interconnection is established from end nodes to the gateway with 1% of duty-cycle. Using 2 dBm of minimum transmission power at gateway, LoRa communication of minimum 5 km based on radio frequency can be accomplished. Due to obstacle created by line of sight from end node to the gateway, RF signal attenuation is raised along with decreases of distance. So the end node should enhance the spreading factor (SF) to reach the desired distance for power optimization.

Liu *et al.* [22] proposed an efficient energy beamforming for "RF energy-harvesting wireless powered sensor networks for smart cities" with time assignment. In this approach, initially sensor nodes harvest the energy from their associated sink node, then they transmit accumulated data into the sink node by time-division multiple access (TDMA) using those harvested energy.

Jabbari *et al.* [23] proposed "energy efficient linear sensor (EELS) infrastructure" for smart city applications. In this structure, they provided different linear sensor nodes with services and support of Internet of Thing (IoT). This infrastructure provides energy optimization in linear sensor while preserving appropriate network delay. Figure 2 illustrates the basic energy-efficient linear sensor network applied in smart city application.

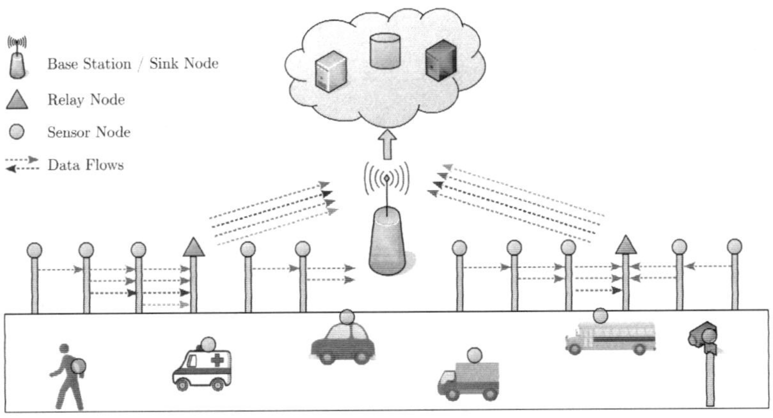

Figure 2. Energy-Efficient Linear Sensor Network [23]

The communication channel of "IR-based wake-up receiver (IR-WuRx)" is implemented over an ordinary WiFi network for improvement of energy efficiency. In multihop communication for linear sensor network topology, the IR pipelined relay method is investigated to verify network delay for end-to-end signal transmission. The energy efficiency and well preserved constant network delay are observed by experimental results.

Moreno *et al.* [24] proposed an IoT enabled location-aware platform which addresses the issue of intelligent data accumulation and processing to optimize the energy by modifying the functioning of related indoor smart devices. The identity and location of the user is the key aspect and essential component of this energy-efficient system which provides customized services and avoids useless power consumption.

Chithaluru *et al.* [25] proposed and implemented "I-AREOR" protocol for IoT enabled green smart city using energy-balanced clustering. For cluster head selection, this method examine proportional node distance, local sensor density and remaining energy of each node in WSN. Various energy factors are considered for selection of cluster head to enhance network lifetime and various impacting factor are applied to manage energy utilization. To forward data from node to cluster head, a dynamic threshold quantity is used in each round. Their outcomes illustrate that "I-AREOR" method enhances the number of rounds earlier to death of first sensor node.

Qureshi *et al.* [26] proposed "interference aware energy efficient transmission protocol (IEETP)" for smart city based healthcare applications using body sensor network (BSN). In this system, next forwarder sensor node is selected based on its residual remaining energy, location, distance and local density. "IEETP" is based on multihop protocol where data is forwarded from source node to the body node coordinator (BNC) of the network. For evaluation of routing parameters, a weighting factor technique is applied in the network, and the next forwarder node is decided based on those values. The "IEETP" protocol presents optimized energy utilization and low loss of packets as compared to other protocols by their experimental results.

3. Proposed System Model

In the smart city based wireless sensor network, the communication between receiver and transmitter nodes depends on the signal strength of electromagnetic wave. The data transmission usually consumes more energy as compared to sensing and processing. In modern samrt city, the base station may be stationary or unmanned aerial vehicle (UAV) based for sensed data collection. The sensor node having higher residual energy will take the responsibility for data collection from other nodes and transmission to the base station. A sensor node sends l bit data using transmission energy E_{Tx} and receiver node consumes E_{Rx} energy to receive that data at distance d, which can be represented as:

$$E_{\mathrm{Tx}}(l,d) = \begin{cases} l \cdot E_{elec} + l \cdot \epsilon_{fs} \cdot d^2, & \text{if } d < d_0 \\ l \cdot E_{elec} + l \cdot \epsilon_{amp} \cdot d^4, & \text{if } d \geq d_0 \end{cases} \tag{1}$$

$$E_{\mathrm{Rx}}(l,d) = l \cdot E_{elec} \tag{2}$$

where ϵ_{fs} is the permittivity of free space, ϵ_{amp} is permittivity using power amplifier losses, E_{elec} is energy loss coefficient of sensor node to send and receive l bit data, whereas d_0 is the threshold distance.

The set of sensor node in this system are $\{S_1, S_2, \dots S_n\}_1^n$. The transmission power E_{Tx} of sensor node completely depends on the residual energy E_{res} of the sensor node, so the linear regression model is considered as the affiliation between residual energy E_{res} (as independent variable) and transmission power E_{Tx} (as dependent variable). The linear relationship residual energy E_{res} and transmission power E_{Tx} of sensor nodes is represented in Figure 3.

The system model can be represented in the mathematical form using linear regression. For total n number of sensor nodes, the mean value for residual energy E_{res} and transmission power

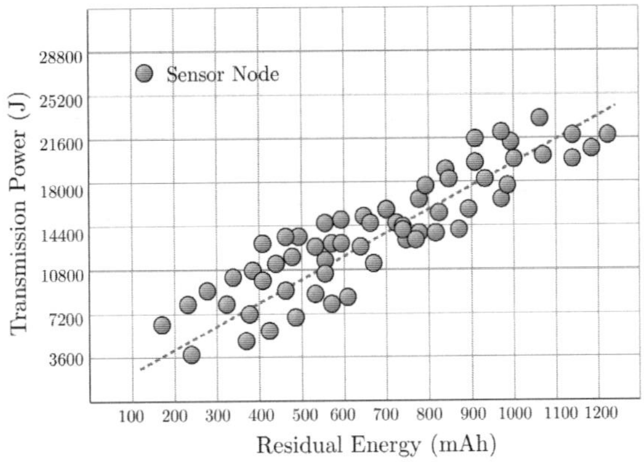

Figure 3. Linear Regression Relationship of Residual Energy and Transmission Power of Sensor Node

E_{Tx} can be calculated as

$$\overline{E_{res}} = \frac{1}{n}\sum_{i=0}^{n} E_{res\ i} \tag{3}$$

$$\overline{E_{\mathrm{Tx}}} = \frac{1}{n}\sum_{i=0}^{n} E_{\mathrm{Tx}\ i} \tag{4}$$

The slop m for the best fit line is calculated as

$$m = \frac{\sum_{i=1}^{n}\left(E_{res\ i} - \overline{E_{res}}\right)\left(E_{\mathrm{Tx}\ i} - \overline{E_{\mathrm{Tx}}}\right)}{\sum_{i=1}^{n}\left(E_{res\ i} - \overline{E_{res}}\right)^2} \tag{5}$$

Finally the prediction model where transmission power can be predicted based on residual energy for selection of best sensor node to transmit data to base station is presented as

$$\overline{E_{\mathrm{Tx}}} = m(\overline{E_{res}}) + c \tag{6}$$

where c represents vertical intercept in this relationship as a constant.

In this proposed system, initially the residual energy and it corresponding transmission power is given to the linear regression model, since transmission power depends on the residual energy of the sensor node. The proposed system is illustrated in Figure 4, the previous values of residual energy and transmission power are given to linear regression model, then it generates a prediction model. Based on this model the transmission power of sensor node can be calculated for data transmission as higher residual node is best suitable for data transmission to the base station. This proposed system creates a balance network for higher energy efficiency in smart city.

4. Result Analysis

The random selection of sensor node for data transmission to the base station can reduce the network lifetime. The random prediction creates higher error rate for node selection to transmit

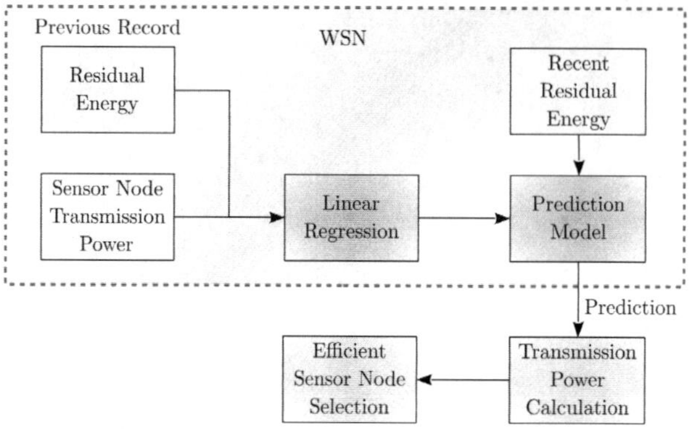

Figure 4. Proposed System Model for Node Selection

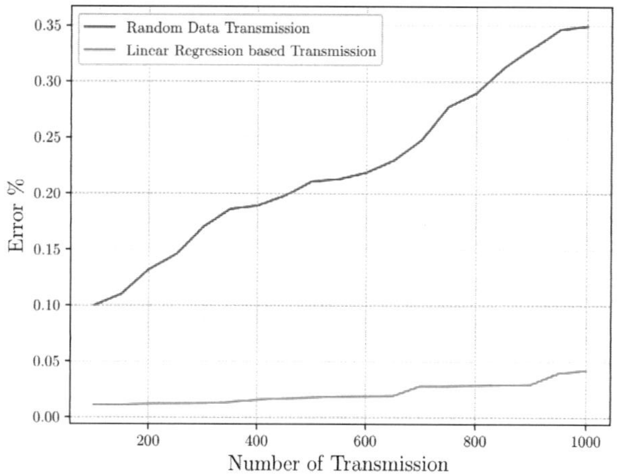

Figure 5. Prediction Error for Sensor Node Selection

data. Figure 5 represents that the random data transmissions have higher error rate and linear regression based prediction transmissions have low error rate.

In the random data transmission, the number of node dead during transmission is higher, whereas during linear regression based transmission less number of node dead occurs as represented in Figure 6. The proposed linear regression based prediction model for sensor node selection significantly improves the network lifetime and reduces the error rate for node selection in smart city based wireless sensor network (WSN).

5. Conclusion and Future Work

Wireless sensor networks (WSNs) are the key component for smart city development. So, the energy efficiency is the central point for research in WSN. This paper presented an energy-efficient approach using linear regression based model for sensor node selection to collect and transmit data to base station. This proposed approach created an efficient method for wireless sensor network which performs far better as compared to random data transmission. The simulation results explain the efficiency of the model concerning prediction error for node selection and network lifetime.

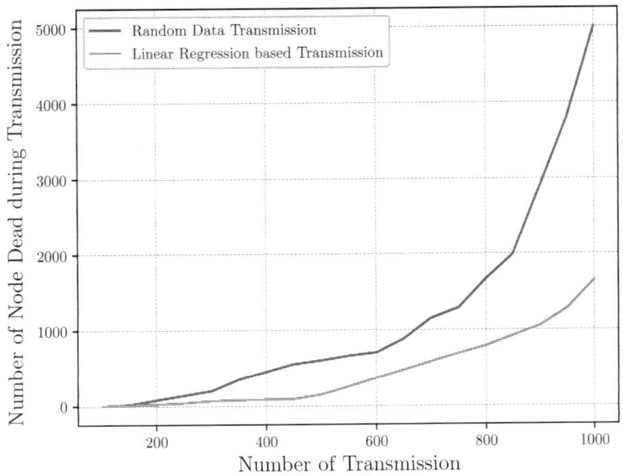

Figure 6. Network Lifetime

The proposed system can be extended using other regression analytical model depend on various conditions. The other factors and parameter can be considered as extended or multiple independent variables to enhance this work. The linear regression analysis can be associated with other modern energy-efficient protocols for smart city based WSNs.

References

[1] J. Jin, J. Gubbi, S. Marusic, and M. Palaniswami, "An information framework for creating a smart city through internet of things," *IEEE Internet of Things Journal*, vol. 1, no. 2, pp. 112–121, 2014. doi: https://doi.org/10.1109/JIOT.2013.2296516

[2] R. Lea and M. Blackstock, "Smart cities: An iot-centric approach," in *Proceedings of the 2014 International Workshop on Web Intelligence and Smart Sensing*, ser. IWWISS '14. New York, NY, USA: Association for Computing Machinery, 2014, pp. 1–2. doi: https://doi.org/10.1145/2637064.2637096

[3] F. Gazzawe and R. Lock, "Smart home," in *Intelligent Computing*, K. Arai, S. Kapoor, and R. Bhatia, Eds. Cham: Springer International Publishing, 2019, pp. 1086–1097. doi: https://doi.org/10.1007/978-3-030-01174-1_82

[4] B. Deepa, C. Anusha, and P. Chaya Devi, "Smart agriculture using iot," in *Intelligent System Design*, S. C. Satapathy, V. Bhateja, B. Janakiramaiah, and Y.-W. Chen, Eds. Singapore: Springer Singapore, 2021, pp. 11–19. doi: https://doi.org/10.1007/978-981-15-5400-1_2

[5] R. R. Patil and V. N. Honmane, "Smart transportation for smart cities," in *Soft Computing Systems*, I. Zelinka, R. Senkerik, G. Panda, and P. S. Lekshmi Kanthan, Eds. Singapore: Springer Singapore, 2018, pp. 53–61. doi: https://doi.org/10.1007/978-981-13-1936-5_6

[6] D. K. Maurya, A. Kumar, S. Kaunoujiya, D. Prasad, and V. Nath, "Study and design of smart industry: A review," in *Nanoelectronics, Circuits and Communication Systems*, V. Nath and J. K. Mandal, Eds. Singapore: Springer Singapore, 2019, pp. 591–598. doi: https://doi.org/10.1007/978-981-13-0776-8_55

[7] Y. Su, "Smart energy for smart built environment: A review for combined objectives of affordable sustainable green," *Sustainable Cities and Society*, vol. 53, p. 101954, 2020. doi: https://doi.org/10.1016/j.scs.2019.101954

[8] J. Al-Dulaimi and J. Cosmas, "Smart safety & health care in cities," *Procedia Computer Science*, vol. 98, pp. 259–266, 2016, the 7th International Conference on Emerging Ubiquitous Systems and Pervasive Networks (EUSPN 2016)/The 6th International Conference on Current and Future Trends of Information and Communication Technologies in Healthcare (ICTH-2016)/Affiliated Workshops. doi: https://doi.org/10.1016/j.procs.2016.09.041

[9] K. W. Al-Ani, A. S. Abdalkafor, and A. M. Nassar, "Smart city applications: A survey," in *Proceedings of*

the 9th International Conference on Information Systems and Technologies, ser. icist 2019. New York, NY, USA: Association for Computing Machinery, 2019. doi: https://doi.org/10.1145/3361570.3361616

[10] A. Giallanza, G. Aiello, G. Marannano, and V. Nigrelli, "Industry 4.0: smart test bench for shipbuilding industry," *International Journal on Interactive Design and Manufacturing (IJIDeM)*, vol. 14, no. 4, pp. 1525–1533, Dec 2020. doi: https://doi.org/10.1007/s12008-020-00739-9

[11] V. S. Parvathy, S. Pothiraj, and J. Sampson, *Automated Internet of Medical Things (IoMT) Based Healthcare Monitoring System*. Cham: Springer International Publishing, 2021, pp. 117–128. doi: https://doi.org/10.1007/978-3-030-55833-8_7

[12] S. Ullah, H. Higgin, M. A. Siddiqui, and K. S. Kwak, "A study of implanted and wearable body sensor networks," in *Agent and Multi-Agent Systems: Technologies and Applications*, N. T. Nguyen, G. S. Jo, R. J. Howlett, and L. C. Jain, Eds. Berlin, Heidelberg: Springer Berlin Heidelberg, 2008, pp. 464–473. doi: https://doi.org/10.1007/978-3-540-78582-8_47

[13] A. Bajaj, M. Bhatnagar, and A. Chauhan, "Recent trends in internet of medical things: A review," in *Advances in Machine Learning and Computational Intelligence*, S. Patnaik, X.-S. Yang, and I. K. Sethi, Eds. Singapore: Springer Singapore, 2021, pp. 645–656. doi: https://doi.org/10.1007/978-981-15-5243-4_61

[14] O. Dehzangi and C. Williams, "Multi-modal biological driver monitoring via ubiquitous wearable body sensor network," in *Proceedings of the 5th International Conference on Digital Health 2015*, ser. DH '15. New York, NY, USA: Association for Computing Machinery, 2015, pp. 65–70. doi: https://doi.org/10.1145/2750511.2750527

[15] K. Bajaj, B. Sharma, and R. Singh, *Integration of WSN with IoT Applications: A Vision, Architecture, and Future Challenges*. Cham: Springer International Publishing, 2020, pp. 79–102. doi: https://doi.org/10.1007/978-3-030-38516-3_5

[16] A. Giri, S. Dutta, S. Neogy, K. Dahal, and Z. Pervez, "Internet of things (iot): A survey on architecture, enabling technologies, applications and challenges," in *Proceedings of the 1st International Conference on Internet of Things and Machine Learning*, ser. IML '17. New York, NY, USA: Association for Computing Machinery, 2017. doi: https://doi.org/10.1145/3109761.3109768

[17] F. Qi, X. Zhu, G. Mang, M. Kadoch, and W. Li, "Uav network and iot in the sky for future smart cities," *IEEE Network*, vol. 33, no. 2, pp. 96–101, 2019. doi: https://doi.org/10.1109/MNET.2019.1800250

[18] D. Marrero, E. Macías, Á. Suárez, J. A. Santana, and V. Mena, "Energy saving in smart city wireless backbone network for environment sensors," *Mobile Networks and Applications*, vol. 24, no. 2, pp. 700–711, Apr 2019. doi: https://doi.org/10.1007/s11036-016-0786-5

[19] H. Shafieirad, R. S. Adve, and S. ShahbazPanahi, "Large scale energy harvesting sensor networks with applications in smart cities," in *Smart City 360°*, A. Leon-Garcia, R. Lenort, D. Holman, D. Staš, V. Krutilova, P. Wicher, D. Cagáňová, D. Špirková, J. Golej, and K. Nguyen, Eds. Cham: Springer International Publishing, 2016, pp. 215–226. doi: https://doi.org/10.1007/978-3-319-33681-7_18

[20] S. Sirsikar and M. Chandak, "Energy-efficient self-organization wireless sensor network for traffic management in smart cities," in *Proceedings of the International Conference on Data Engineering and Communication Technology*, S. C. Satapathy, V. Bhateja, and A. Joshi, Eds. Singapore: Springer Singapore, 2017, pp. 681–690. doi: https://doi.org/10.1007/978-981-10-1675-2_67

[21] V. Fialho and F. Fortes, "Low power iot network sensors optimization for smart cities applications," in *2019 International Conference on Smart Energy Systems and Technologies (SEST)*, 2019, pp. 1–6. doi: https://doi.org/10.1109/SEST.2019.8849071

[22] J. Liu, K. Xiong, P. Fan, and Z. Zhong, "Rf energy harvesting wireless powered sensor networks for smart cities," *IEEE Access*, vol. 5, pp. 9348–9358, 2017. doi: https://doi.org/10.1109/ACCESS.2017.2703847

[23] A. Jabbari, K. Ayinala, K. J. Almalki, S. Sung, B. Choi, and S. Song, "Building an energy efficient linear sensor (eels) infrastructure for smart cities," in *2019 IEEE International Smart Cities Conference (ISC2)*, 2019, pp. 665–670. doi: https://doi.org/10.1109/ISC246665.2019.9071656

[24] M. V. Moreno, M. A. Zamora, and A. F. Skarmeta, "User-centric smart buildings for energy sustainable smart cities," *Transactions on Emerging Telecommunications Technologies*, vol. 25, no. 1, pp. 41–55, 2014. doi: https://doi.org/10.1002/ett.2771

[25] P. Chithaluru, F. Al-Turjman, M. Kumar, and T. Stephan, "I-areor: An energy-balanced clustering protocol for implementing green iot in smart cities," *Sustainable Cities and Society*, vol. 61, p. 102254, 2020. doi: https://doi.org/10.1016/j.scs.2020.102254

[26] K. N. Qureshi, M. Q. Tayyab, S. U. Rehman, and G. Jeon, "An interference aware energy efficient data transmission approach for smart cities healthcare systems," *Sustainable Cities and Society*, vol. 62, p. 102392, 2020. doi: https://doi.org/10.1016/j.scs.2020.102392

CICERONE- A Real Time Object Detection for Visually Impaired People

Therese Yamuna Mahesh[1], Parvathy S S [2], Shibin Thomas[3], Shilpa Rachel Thomas [4] and Thomas Sebastian[5]

[1]Associate Professor, Dept. of Electronics & Communication Engineering, Amal Jyothi College of Engineering, Kanjirappally, Kerala, India
[2,3,4,5]B-Tech Electronics and Communication Engineering, Amal Jyothi College of Engineering, Kanjirappally, Kerala, India
thereseyamunamahesh@amaljyothi.ac.in
parvathyss2020@ec.ajce.in
shibin0598@gmail.com
shilparachelthomas2020@ec.ajce.in
thomassebastian2020@ec.ajce.in

Abstract. Our work provides a solution to people who are suffering from partial blindness due to diseases of the eye or due to accidents. Unlike people who are born blind, they depend a lot on other people for doing their day to day activities. Our project is a boon to such people as our end product; a smart walking stick detects the trained objects that are used daily by the person and intimates the person using audio messages. This project changes the visual world into an audio world by informing the visually impaired of the objects in their environment. These people can use this particular prototype for self -navigating their way. A YOLO (You look only once) algorithm is used in our project. Real-time objects in an image are detected with their names represented on a bounding box and these names are converted to speech signals. The conversion to audio signals is done by using an e-Speak tool which forms Googles Text to Speech (gTTS) system. The prototype consists of several modules. The Raspberry Pi camera module takes the image and transfers it to the Raspberry Pi desktop. Then, real-time object detection is carried out by using YOLO network. The detected image is converted to speech by using the gTTS module and the audio result is provided to the user through a headset. For achieving the required portability, the battery backup is being used.

1. INTRODUCTION

Millions of people living in the world today suffer from visual impairment, and the main reasons are, age-related issues, diabetes and accidents.Blindness or vision impairment, one of the top ten disabilities among men and women. Challenges faced by blind and partially sighted people include loss of flexibility & spontaneity, loss of employment, inability to carry out household task, inability to shop alone, time & effort required to accomplish simple task.The new technologies in the market are now trying to help them lead a normal life. Even though new products are rapidly being developed, they suffer from several self-navigation issues due to which social awkwardness follows. Some of the technologies before are **NAVI**, Navigation Assistance for the Visually Impaired could differentiate between background and the obstacle, **VOICE**, the system converts the image into audio, so that the user can identify an obstacle, **TVS** (Tactile Vision System) and **Tyflos** systems have a device at the user's abdomen. It converts the depth into vibrations. These systems are quite slow and not practically viable.

Here, the proposed prototype is connected on top of a walking stick and an image is captured by using the pi camera connected to a Raspberry Pi mounted onto the stick. Intelligent object detection is performed by using YOLO (You Look Only Once) [9], [10] algorithm, a new and state-of-the-art method to detect objects by regression and classification. There are different object detection techniques but are not as accurate as YOLO. The image passes through a neural network only once and provides the output of all detected objects [11]. The next phase involves text to speech conversion which is implemented using Google's gTTS module (Google text-to-speech) and thus artificially producing human speech. With the aid of a headset the visually impaired person receives the required information. As a result, a camera based effective self-navigation can be provided.

2. LITERATURE SURVEY

Much like guide dogs, that are trained to guide visually impaired people in a particular locality, use of a smart walking stick would help them to avoid obstacles in their path. The camera-based system in [1] helps to recognize text on the hand held objects. The proposed system consists of camera, to capture the object of interest and an algorithm to extract the text from the background. The optical character recognition (OCR) separates each letters of the detected text. The detected text is given as an audio output with the help of software development kit. The wearable device in [2] is a system that accepts input from the user and detects the objects. The device contains ultrasonic sensor that helps to alert the user about the object in his/her path. Haar cascade algorithm is used to detect the objects. It is a wearable device and placed on the chest of the user. The detected object is given back to the user as audio.

The object detection system in [3] is used to detect objects in the traffic scenes. Here they have used the combination of optimized you only look once (OYOLO), which is 1.18 times faster than YOLO and R-FCN (Regression based Full Convolution Network). It is used to detect and classify the images such as cars, cyclist and pedestrian. Use of YOLO makes location errors, to avoid that we use OYOLO. Paper [4] presents a prototype that extracts the text from image and is converted to speech. Extraction of text is done by using the Tesseract Optical Character Recognition (OCR). This method is carried out by using Raspberry Pi. Text recognition is done by using Open Computer Vision (Open CV), considering the large library of functions when compared to MATLAB. Capturing of image is done by using a portable camera and the image is converted to gray scale and filtered by Gaussian filtering. Then it is binarized and cropped. The cropped image is given to Tesseract OCR. The e-Speak creates an analog signal of the text and is given to the headset.

3. METHODOLOGY

The aim of our project is to design a system at a low cost to enable a person who's vision is not clear to lead a normal life without depending on others in a selected environment like our house, workplace etc. All the things that are used for the day to day activities can be detected using a walking stick with the raspberry pi attached. The process is as shown in the flowchart in Figure 1.

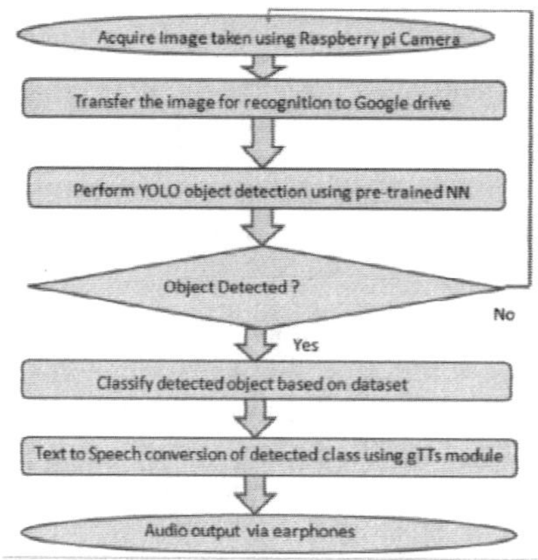

Figure 1. Flow chart of the proposed system.

Raspberry pi 4 model B is used for acquiring the images. The acquired images are transferred to the Google drive using rclone. YOLO network is used for object detection and classification. The test to speech conversion of the detected class is done using gTTS (Google Text to Speech module) to be sent to the users earphones.

4. RASPBERRY PI CONFIGURATION

Initially, to get started with the raspberry pi we need the operating system. NOOBS (New Out of the Box Software) is a preconfigured and an easy OS installer manager for the Raspberry Pi. NOOBS include the

following operating system
- Raspbian,
- Pidora,
- OpenELEC,
- RaspBMC,
- RISC OS

Installation procedures for NOOBS are as follows:
1. NOOBS is available for download on the official Raspberry Pi website [5].
2. Next, we have to download and install SD card formatter software into windows OS in order to make the SD card compatible for installing NOOBS.
3. Format the SD card using the above software in FA format. Now, micro SD card is ready for use with the Raspberry Pi.
4. Next, we have to upgrade and update the operating system with the latest version of Raspberry Pi. For that the following two commands are initiated:
    i. sudo apt-get update;
    ii. sudo apt-get upgrade
5. After the above procedures, we have successfully configured the Raspbian operating system

Figure2. Raspberry pi camera setup.

5. YOLO ALGORITHM

YOLO algorithm [6],[8] is primarily used for the prediction of bounding boxes accurately from an image. Images are divided into N x N grids and for each grid the prediction of the bounding boxes are done as well as the class probabilities. After performing object localization and image classification for each grid of the input image, each grid is given a different label. The following project is designed as a YOLO algorithm applied separately on each grid and the objects in it are marked with their particular label and corresponding bounding boxes are also highlighted. The grids having no object are labeled as 0.

Initially, YOLO algorithm is applied to the received image. In our project, the real time image is divided into grids of matrices. As the image complexity varies the image can be split into any number of grids. After division of the images, both classification and the localization process are done on each grid containing the object. The confidence score is computed for all the grids. The confidence score and also the bounding box for each of the grid will differ based on whether the object is detected or not. For no object it displays the value as 0 and displays the value 1 if object exist. Bounding box value will show how confident the network is, that is, how much the detected object matches to the object under observation. The prediction of bounding box is illustrated below.

6. EXPERIMENTAL PROCEDURE

The experimental procedure started with collecting the database, training the YOLO network and then testing the output by giving test images. The network was trained to detect 5 classes namely C1 (class 1) for bottles, C2 (class 2) for watches, C3 (class 3) for keys, C4 (class 4) for pens and C5 (class 5) for glasses.

6.1. Collection of Data Base

The example of dataset collected for class 4 pen is shown in figure 3. Similar datasets were collected for other classes.

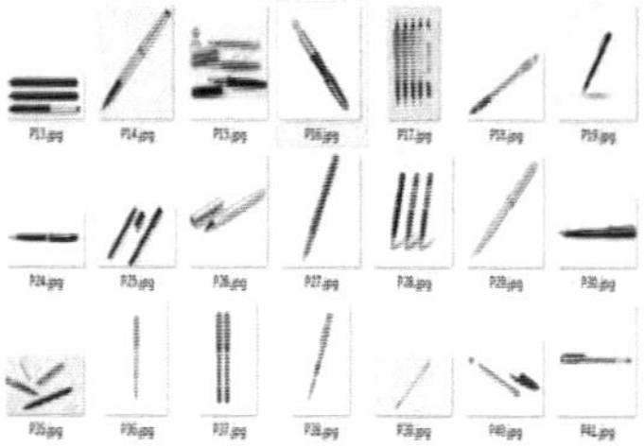

Figure 3. Dataset for the class-pen.

6.2. Network Output Analysis

Consider for example, input images being processed in batches. The output must be considered as a vector onto which we are mapping the features or simply a feature map. If N is the number of bounding boxes and C is the number of classes the detector is detecting on, then, these N bounding boxes are responsible for detecting different objects. Several bounding boxes are output with corresponding class names and each bounding box consists of six parameters (*pc*,bx,by,bz,*bw*,*c*). Therefore, each of these bounding boxes contain (C+5) characteristics that provide relevant information about the bounding boxes such as presence/absence of objects, dimensions of anchors, detected class and corresponding confidence scores also. Since the custom object detector in the following project is trained on only 5 classes, each bounding box is represented with 10 parameters as (C=5 classes).

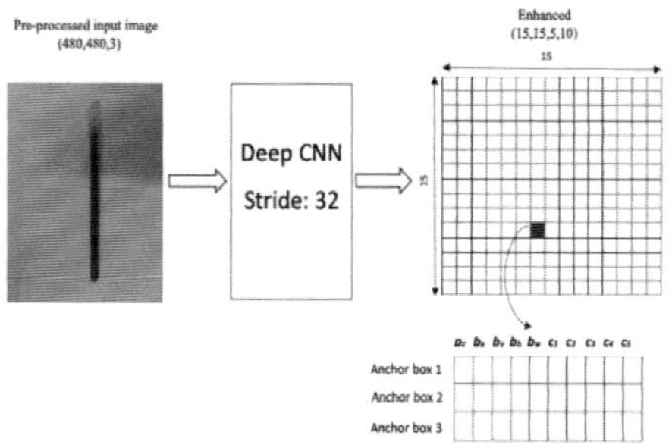

Figure 4. Image processing analysis.

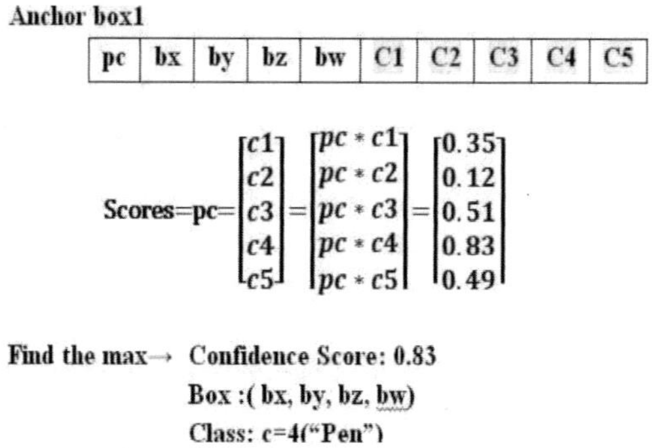

Figure 5. Bounding box evaluation.

In YOLO v3, 3 anchor boxes are computed for every cell. The expectation behind feature maps is to predict the desired objects using anchor boxes provided. The object center point coincides with that particular cell. The defining concept of YOLO v3 is that only a single bounding box must be responsible for detecting a particular object. For detection, the center point of the object is evaluated and is mapped to the feature map. The grid cell to which the center point falls on is responsible for detecting the object. Next, for each anchor box a probability is computed that indicates the certainty that the box contains a certain class. It can be done as shown in figure 5.

The generated bounding box dimensions are evaluated using simple mathematical formulas. The anchor box width and height are predicted from neighboring clustering centroids and the center coordinates can be obtained using a sigmoid function. Transition from network output to the bounding box generation is done as follows:

$$b_x = \sigma(\mathrm{a_x}) + \mathrm{i_x} ---------- (1)$$

$$b_y = \sigma(\mathrm{a_y}) + \mathrm{i_y} ---------- (2)$$

$$b_w = (\mathrm{q_w}) * \mathrm{e^{aw}} ---------- (3)$$

$$b_z = (\mathrm{q_z}) * \mathrm{e^{az}} ---------- (4)$$

Here, $\mathbf{b_x}$ and $\mathbf{b_y}$ are the (x, y) center coordinates and $\mathbf{b_w}$ and $\mathbf{b_z}$ are the bounding box width and height respectively. The parameters $\mathbf{a_x}$, $\mathbf{a_y}$, $\mathbf{a_z}$, $\mathbf{a_w}$ are the output of the network. The top left most cell coordinates are denoted as $\mathbf{i_x}$ and $\mathbf{i_y}$. $\mathbf{q_w}$ and $\mathbf{q_z}$ are the anchor box dimensions

6.3. Bounding Box Dimensions

The sigmoid function is deployed to normalize the output values between 0 and 1. The actual coordinates of bounding box are not computed by YOLO, only the offset values are determined relative to the top left most corner of the grid cell responsible for detecting the object.

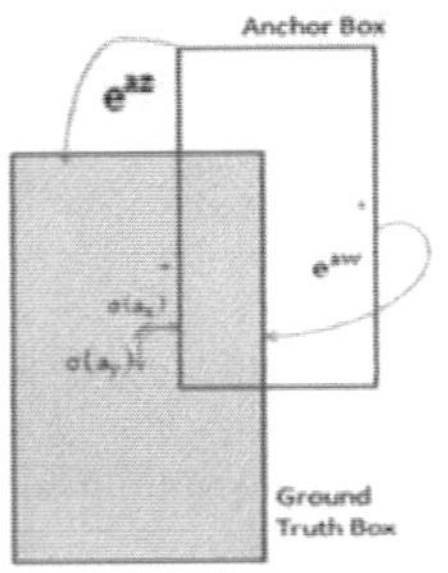

Figure 6. Mapping of anchor box and ground truth box

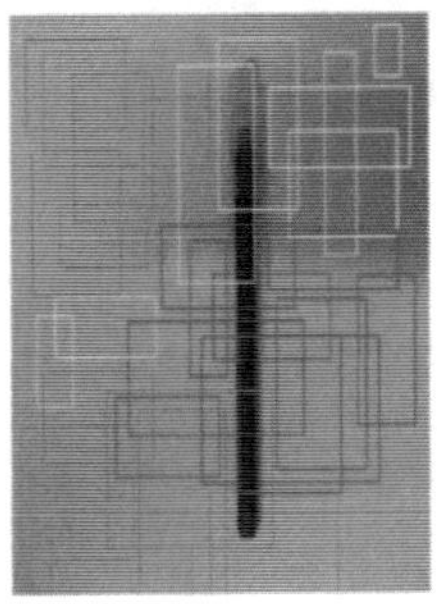

Figure 7. Interpretation of network output

Interpretations of the YOLO results are as follows:
For each cell, the maximum of the probability scores is selected (ie taking the maximum value of both, the 3 anchor boxes as well as the various classes. The above process generates a large number of bounding boxes and the generated anchor boxes can be illustrated in figure 7.

6.4. Output Filtering

The process described above yields a large number of bounding boxes which in turn may cause multiple detections of the same objects in an image. To overcome such issues, the technique of Non-Maximum Suppression (NMS) is deployed which neglects those generated anchor boxes having low confidence scores based upon a pre-defined threshold value (cut off value). That is, if the threshold value is chosen as 0.5 then all generated anchor boxes having a confidence score below 0.5 are discarded. Despite filtering the bounding boxes using threshold values, the network might still end up with the condition of overlapping boxes, as a result, an additional function of NMS known as Intersection over Union (IoU) is used which acts as a second filter for selecting the right boxes. IoU can be implemented as follows:

- An anchor box is defined with two corners, that is the top left most corner and the bottom right most corner (a_1, b_1, a_2, b_2)
- Area of the box is computed by multiplying the height (b_2- b_1) and the width of the box (a_2- a_1).
- Next step is locating the coordinates of the intersection of two anchor boxes. (a_1', b_1', a_2', b_2') as follows:

  a_1'= Min {a_1 coordinates of box1 and box2}
  b_1'= Min {b_1 coordinates of box1 and box2}
  a_2'= Min {a_2 coordinates of box1 and box2}
  b_2'= Min {b_2 coordinates of box1 and box2}
- IoU can be illustrated as follows:

$$\textbf{Intersection Area} = (a_2' - a_1') * (b_2' - b_1') - - - -(5)$$

$$\textbf{Union Area} = [\text{box1 area} - \text{box2 area}] - \text{Intersection Area} - - - -(6)$$

Therefore,

$$\textbf{IoU} = (\text{Intersection Area})/(\text{Union Area}) - - - (7)$$

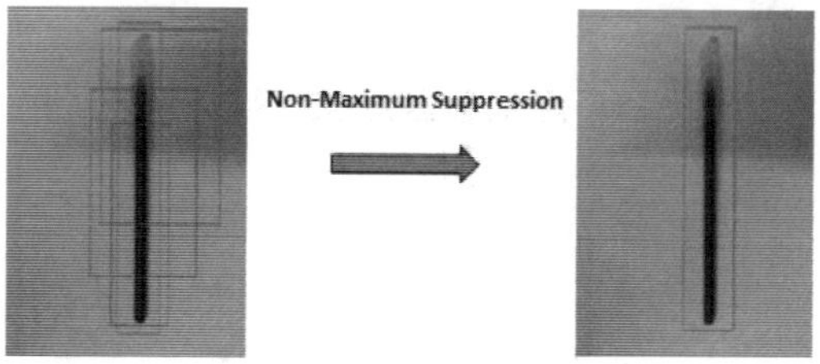

Figure 8. Input image before NMS and after NMS

6.5. Training

- The resolution of the training images was pre-adjusted in order to keep a constant stride value.
- The entire network was trained for 10000 epochs (2000*no. of classes) with the following specifications:

Table 1. Training network specifications and values.

Training Specifications	Value
Batch size	64
Subdivisions	16
Channels	3
Momentum	0.9
Decay	0.0005
Angle	0
Saturation	1.5
Exposure	1.5
Hue	1
learning rate	.001
Policy	steps
Steps	8000,9000

Training was performed on the collected dataset by using open source software known as LabelImg [11] which is a graphical image annotation tool. The software is written in Python and uses python libraries such as Qt for its graphical interface. The annotations are saved as XML files in both the PASCAL VOC format and YOLO format. For this project, we have opted it to be saved in the YOLO format.

7. TEXT TO SPEECH

The main aim of this project is to get the audio output of the detected text. For recognition of the text an optical character recognition (OCR) engine is used which is called the Tesseract-OCR. So, the text file extracted from the Tesseract, needs to be converted to audio which is performed by Python gTTS module, which is a text to speech engine for converting the text words that is extracted by using Tesseract-OCR to speech form. The complete process is executed on the Raspberry Pi as a compact hardware design. The detection of text from the image and text to audio conversion has an accuracy of 85 % by taking random samples. The Python interface for Google's Text to Speech API (Application Package Interface), Google Text to Speech (gTTS) is used. gTTS module produces an mp3 file.. The audio conversion by making use of gTTS is much faster. The mp3 file obtained using gTTS can be heard using the play sound, which is an open source - based sound player compatible with python. Play sound is a sound player for the Raspberry Pi which is command line-based.

8. RESULT

The YOLO network was trained to detect five customized objects .The input scenario is as shown in the figure 9 and the objects detected by the YOLO network are shown in figure 10.

Figure 9. Input image containing several objects

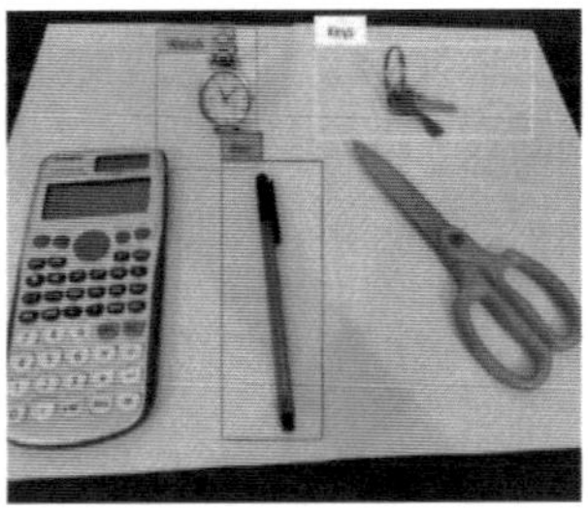

Figure 10. Desired objects detected from several objects

As seen from the figures 10, the custom Yolo v3 architecture was able to detect the 5 classes it had been trained on. Since the network had been trained for an epoch of 10000, the confidence score of the detected object is relatively high.

Table 2. Confidence score for figure 19 (three objects detected).

Classes	Confidence score
Pen	99%
Watch	98%
Key	96%

Higher confidence value resulted from exponential decrease of the loss curve where an ideal detector model is required to have a loss value of less than two 2.

After running the code for gTTS we obtain the audio file which is given to the user.

9. CONCLUSION

The main objective of this project is to develop an object detector that is capable of detecting customized classes. This is implemented for five classes of different objects which the user is likely to use daily. The proposed system was implemented using a camera, which captures image, raspberry pi for interface and Bluetooth for the audio output. Ultrasonic sensor is connected to the walking stick for the detection of obstacles. YOLO leads to faster run time [7] and the performance measure using IoU is above 90%. The future scope includes complete independence to the partially blind person both indoors and outdoors by incorporating cards with GPU in the hardware and cloud based execution of the network.

IOP Conf. Series: Materials Science and Engineering 1085 (2021) 012006 doi:10.1088/1757-899X/1085/1/012006

REFERENCES

[1] Samruddhi Deshpande and Revati Shriram "Real time text detection and recognition on hand held objects to assist blind people" International Conference on Automatic Control and Dynamic Optimization Techniques (ICACDOT) International Institute of Information Technology (I^2IT), Pune, 2016

[2] Deepthi Jain B, Shwetha M Thakur and K V Suresh "Visual assistance for blind using image processing" *International Conference on Communication and Signal Processing*, April 3-5, India, 2018

[3] Jing Tao, Hongbo Wang, Xinyu Zhang, Xiaoyu Li and Huawei Yang,"An object detection system based on YOLO in traffic scene" , *6th International Conference on Computer Science and Network Technology (ICCSNT)*, 2017

[4] Rajesh M, Bindhu K Rajan, Ajay Roy, Almaria Thomas K, Ancy Thomas, Bincy Tharakan T and Dinesh C, "Text recognition and face detection aid for visually impaired person using raspberry pi ", *International Conference on Circuits Power and Computing Technologies (ICCPCT)*, 2017

[5] Installing operating system in Raspberry pi available at https://www. raspberrypi.org/downloads

[6] Geethapriya. S, N. Duraimurugan and S.P. Chokkalingam "Real-Time Object Detection with Yolo", *International Journal of Engineering and Advanced Technology (IJEAT)* ISSN: 2249 – 8958, Volume-**8**, Issue-3S, February 2019

[7] Deepa,Tamilselvan, Abrar and Shrinivas Sampath"Comparison of Yolo, SSD, Faster RCNN for Real Time Tennis Ball Tracking for Action Decision Networks", *International Conference on Advances in Computing and Communication Engineering (ICACCE)*, 2019

[8] Joseph Redmon∗, Santosh Divvala∗†, Ross Girshick and Ali Farhadi "You Only Look Once: Unified, Real-Time Object Detection", *IEEE Conference on Computer Vision and Pattern Recognition,* 2016

[9] R.Rajalakshmi,K.Vishnupriy,M.S.Sathyapriya and G.R.Vishva-ardhini "Smart Navigation System for the Visually Impaired Using Tensorflow", *IJARIIE*-ISSN(O)-2395-4396, Vol-**4** Issue-2 2018

[10] Prof. Seema Udgir , Shivaji Sarokar2 , Sujit Gore , Dinesh Kakuste and Suraj Chaskar "Object Detection System for Blind People", *International Journal of Innovative Research in Computer and Communication Engineering*, Vol. **4**, Issue 9, September 2016

[11] Installation process of labelImg is available at https://github.com/tzutalin/labelImg

Analysis of image inpainting and object removal methodologies

Shiney Thomas[1*], **Dr. Juby Mathew**[2*]

[1]Research Scholar, [2]Associate Professor,
[*]Department of Computer Science and Engineering,Amal Jyothi College of Engineering, Kanjirappally, Kottayam, Kerala, India

Email: shineythomas@amaljyothi.ac.in

Abstract: Image inpainting is a process that tries to fill in missing parts of a degraded image or remove objects from an image, but maintain the realistic content in the image. Various methods can be employed to perform inpainting. Here three benchmark methods of which two are conventional and other GAN based are evaluated to check the effectives of the inpainting process, for different class of images. The visual quality evaluation as well as quantification parameter namely the SSIM score was computed. Based on the test score it was seen that the GAN based method was capable of providing visually good restored images with an average SSIM score of 0.9516, which is an indicator on the quality of the image. The GAN based methods can be further enhanced to provide more realistic images by incorporating functions to extract both local and global contextual content.

1. Introduction

Image is one of the most common forms of information source that is used in everyday life with the help of many application/tools. The information content will be critically affected if these images are corrupted, missing or damaged. Image inpainting or restoration models include operations that try to alter an image by filling in or removing details from the targets regions in the image. The process tries to remove object like text, date, or broken regions as well as add objects into the image but still provide a realistic image. Image restoration is a very active research content in the field of image processing [1].

A generic architecture of any image inpainting models can include various processes. The first process is the acquisition of the image that needs to be processed. This is followed by image pre-processing; these can incorporate operations enhancement operation like brightness or contrast, colour to greyscale conversion or noise removal. The next process would involve identification of the area of interest where the inpainting operation needs to be performed. Once this has been identified an appropriate mask has to be created so that image inpainting process can be initiated.

There are numerous papers in image inpainting focusing on different aspects. Image inpainting can be done by two methods- patch matching using low level image features and deep neural networks based on feed forward generative models. Patch based methods detects plausible fixed textures but are not successful in detecting complex scenes, human faces and some objects. Deep generative models learns from large scale datasets to capture semantics of the complex images and apply feature continuity of the generated contents around the mask region.[12]. When a key region missing (it is the only information in the image, which cannot be obtained from existing information, such as the nose loss of a single face picture), the target regions cannot be filled from existing information [1]. In [9] authors elaborate on an inpainting method based on a fast non-iterative nonlinear analysis of stationary first order transport equations using higher-order partial differential equations. Structure tensors are employed to estimate the image coefficients in the coherence direction robustly using fast marching method which is run only once while transporting. Based on the coherence parameter the method switches between two operations namely diffusion and directional support. Another benchmark method is discussed in [10][11] that employ exemplar-based image inpainting, where the process is divided into have two phases namely, deciding the filling-in order and selecting good exemplars. Therefore, the traditional method does not have a semantic inpainting. In [3], a framework for coarse-to-fine generative image inpainting was introduced along with a novel contextual attention module. The addition of contextual attention module considerably improves image inpainting results by learning feature representations for explicitly matching and attending to relevant background patches. The input to the network is a 256 × 256 image with a rectangle missing region sampled randomly during training, and the trained model can take an image of different sizes with multiple holes in it. The authors in [4], proposes a gated convolution for free-form image inpainting where a dynamic feature gating mechanism is computed for both RGB channel and spatial location. Initially the features are computed based on the gating function defined by $g = \sigma(w_g x)$ where σ and w_g represents the sigmoid function and learnable variable respectively. The final output operation is defined by the operation $y = \Phi(wx)\Theta g$ where Φ can be any activation function and Θ represents the gating function. In other words the operation is simply a multiplication of learned feature and gating values. The main advantage of gated convolution is its ease of computation and that it can be effective used on arbitrary shaped masks. The gated convolution modules are stacked for designing a encoder-decoder network as discussed in [3]. For previous inpainting networks which try to fill a single rectangular hole, an additional local GAN is used on the masked rectangular region to improve results [3][5]. The task of free-form image inpainting where there may be multiple holes with any shape at any location employing SN-Patch GAN has been discussed in [4]. More recent work [13] focuses on encoder combined with a context loss function. In order to obtain clear repaired images and ensure that the semantic features of images are fully learned with the generator and discriminator operating on a fusion of squeeze-and-excitation networks deep residual networks.

Different GAN models vary in their implantation strategies in terms of number of layers, types of layers, parameters used in the layers and so forth. In this papers a GAN model is explored that can be used for both filling in missing regions as well as removing objects from the scene. The efficacy of the model is compared with existing conventional model. Section 2 provides a crisp discussion on the methodology employed and the results and findings are elaborated in sections 3. The manuscript is concluded in section 4.

2. Methodology

The GAN has basically two modules namely the generator and discriminator. Both the networks are trained simultaneously to achieve its required level namely: Generator is trained to generate data that is capable of fooling the discriminator and the discriminator is trained such that is capable to distinguish between real and generated data. A GAN generator and discriminator that can be employed for inpainting application is illustrated in figure 1a and 1b. The choice of the layers and filter kernels determine the effectiveness of the model as well as the novelty.

From the figure, it can be seen that the common functional layers in the generator and discriminator are Convolution, Batch Normalization and Leaky ReLU. Some other design constraints in the generator include the embedding dimension of 50; each convolution layer consist of 5x5 filter kernel and the number of filters are reduced at subsequent layers, with stride of 2. The last convolution layer has 3 filters of size 5x5 corresponding to the RGB color space. This is followed by a *tanh* activation function. In the discriminator the each convolution layer consist of 5x5 filter kernel and the number of filters is increased at subsequent layers, with stride of 2. Here the scale factor of the Leaky ReLU layer is taken as 0.2

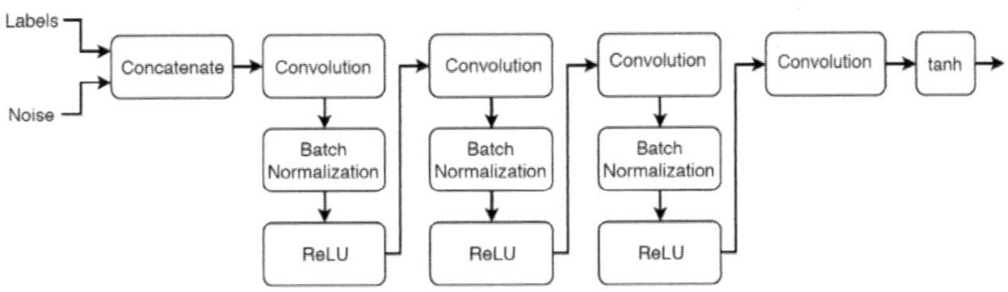

a. Generator

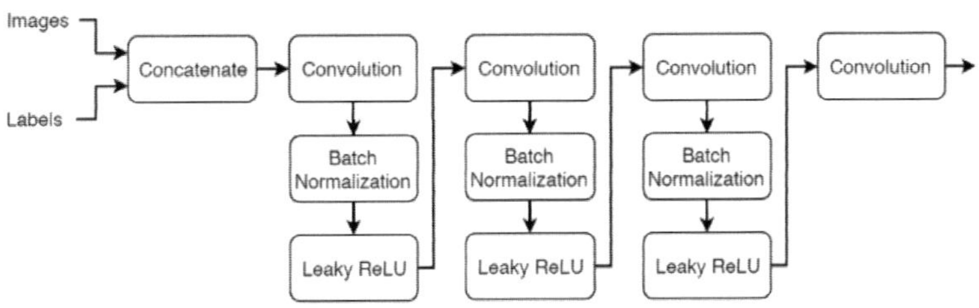

b. Discriminator

Figure 1: GAN model for Inpainting a. Generator b. Discriminator

3. Results and Finding

To get a clearer understanding on efficacy of the proposed GAN based inpainting process, it is compared with two conventional benchmark models based on [9] and [11]. For the purpose of training and testing of the image inpainting models images from various publically available datasets were taken these include CelebA [6], CelebA-HQ [7] and Places2 [8]. For the purpose of evaluation illustration few critical outputs are illustrated in Figure 2. From the figure following observations were made

1. All the methods perform equally well in removing an object from a scene and is able to maintain good realistic content.
2. In the case of text removal from a scene all the methods perform equally well, but with slight reconstruction error in method using coherence transport.
3. When multiple patch removal is involved the coherence transport method provides a poor quality scene reproduction when with the other two methods performing equally well.
4. Considering the task of filling the missing region in a scene, celebrity faces data is used as an illustration. The main difference between the two input images is the size of missing patch. It can be seen from first celebrity face that both the conventional methods fail to create a realistic image. The GAN model gave a far superior image, but still sharper features are still missing. In the case of the second celebrity image all the methods gave an equally good result. Reason why these methods gave goods result can be attributed to two parameters namely the size of patch and area of the patch. The patch is located in the region of least texture transition.

5. Finally scene filling scenario is considered next. In the first scenario the window pane was perfectly reconstructed by the GAN model and the exemplar method provided a marginally good reconstruction followed by coherence transport method. The same result is observed in the last scenario, but realistic or the contextual content can still be improved in the GAN model.

To obtain a quantitative analysis on the performance of the algorithm structural similarity index measure (SSIM) has been employed which is given equation (1). SSIM takes on values between 0 and 1, with the values greater than 0.95 representing good quality images.

$$SSIM(x,y) = \frac{2\mu_x\mu_y+C_1+(2\sigma_{xy}+C_2)}{(\mu_x^2+\mu_y^2+C_1)(\sigma_x^2+\sigma_y^2+C_2)} \quad (1)$$

SSIM has been evaluated on all the reconstructed images and average scores of the methods are quantified in Table 1.

a. Input Image	b. Using Coherence Transport	c. Exemplar Based	d. GAN Based

Figure 2: Generic Inpainting Model

Table 1: SSIM Score Average Across Different Images

Method	Average SSIM Score
Coherence Transport	0.8215
Exemplar Based	0.9111
GAN	0.9516

4. Conclusion

Image inpainting is a process used to fill in missing part or remove objects from an image, while maintaining the image integrity (realistic content). Benchmark conventional and GAN based methods has been evaluated on different class of images and their performances were evaluated on both visually and quantitatively using the SSIM score. In all the evaluations across different class of images the GAN based methods performed better than the conventional methods with higher average SSIM score (0.9516). Furthermore GAN based methods fail to address the contextual content perfectly which has lead to smaller SSIM scores for certain images. Hence use of contextual methods on both local and global level can improve the SSIM score as well as realistic content.

5. References

1. H. Liu, G. Lu, X. Bi, J. Yan and W. Wang 2018 Image Inpainting Based on Generative Adversarial Networks *14th International Conference on Natural Computation, Fuzzy Systems and Knowledge Discovery (ICNC-FSKD), Huangshan, China*, pp. 373-378, doi: 10.1109/FSKD.2018.8686914.
2. Marcelo Bertalmio, Guillermo Sapiro, Vincent Caselles, and Coloma Ballester 2000 Image inpainting *Proceedings of the 27th annual conference on Computer graphics and interactive techniques (SIGGRAPH '00)* ACM Press/Addison-Wesley Publishing Co., USA, pp. 417–424, doi:10.1145/344779.344972.
3. J. Yu, Z. Lin, J. Yang, X. Shen, X. Lu and T. S. Huang 2018 Generative Image Inpainting with Contextual Attention *IEEE/CVF Conference on Computer Vision and Pattern Recognition, Salt Lake City, UT,* pp. 5505-5514, doi: 10.1109/CVPR.2018.00577.
4. J. Yu, Z. Lin, J. Yang, X. Shen, X. Lu and T. Huang 2019 Free-Form Image Inpainting With Gated Convolution *IEEE/CVF International Conference on Computer Vision (ICCV), Seoul, Korea (South)*, pp. 4470-4479, doi: 10.1109/ICCV.2019.00457.
5. Satoshi Iizuka, Edgar Simo-Serra, and Hiroshi Ishikawa 2017 Globally and locally consistent image completion *ACM Trans.on Grapic* . **Vol.36 Issue 4** Article 107 (July 2017), 14 pages. doi: 10.1145/3072959.3073659.
6. Ziwei Liu, Ping Luo, Xiaogang Wang, and Xiaoou Tang December 2015 Deep Learning Face Attributes in the Wild *Proceedings of the 2015 IEEE International Conference on Computer Vision (ICCV '15)* IEEE Computer Society, USA pp.3730–3738. doi:10.1109/ICCV.2015.425.
7. C. -H. Lee, Z. Liu, L. Wu and P. Luo 2020 MaskGAN: Towards Diverse and Interactive Facial Image Manipulation *IEEE/CVF Conference on Computer Vision and Pattern Recognition (CVPR), Seattle, WA, USA*,pp.5548-5557,doi: 10.1109/CVPR42600.2020.00559.
8. Zhou B, Lapedriza A, Khosla A, Oliva A, Torralba A. June 2018 Places: A 10 Million Image Database for Scene Recognition *IEEE Trans Pattern Anal Mach Intell.* **40(6)** pp.1452-1464 doi: 10.1109/TPAMI.2017.2723009. Epub 2017 Jul 4. PMID: 28692961.
9. Bornemann, F., März, T. 2007 Fast Image Inpainting Based on Coherence Transport. *J Math Imaging Vis* **28**, pp. 259–278 .doi: 10.1007/s10851-007-0017-6.

10. Criminisi, A., P. Perez, and K. Toyama 2004 Region Filling and Object Removal by Exemplar-Based Image Inpainting *IEEE Transactions on Image Processing* **Vol. 13, No. 9**, pp. 1200–1212.
11. Le Meur, O., M. Ebdelli, and C. Guillemot 2013 Hierarchical Super-Resolution-Based-Inpainting *IEEE Transactions on Image Processing* **Vol. 22, No. 10**, pp. 3779–3790.
12. R. A. Yeh, C. Chen, T. Y. Lim, A. G. Schwing, M. Hasegawa-Johnson and M. N. Do 2017 Semantic Image Inpainting with Deep Generative Models *IEEE Conference on Computer Vision and Pattern Recognition (CVPR), Honolulu, HI,* pp. 6882-6890, doi: 10.1109/CVPR.2017.728.
13. Chen, Y., Liu, L., Tao, J. et al. 2020 The improved image inpainting algorithm via encoder and similarity constraint. *Vis Comput* . doi:10.1007/s00371-020-01932-3.

Performance analysis of Word Embeddings for Cyberbullying Detection

Subbaraju Pericherla[1], E Ilavarasan[2]

[1] Research Scholar, Dept. of CSE, Pondicherry Engineering college, Pondicherry, India-605014.
[2] Professor, Dept. of CSE, Pondicherry Engineering college, Pondicherry, India- 605014.
*Email: raju.pericherla74@gmail.com

Abstract. Cyber bullying activities are increasing day by day with the increase of Social Media Platforms such as Face book, Twitter, Instagram etc. Bullies take the advantage of these large online connected platforms due to which it became as a big challenging task in Natural Language Processing (NLP). In this paper, we compare the performance of various word embedding methods from basic word embedding methods to recent advanced language models such as RoBERTa, XLNET, ALBERT, etc. for cyberbullying detection. We used LightGBM and Logistic regression classifiers for the classification of bullying and non-bullying tweets. Among all the models, RoBERTa is outperformed as compared to state-of-the-art models.

1. Introduction

The usage of social media is much more increasing from day to day with increase of digital devices like smartphones, tabs, and electronic gadgets. The most common social media platforms like Facebook, Twitter, Instagram, WhatsApp etc. are majorly used to share the messages and express their opinions. According to Statista* website information, the number of social media users increased from 10 million to 260 million. These social media platforms give more freedom to users to express their ideas and feelings in different forms like text message, sharing image, audio, and graphics. Some of the people using these social media networks for anti-social activities like online trolling, cyberbullying etc.

In recent years, cyberbullying is a big challenging task in social media era. Cyberbullying is a kind of online hate speech by intentionally posting hate kind of text messages, images, audios and videos[1]. The effect of cyberbullying is more dangerous than traditional bullying because, it reaches very large number of people in short time. The major victims of cyberbullying activities are school children and adolescents. The victims are psychologically affected, and sometimes lead to suicide attempts[2]. To save the bullying victims from these activities, some researchers developed machine learning algorithms that can predict cyberbullying text messages which are having humiliating content. In natural language processing tasks, the machine learning algorithms depends on word embeddings. Each text message must convert into numeric form by different word embedding techniques.

In this paper, we study the performance of word embedding techniques on cyberbullying detection on two classifier algorithms called Logistic regression and LightGBM. The major contributions of the paper as follows

- We conducted experiments on most advanced word embedding techniques in natural language processing research area.
- Compared the performance of word embeddings in three different sampling modes.
- Demonstrated the mis-classification rate of word embedding techniques.

2. Related work

In this section, we discussed about previous works on cyberbullying detection using various word embedding techniques.

M. Karan and J. Snajder [3] proposed a pre-emptive detection system for toxic language. They used support vector machine as classifier and tf-idf used for word embeddings. Xu et al.[4] used bag of words word embeddings and studied different roles of participants in cyberbullying activities such as victims, bullies, accuser. They achieved f1-score of 0.77 using support vector machine classifier. Nahar et al.[5] implemented graph model to estimates victim scores on Twitter dataset. They used tf-idf to extract features from the text. Pang et al.[6] employed three machine learning algorithms SVM, Navie Bayes, Maximum Entropy for sentiment classification on IMDb dataset to classify the movie review is positive or negative. They used Bag of words word embeddings technique. Zhao and Mao[7] proposed a method called Semantic-enhanced Marginalized Stacked Denoising Autoencoder (smSDA) for cyberbullying detection. This smSDA capable of reducing noise and able to identify best features from Bag of Words representation in effective and efficient way. Yen et.al [8] employed a supervised learning approach to identify online harassment in web 2.0. They combine different features like contextual features, sentiment features and content features of documents and trained on support vector machine classifier. The proposed technique outperforms the traditional tf-idf. Vikas et al.[9] included two additional features called pronoun capturing and skip-grams to the features extracted by tf-idf and n-gram. This method increases the accuracy of the detection system up to 86% using logistic regression classifier. Kasture [10] considered Linguistic Inquiry and Word count(LIWC) for feature extraction. They train with different machine leaning classifiers and found that random forest is the best among the twitter dataset on recall of 96.3% and precision of 98.4%. Ghasem et al[11], chosen the top most 500 informative words as the feature vector and train on support vector machine(SVM) and achieved f1-score of 95%. Jiale Wu et.al[12], PROPOSED ITFT (improved tf-idf based fastText) model for bullying detection by considering keywords, position weight to traditional tf-idf word embedding method. The proposed model able to filter the noise and improved the accuracy. B Sri Nandini et al.,[13] proposed a model using Fuzzy logic and genetic algorithms to classify the bullying actions such as Terrorism, Racism, Harassment. They used to genetic algorithm for optimization of features and fuzzy rule set to retrieve the data for classification task. Charkrabarthy N[14] used Bag of Words and tf-idf word embedding techniques for toxic comments classification from Wikipedia dataset. Salminen et al.[15] proposed online hate content detection approach for various social media networks. The authors collected 197,566 comments from YouTube, Reddit, Wikipedia and Twitter.

3. Materials and Methods

In this section we discuss about data collection and proposed methodology for cyberbullying detection and evaluation metrics. Fig 1 shows the key components in the proposed methodology. We collected the dataset from the source Waseem and Hovy[16]. The Waseem and Hovy collected tweets from popular social networking site called Twitter over a period of two months with the help of Twitter API. They retrieved 136,052 tweets totally and assigned the labelling for 13,762 tweets. The 9,533 tweets neither sexist nor racist labelled as 0, and 4,229 tweets labelled as 1 which contains either sexism or racism. We assumed labelled '0' tweets are non-bullying comments and labelled '1' tweets are bullying comments. Table 1 shows the example labelling of bullying and non- bullying tweets.

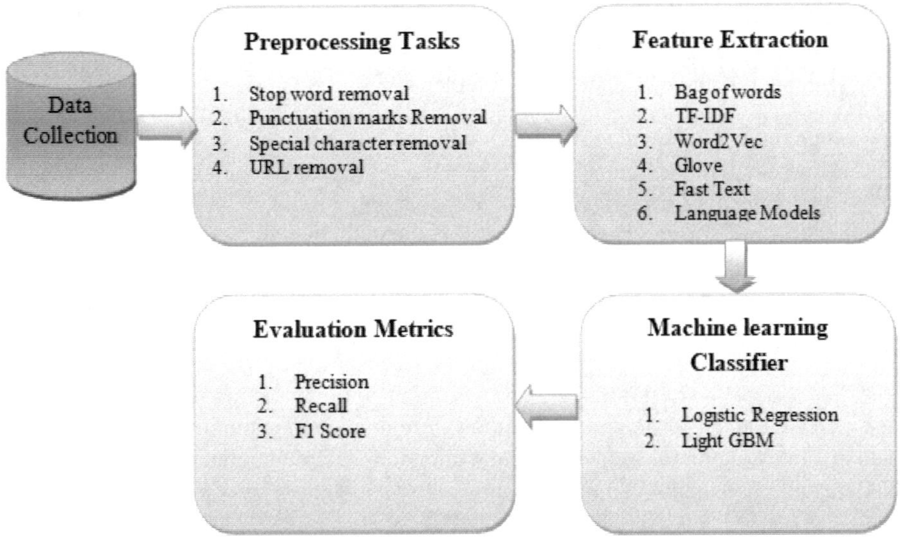

Figure 1. Flow chart of the proposed work.

3.1 *Data collection*:

Table 1. Labelling of tweets. Label 0 represents non bullying and label 1 is for bullying.

Example Tweet	Label
user if we're lucky. <hashtag>mkr	0
i think i just threw up in my mouth a little <hashtag>mkr	0
trapped in a pole-dancing venue with that hotch potch and sausage impersonating food <hashtag>mkr	0
rt user **violence** censorship & **intolerance** of dissent. it's why we call **feminism a hate movement**. User	1
self described **"radical feminist"** makes death threats against <hashtag>gamergate woc for not doing/thinking as she was told.	1
rt user it's absolutely crystal clear now that with the amount of sjw **feminists** on campuses that guys must tape every **sexual** enemoticon	1

Table 2. Example of Noisy Data.

Example Tweets	**Noisy Data**
â€œAnd so weâ€™ll have whittled **18** down to **14** in just a few short monthsâ€	Numbers
#MKR that was useless from nikki	Hash Tags
#MKR2015#MKR**http://t.co/mD9TyU1MOn**	URL links
Twitter makes me laugh....Better than the series 😊 ☹	Emojis
4 rounds of instant restaurants???? **Craaaazy**	Spelling corrections

The input raw data is first pre-processed to avoid inconsistencies and then different feature extraction techniques are used to extract the robust features. Then the classification methods are used to compare the effectiveness of each feature in terms of evaluation metrics.

3.2 *Data pre-processing*
The pre-processing is very important step for any natural language processing (NLP) tasks. The data collected from various social media consists of noise data such as stop words, URL, unwanted numbers. Table 2 shows the example of noise in social media tweets. To remove noise data from the tweet, we applied various natural language pre-processing steps such as tokenization, lemmatization [17] to get the original root words and stemming [18] to reduce inflections words to root form.

3.3 *Feature Extraction*
For any natural language processing tasks, feature extraction methods play important role. Machine learning classifier accepts only numerical representation as input data to get trained. So every text information needs to convert into numeral form by applying various feature extraction techniques. In this proposed methodology, we used different feature extraction methods such as Bag of Words (BoW), TF-IDF, word2vec, glove, FastText and various language models such as ALBERT [19], ELCTRA [20], GPT-2[21], XL-NET[22] and RoBERTa[23].

3.3.1 *Feature Extraction Techniques*
In this section, we give outline of the feature extraction techniques that are used for comparison in our work.

ALBERT:
ALEBERT is a language model stands for A LITE BERT. It is self-supervised learning language model of language representations. ALBERT incorporates two-parameter reduction techniques which are factorized embedding parameterization and cross-layer parameter sharing in order for lifting the major obstacles in scaling pre-trained models in NLP. In this model, number of parameters are reduced 89% compared to BERT [24] model and it achieves respectable performance on benchmark datasets. The major advantage of ALBERT is parameter efficiency, because it can fit much larger batches into memory for quick inference.

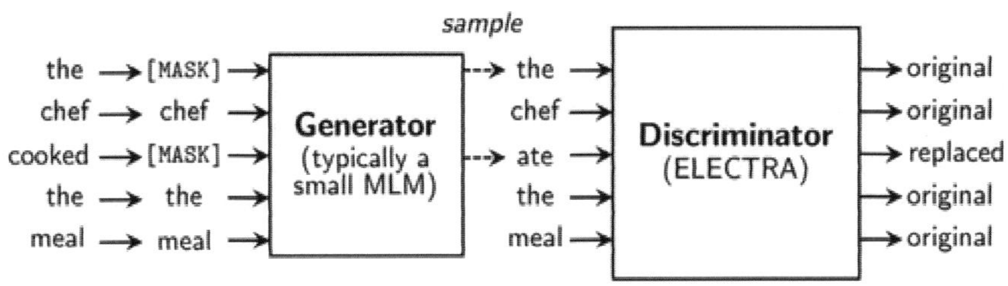

Figure 2.: Replaced Token Detection

ELECTRA
It is novel pre-training method that uses a technique is called Replaced Token Detection (RTD). Fig 2 shows the Replaced Token Detection system in ELECTRA. The model returns the efficient performance as same as masked language model such as BERT. It overcomes the drawback of masking approach in BERT models.

GPT-2
GPT-2 is a pre-trained language model which is useful for various natural language processing tasks such as text generation, question and answer system and language translation. It is a transformed-based model trained on a dataset eight million web documents. It uses huge vocabulary size in training. It can predict the next occurring word in a sentence. The GPT-2 architecture based on Transformers concept which provides mechanism of encoder-decoders to identify input and output dependencies. It was evaluated through zero-shot learning.

XL-NET

It is a general auto regressive pretraining method that enables learning bidirectional contexts by maximizing the expected likelihood over all permutations of the factorization order. It integrates two methods called autoregressive and auto encoding. In a given text sequence k = (k_1,…,k_T), autoregressive models use probability of $p(k) = \prod_{t=1}^{T} p(k_t \mid k_{<t})$ for forward prediction of the word and use probability of $p(k) = \prod_{t=T}^{1} p(k_t \mid k_{>t})$ backward prediction of the word. XL-Net outperforms BERT on 20 tasks, including question answering, natural language inference, sentiment analysis, and document ranking.

RoBERTa

RoBERTa is an enhanced version of BERT model by improved training methodology. Dynamic masking pattern is introduced whereas BERT follows static masking pattern. It removes Next Sentence Prediction(NSP) task from BERT's pre-training. It uses 160GB of text for pre-training with various data such as BookCorpus, English Wikipedia, OpenWebText whereas BERT trained against BookCorpus only. This model outperforms both XL-NET and BERT on GLUE benchmark results.

3.4 *Classifiers*

Classification is supervised learning approach in which the model predicts discrete output variable based on input variables. Machine learning classifier plays vital role for after generating word embeddings. These classifiers are used in many applications such as speech recognition, handwriting recognition, face detection, document classification etc.... A bunch of machine learning classifiers such as Support Vector Machine (SVM), Navie Bayes(NB), Radom Forest(RF), K-nearest neighbor(kNN) used for these tasks. We used simple baseline linear classifier Logistic Regression (LR)[25] and non-linear classifier called LightGBM[26] in the experiments for classification task.

3.4.1 Logistic Regression

It is a supervised machine learning algorithm and intended for binary classification tasks. It uses logistic function called sigmoid function which is in the range of 0 and 1. Eq.1 is sigmoid equation for logistic regression. The sigmoid function generates probability scores based on input features. For example, the probability score is ≥ 0.5, it belongs to bullying tweet; otherwise, the prediction will be for the other class (non-bullying).

$$\text{Sigmoid}(z) = \frac{1}{1+\exp(-z)} \quad (1)$$

3.4.2 Light GBM

LightGBM(Light Gradient Boosting Machine) improved many demerits of Decision Tree and Gradient Boosting Decision Tree algorithms and hence is very useful in bringing great insights from Cyberbullying Data. The LightGBM classifier follows leaf-wise tree growth which is big advantage for this classifier. Earlier boosting algorithms unable to handle large datasets while LightGBM able to handle a huge amount of data with many advantages such as faster training speed, good accuracy, parallel learning, and less memory usage. Fig 3 shows the leaf-wise growth in LightGBM classifier. Each node represent a decision of classify the tasks.

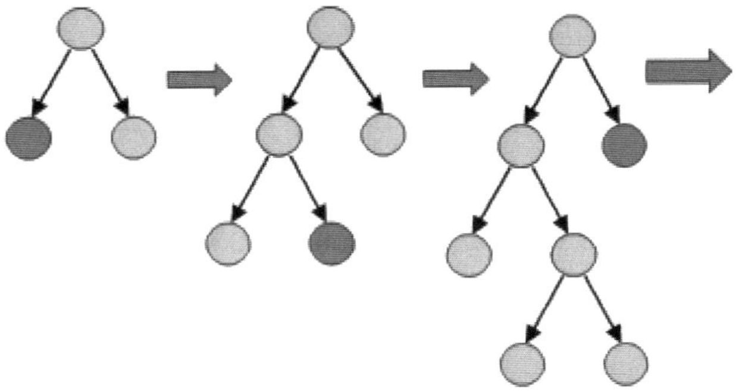

Figure 3.: Leafwise growth of LightGBM classifier.

3.5 Evaluation Metrics
We considered precision, recall and f1-score as evaluation metrics.
Precision: The precision can be defined as the ratio of correctly predicted as bullying tweets to total number of predictions as bullying tweets. Eq.2 shows precision formula.

$$P_r = \frac{C}{C+B} \quad (2)$$

Where C indicates correctly predicted as bullying tweets, B indicates non bullying tweets predicted as bullying tweets.
Recall : The recall can be defined as the ratio of correctly predicted as bullying tweets to total number of actual bullying tweets. Equation-3 shows recall formula. Here, NB indicates non bullying tweets.

$$R_c = \frac{C}{C+NB} \quad (3)$$

F1-score **:** f1-score is the weighted average of recall and precision. Equation-4 gives f1-score formula.

$$\text{f1-score} = \frac{2*Pr*Rc}{(Pr+Rc)} \quad (4)$$

4. Results and Discussion
The proposed approach is implemented in Python programming language with help of several packages called Pandas, Numpy, matplotlib, Scikit-Learn, LightGBM, Tensorflow in Linux operating system. We ran our codes on Intel i7 8th Gen 12core CPU processor and Nvidia Max-Q 1070 8GB RAM.

We conducted several experiments to check the performance of various Word Embedding methods for cyberbullying detection data. We performed 5-fold validation of the dataset. We consider macro average of precision, recall, f1-scores of class-0 and class-1. Fig 4 shows the f1-scores of Waseem dataset with No sampling mode. We observed that most of the language models scored high f1-scores as compared to traditional word embedding such as BoW and T-IDF and semantic word embeddings such as word2vec, glove , FastText. Over sampling and under sampling are two techniques used for imbalanced datasets to avoid class bias problems. We conducted experiments in Over sampling and Under sampling. Table 3 shows precision, recall and f1-score of Waseem dataset with over sampling and under sampling modes. We noticed that the performance of language models is better than traditional word embedding techniques in terms of precision, recall and F1-score.

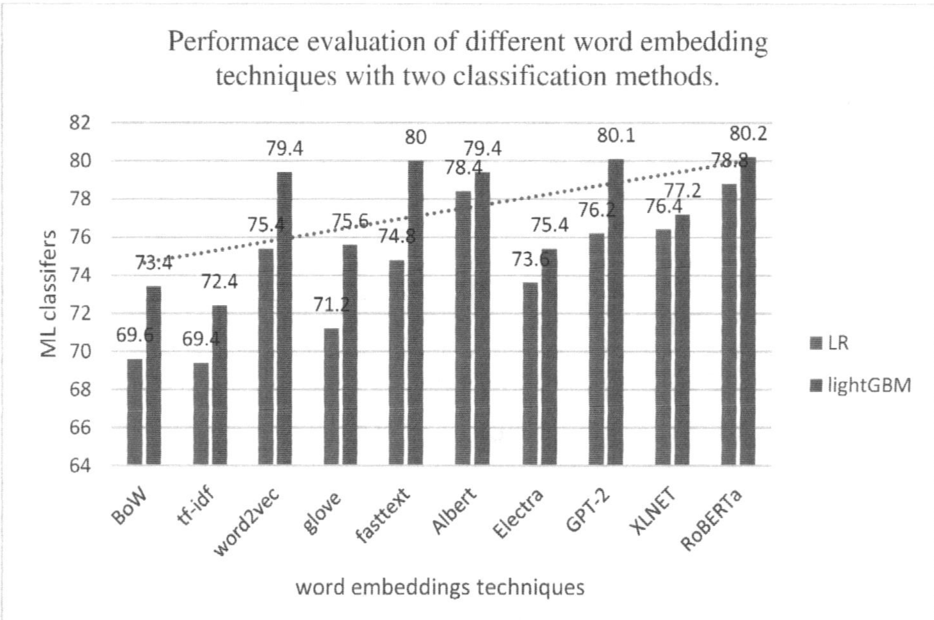

Figure 4. f1 score on Waseem dataset with No sampling.

Table 3. Precision, Recall and f1-scores of under sampling and over sampling of various word embeddings.

Word Embeddings + classifier	Under Sampling			Over Sampling		
	Precision	Recall	F1-score	Precision	Recall	F1-score
BoW+ LR	71.2	72.2	71.4	71.2	72.4	71.4
BoW+ LightGBM	72.0	74.4	72.8	73.8	74.2	74.0
TFIDF+LR	70.0	71.8	70.8	70.2	71.8	70.8
TFIDF+LIGHTGBM	70.6	73.0	71.0	73.4	73.4	73.4
WORD2VEC+LR	75.2	77.8	75.8	75.6	78.0	76.6
WORD2VEC+lightGBM	77.6	80.0	78.6	81.0	78.4	79.4
GLOVE+ LR	71.8	74.2	72.6	72.2	74.2	72.6
GLOVE+ light GBM	73.8	76.8	74.8	78.6	76.0	77.0
FastText + LR	75.6	77.2	76.2	76.0	78.2	76.2
FastText + light GBM	78.0	80.4	78.6	80.2	79.2	80.2
ALBERT+LR	77.0	79.4	77.4	76.4	79.0	77.2
ALBERT+ light GBM	78.0	80.6	78.6	81.4	78.2	79.8
ELECTRA+LR	71.8	74.6	72.0	71.8	74.6	72.6
ELECTRA+ light GBM	73.2	75.6	73.8	77.4	73.8	75.2
GPT-2 + LR	74.4	77.0	75.2	74.2	76.8	75.0
GPT-2 + light GBM	78.2	80.6	**79.2**	83.2	79.4	**81.2**
XL NET + LR	73.6	76.4	74.6	74.4	76.6	75.4
XL NET + light GBM	75.2	77.6	76.0	80.4	75.4	77.4
RoBERTa + LR	77.2	79.2	78.2	77.8	80.0	78.8
RoBERTa + light GBM	78.0	80.8	**79.0**	82.8	79.4	**80.6**

Error Analysis:

Mis-classification rate(MCR) is the ratio of false positive(FP) and false negatives(FN) to the total number of records.Eq.5 shows mis-classification rate. In this experiment we applied five fold validation. The total number of tweets in each fold is 2753. Table 4 shows the total number of misclassifications i.e both false positives and false negatives in each fold in terms of percentage. The least number of misclassifications found in language models such as RoBERTa, Open-AI, GPT-2. We noticed that LightGBM classifies better than logistic regression classifier in all the word embedding methods. Fig 5 show the misclassification rate of word embedding methods in terms of percentage. The basic word embeddings rate is 21% which is reduced to 16% with language models.

$$\text{MCR} = \frac{FP+FN}{Total\ number\ of\ Records} \tag{5}$$

Table 4. Mis-classification rate in each fold in percentage.

Word Embedding Techniques + Classifier	Fold-1(%)	Fold-2(%)	Fold-3(%)	Fold-4(%)	Fold-5(%)
BoW+ LR	22.23	21.76	22.96	22.96	22.12
BoW+ LightGBM	20.81	20.41	22.3	21.32	19.69
TF-IDF+LR	23.03	23.1	21.79	22.34	21.5
TF-IDF+LIGHTGBM	21.32	21.72	20.96	22.12	21.25
WORD2VEC+LR	18.85	19.22	20.09	18.63	20.16
WORD2VEC+lightGBM	15.66	15.76	16.64	16.6	16.78
GLOVE+ LR	23.14	21.61	22.12	22.63	20.09
GLOVE+ light GBM	19.40	19.87	18.38	19.29	18.71
FastText + LR	19.83	20.27	18.82	19.22	19.25
FastText + light GBM	16.16	15.98	15.04	16.6	16.31
ALBERT+LR	18.31	17.18	16.06	16.89	18.96
ALBERT+ light GBM	16.78	16.27	15.51	16.02	17.14
ELECTRA+LR	20.74	20.01	20.52	20.6	21.21
ELECTRA+ light GBM	18.82	19.03	20.12	18.63	19.4
GPT-2 + LR	20.27	17.14	18.63	19.07	19.8
GPT-2 + light GBM	16.13	14.89	16.24	15.66	16.02
XL NET + LR	19.03	18.92	19.72	19.25	20.12
XL NET + light GBM	17.22	17.58	18.09	17.62	17.87
RoBERTa + LR	17.14	16.67	16.82	17.4	17.47
RoBERTa + light GBM	16.27	15.76	15.51	16.6	15.29

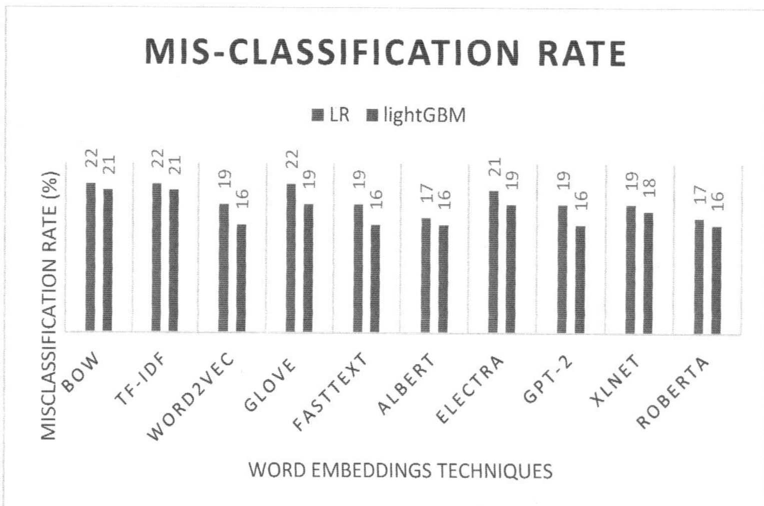

Figure 5.: Mis-classification rate of Word Embeddings.

5. Conclusion and Future works

With the increase of social media platforms, the bullying activities are increasing significantly. This paper proposed a methodology for cyberbullying detection by experimenting various word embeddings techniques with two classifiers. We conducted experiments in three types of sampling modes on the Twitter dataset. 1) no sampling 2) under sampling 3) oversampling. The f1-score of language models in sampling mode almost close to 80 percentage whereas basic word embeddings methods are around 75%. The mis-classification rate of basic word embeddings is 21%, whereas mis-classification rate of language models reduced to 16%. We observed that the drawbacks of basic word embedding techniques can overcome with these language models for many downstream tasks in Natural language processing. As part of the future work for cyberbullying detection system, language models have to be considered to build the robust features.

6. References

[1] Cohen A and Raphael 2018 Social responsibility on the internet: Addressing the challenge of cyberbullying. *Aggression and Violent Behavior.* DOI: 10.1016/j.avb.2018.01.001
[2] Nikolaou and Dimitrios 2017 Does cyberbullying impact youth suicidal behaviors? Journal of health economics 56 30-46
[1] Karan M and Snajder J 2019 Preemptive toxic language detection in Wikipedia comments using thread-level context in Proceedings of the Third Workshop on Abusive Language Online. Florence Italy: Association for Computational Linguistics 129–134
[2] Xu. X Zhu J M and Bellmore A 2012 Fast learning for sentiment analysis on bullying proceedings of the First International Workshop on Issues of Sentiment Discovery and Opinion Mining - WISDOM 12 1–6
[3] Nahar V Li X and Pang C 2014 An Effective Approach for Cyberbullying Detection Communications in Information Science and Management Engineering 3 5 238-247
[4] Pang B Lee L and Vaithyanathan S 2002 sentiment classification using machine learning techniques In: Proceedings of the ACL-02 Conference on Empirical Methods in Natural Language Processing 10 79–86
[5] Zhao R and Mao K 2016 Cyberbullying detection based on semantic-enhanced marginalized denoising auto-encoder IEEE Transactions on Affective Computing 8 328–339
[6] Yin D Xue Z Hong L Davison B D Kontostathis A and Edwards L 2009 Detection of harassment on web 2.0 Proceedings of the Content Analysis in the WEB 2 1–7
[7] Chavan V S and Shylaja S S 2015 Machine learning approach for detection of cyber-aggressive comments by peers on social media network International Conference on In Advances in computing communications and informatics (ICACCI) 2354–2358
[8] Kasture A S2015 A predictive model to detect online cyberbullying Auckland University of Technology
[9] Ghasem Z Frommholz I and Maple C 2015Machine learning solutions for controlling cyberbullying and cyberstalking J Inf Secur Res 6 55-64
[10] Jiale Wu Wen M Lu R Li B and Jinguo Li 2020 Toward efficient and effective bullying detection in online social network Peer-to-Peer networking and Applications Springer 13 1567–1576
[11] Sri Nandhinia B and Sheeba J I 2015 Online Social Network Bullying Detection Using Intelligence Techniques International Conference on Advanced Computing Technologies and Applications (ICACTA-2015), Elsevier 45 485 – 492
[12] Chakrabarty N 2019 A Machine Learning Approach to Comment Toxicity Classification Computational Intelligence in Pattern Recognition Advances in Intelligent Systems and Computing Springer Singapore 999 183-193

[13] Salminen J Hopf M Chowdhury S A Jung S G Almerekhi H and Jansen B J 2020 Developing an online hate classifier for multiple social media platforms Human-centric Computing and Information Sciences 10 15-20
[14] Skorkovská L 2012 Application of Lemmatization and Summarization Methods in Topic Identification Module for Large Scale Language Modeling Data Filtering 7499 Springer
[15] Waseem Z and Hovy D 2016 Hateful symbols or hateful people? Predictive features for hate Speech detection on twitter Proceedings of the NAACL Student Research Workshop Association for Computational Linguistics 88-93
[16] Pande B P and Dhami H S 2011 Application of natural language processing tools in stemming International Journal of Computer Applications 27(6) 14-19
[17] Lan Z Chen M Goodman S Gimpel K Sharma P and Soricut R 2020 ALBERT: A Lite BERT for Self-supervised Learning of Language representations ICLR
[18] Clark K Luong M Le QV and Manning C 2020 ELECTRA: Pre-training Text Encoders as Discriminators Rather Than Generators ICLR
[19] Radford A Wu J Child R Luan D Amodei D and Sutskever I 2019 Language Models are Unsupervised Multitask Learners
[20] Yang Z Dai Z Yang Y Carbonell J Salakhutdinov R Le and Quoc V 2019 XLNet: Generalized Autoregressive Pretraining for Language Understanding NeurIPS
[21] Liu Y Ott M Goyal N Du J Joshi M Chen D Levy O Lewis M Zettlemoyer L and Stoyanov V 2019 RoBERTa: A Robustly Optimized BERT Pretraining Approach ICLR
[22] Devlin J Chang M W Lee K and Toutanova K 2019 BERT: Pre-training of Deep Bidirectional Transformers for Language Understanding Association for Computational Linguistics DOI:10.18653/v1/N19-1423
[23] Park H 2013 An introduction to logistic regression: from basic concepts to interpretation with particular attention to nursing domain Journal of Korean Academy of Nursing
[24] Guolin K Meng Q Finely T Wang T Chen W Ma W Ye Q and Liu T 2017 LightGBM: A Highly Efficient Gradient Boosting Decision Tree Advances in Neural Information Processing Systems 3149–3157

Robust Segmentation and Classification of Moving Objects from Surveillance Video

Chandrajit M[1*], Shobha Rani N[2] and Manohar N[2]

[1] Maharaja Research Foundation, Maharaja Institute of Technology Mysore, India;
[2] Department of Computer Science, Amrita School of Arts & Science, Mysuru Campus, Amrita, Vishwa Vidyapeetham, India.

*Email: chandrajithmmca@mitmysore.in

Abstract. Object segmentation and classification in the video sequence is a classical and critical problem that is constantly addressed due to vast real-time applications such as autonomous vehicles, smart surveillance, etc. Segmentation and classification of moving objects in video sequences captured from real-time non-constraint sequences is a challenging task. This paper presents a moving object segmentation and classification method for video sequences captured in a real-time environment. The key contributions of the paper include a method to detect and segment motion regions by applying the non-parametric Kolmogorov–Smirnov statistical test in the Spatio-temporal domain and a probabilistic neural network-based classification method to classify the moving objects into various classes. Promising results are obtained by experimentation using challenging PETS and Change Detection datasets. To corroborate the efficacy, a comparative analysis with contemporary method is also performed.

1. Introduction

Enormous amount of work in the domain of visual analysis has been reported in the last two decades due to the huge demand in applications such as robotics, video surveillance, driver assist system, video indexing, traffic video monitoring etc. Nevertheless, success of any visual analysis application directly depends on accurate segmentation and classification of moving objects [1, 2]. Surveillance video feed generally suffers from illumination variations, object occlusion, noise, cluttered environment etc. These problems in turn increase the complexity for devising an accurate solution to the problem. Therefore, research in this direction focus to contribute substantial evidence based solutions.

In this paper emphasis is given to accurately extract moving objects from surveillance video sequence and further classify the objects into predefined classes.

The article is organized as follows. Section 2 reports the related work. Section 3 and 4 elaborates the proposed motion segmentation and object classification method respectively, Section 5 presents the experimental and comparative results and Section 6 presents the conclusion.

2. Related Work

Background subtraction, optical flow and temporal differencing are the major category of techniques for motion segmentation. Background subtraction technique constructs the background model for motion segmentation. Optical flow technique computes the direction vectors for every pixel using information from successive frames to achieve motion segmentation. Temporal differencing based technique achieves motion segmentation on the basis of pixel difference between frames using a threshold. Statistical tools are also used in these techniques for correlating the pixel values from consecutive frames.

Stuaffer and Grimson [3] proposed GMM based background subtraction method for motion segmentation. Chen et al. [4] combined pixel-level and block-level features to construct background model for motion segmentation. Armanfard et al. [5] proposed texture and edge feature based background subtraction for pedestrian segmentation. Elgammal et al. [6] proposed kernel density estimation based background modeling for moving object segmentation. Sadaf et al. [7] proposed a temporal domain based background subtraction method. Subudhi et al. [8] proposed Wronskian framework based background construction technique. Sergi et. al. [9] proposed convolution network based semi-supervised method for motion segmentation. Variation of convolution network based motion segmentation work is reported in [10-12].

Denman et al. [13] proposed a optical flow based method for motion segmentation. Variation of optical flow based technique for motion segmentation can be found in the work reported by [14, 15].

Murali and Girisha [16] proposed temporal differencing based method for motion segmentation. Variation of temporal differencing based techniques can be found in the work reported by [17-19]. Cheng and Chen [20] proposed achieved temporal based motion segmentation in frequency transformed domain. Liu et al. [21] proposed information theory based temporal differencing motion segmentation method.

The methods reported can be classified as online or learning based methods. Online based methods use geometric properties of objects for classification. Whereas the learning based method uses a trained model for classification. Learning based method is more suitable due to increased accuracy. Javed and Shah [22] proposed a method to classify objects into three classes by using temporal motion features. Liang and Juang [23] proposed a method to classify the objects into three classes using histogram feature with SVM classifier. Zang and Klette [24] proposed a method for two class classification using geometric features. Asaidi et al. [25] proposed a method for vehicle classification using shape features. Zhang et al. [26] proposed a three class classification using shape and motion features. Zhang et al. [27] proposed adaptive boosting based method for four class classification using block based local binary pattern. Vijaya and Venkata [28] proposed war filed object classification using SIFT features. Mahalingam and Subramoniam [29] proposed texture based decision tree classification method for object classification.

In summary, a great amount of work has been dedicated for motion segmentation using background subtraction. Generally, the distribution of data is assumed to be normal distribution in background subtraction based methods. Nevertheless, in real-time environment, this assumption need not be valid [6]. Further, the background subtraction based methods cannot handle dynamic and cluttered environment. Moreover, the background subtraction technique cannot be applied for moving and Pan Tilt and Zoom (PTZ) cameras. Optical flow technique based methods accurately segments moving objects. However, these methods are not suitable for illumination variation and cluttered environment sequences. Temporal differencing based methods can extract motion objects satisfactorily in the case of dynamic and cluttered environment. However, these methods fail to accurately extract the objects.

Further, the methods reported on moving objects classification are limited in numbers. Most of them classify objects into few categories. However, in real-time scenario the scene will be composed of many kinds of objects. Therefore, effort is made in this work for multi-class classification.

3. The Proposed Motion Segmentation Method

Figure 1 presents the schematic representation of proposed method devised for moving object segmentation from video sequence. The input frames I_t of video are smoothened by 2D 3 x 3 Gaussian filter [30] to reduce the noise caused by sensors and illumination variations. Consequently, the smoothened frame F_t is obtained, where, t is the temporal frame. Subsequently, RGB mean value for each pixel $P(w,h)$ is considered as pixel value.

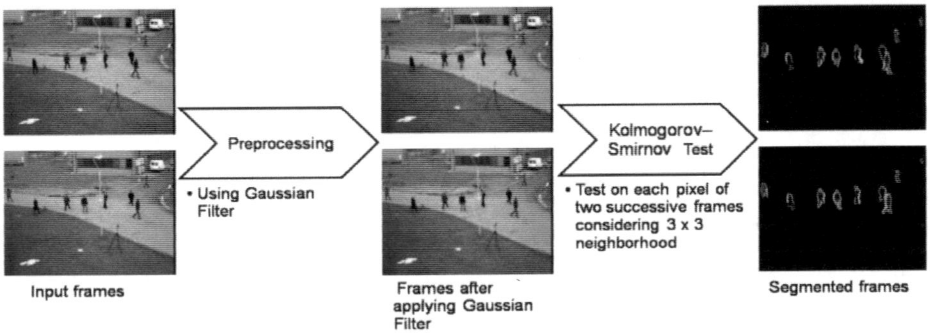

Figure 1. Schematic representation of proposed method for segmentation

The detection of motion pixels can be determined by capturing the variations among pixel values from successive frames. However, capturing the variations using single pixel value from successive frame will not lead to accurate motion segmentation. Therefore, for each pixel its 3 x 3 neighbors are considered in this work.

The displacement of object in frames of the video sequence in the case of fast moving object will be very high. Therefore, successive temporal frames are used for motion segmentation.

Kolmogorov–Smirnov (KS) test [31] is a popular nonparametric test in statistics used to compare a sample with a reference probability distribution, or it can also be used to compare two samples. The variation in pixel values from successive temporal frames for moving objects segmentation is determined by employing the KS test. In the case of dynamic environment, pixel data obtained from video frame cannot be assumed to follow any specific distribution. Therefore, this test has been chosen for moving object segmentation in dynamic environment.

The KS test requires sample values for establishing the variation. The neighborhood pixel values in smoothened frames F_t and F_{t+1} which is denoted as $F_{t_{P_{N9(RGB)}}}$ and $F_{t+1_{P_{N9(RGB)}}}$ respectively are used as sample values.

The KS test is applied on the pixel values in smoothened frames $F_t$and F_{t+1} using Eq.(1) for the segmentation of moving objects.

$$D_{F_t,F_{t+1}} = \underbrace{sup}_{x} \left| F_{t_{P_{N9(RGB)}}} - F_{t+1_{P_{N9(RGB)}}} \right| \quad (1)$$

where, D is KS test statistic for each pixel and sup is supremum function.

The variation among pixel values is captured by comparing the test statistic $D_{F_t,F_{t+1}}$ with threshold value τ chosen from table for $\alpha = 0.5$. The null hypothesis is rejected at level α based on Eq.(2).

$$D_{F_t,F_{t+1}} > c(\alpha)\sqrt{\frac{n+m}{nm}} \tag{2}$$

where, (n, m) are size of two samples and

$$c(\alpha) = \sqrt{-\frac{1}{2} ln \frac{\alpha}{2}} \tag{3}$$

Consequently, the motion pixels are detected using KS test. The output frames O_t and O_{t+1} that contain extracted foreground objects from input frames I_t and I_{t+1} respectively are generated using Eq.(4).

$$O_{m_{P(w,h)}} = \begin{cases} RGB\ of\ I_{m_{P(w,h)}} & if\ (D_{F_t,F_{t+1}} \geq \tau) \\ 0 & (otherwise) \end{cases} \tag{4}$$

where, m is the temporal frame and $I_{m_{P(w,h)}}$ is input frame pixel.

4. The Proposed Classification Method

Schematic representation of proposed method to classify objects is shown in figure 2. Objects are classified into predefined classes viz., single person, group of people, two wheeler, small vehicle or big vehicle using PNN classifier [32].

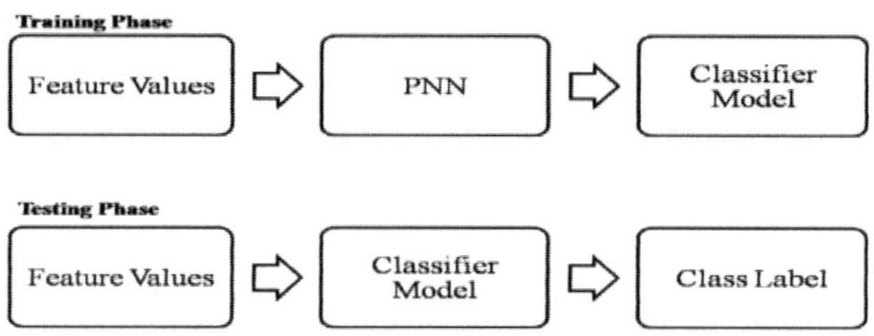

Figure 2. Schematic representation of proposed method for classification

PNN is feed forward neural network based on Bayesian framework with Kernel Fisher discriminant analysis. In this work, PNN consists of four layered network which estimate PDF for each of the classes. The first layer is the input feature values. The second layer consists of the Gaussian functions estimated by given set of feature values. Subsequently, averaging of values is performed in third layer for each class. Finally, class label is determined for maximum value among the distribution value output in the fourth layer.

In general, a PNN for M classes is defined as;

$$y_j(X) = \frac{1}{n_j}\sum_{i=1}^{n_j} \exp(-\frac{(\|x_{j,i}-x\|)^2}{2\sigma^2}) \tag{5}$$

$j = 1, 2, \dots, M$

n_j is the number of features.

The PNN classifier assigns X to class j having maximum $y_j(X)$.

Object representation using features plays a pivot role in classification. In this work, object is represented by the following features; Seven invariant Hu moments features [33], Gabor Features, geometric features such as, centroid, major axis, minor axis, orientation.

5. Experiments

Experiments on the challenging benchmark IEEE PETS and IEEE CD [34] and DARPA Neovision datasets have been conducted to evaluate the proposed method. The dataset contains outdoor and indoor sequences captured from normal, thermal, night vision and PTZ sensors. The result of motion segmentation is shown in figure 3 and figure 4. Similarly, the result of classification is shown in figure 5 and figure 6.

Quantitative evaluation has been performed using metrics viz. precision, recall and F-measure [34]. Further, sample groundtruth frames from the same dataset have been used for evaluation. Table 1 presents the obtained results.

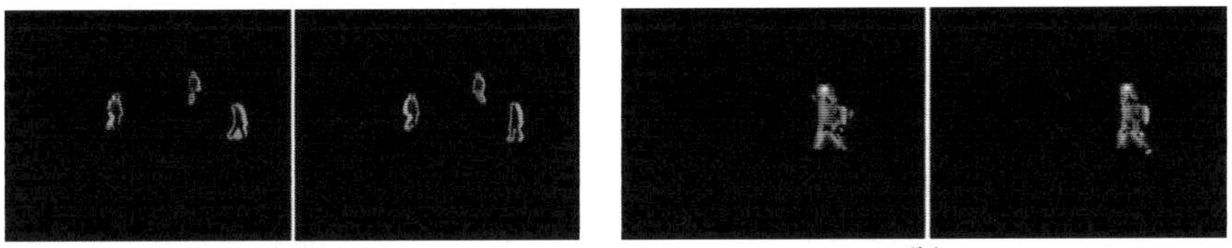

(a) (b)

Figure 3. Motion Segmentation result for (a) PETS 2013 People Tracking (b) Thermal

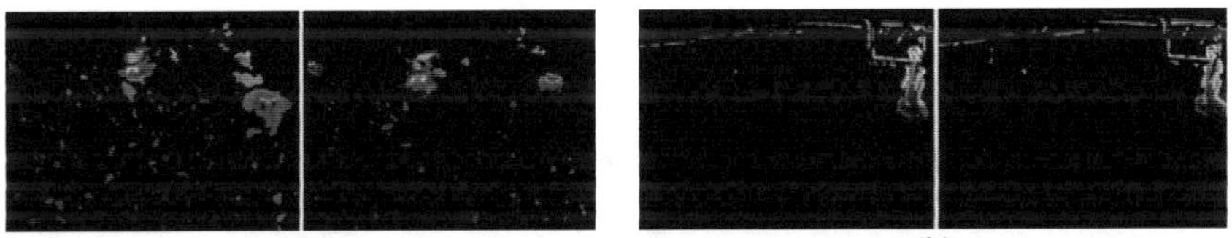

(a) (b)

Figure 4. Motion Segmentation result for (a) Night vision Fluid Highway (b) PTZ frames

(a) (b)

Figure 5. Results of Classification for (a) PETS 2001 (b) PETS 2013

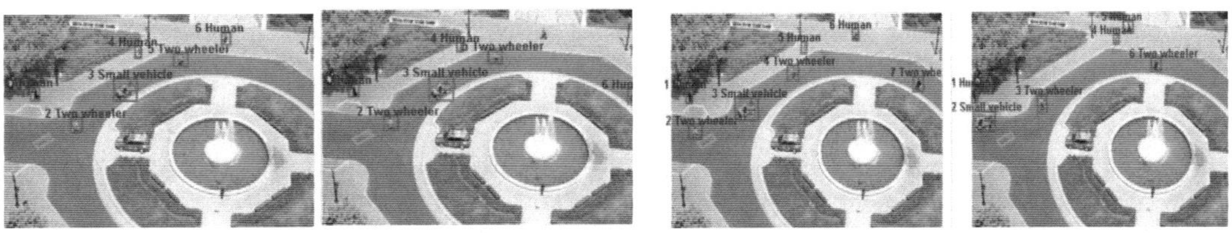

Figure 6. Results of classification for DARPA Neovision2

Evaluation results in table 1 indicate the accurate extraction of foreground object blobs from sequences captured from variety of sensors. Further, the results presented are raw and no post processing has been incorporated for enhancement.

Comparative results from the SOBS [35], Chi-Square test based [36] and proposed methods are presented. Table 2 presents the comparative performance metrics. It indicates the significant performance of proposed KS test based segmentation. This corroborates the suitability of proposed method for video sequences captured using various kinds of sensors.

A quantitative result of classification is presented in table 3 which clearly demonstrates the promising results obtained by the proposed method for real-time challenging surveillance video feed. The combination of geometric, shape and texture features has proven in retaining intra and inter class variations. An F-measure of 96% and 71% is the maximum and minimum score obtained.

Table 1. Performance metrics values obtained by the proposed KS test based motion segmentation

Dataset Sequence	Precision	Recall	F-measure
PETS 2013	0.79	0.75	0.77
PETS 2006	0.86	0.72	0.78
CD night vision	0.54	0.63	0.58
CD PTZ	0.63	0.6	0.61
CD thermal sensor	0.81	0.69	0.75
Average	0.73	0.68	0.70

Table 2. Comparative performance metrics values

Dataset Sequence	Precision			Recall			F-measure		
	Proposed	SOBS	Chi-Square Test	Proposed	SOBS	Chi-Square Test	Proposed	SOBS	Chi-Square Test
PETS 2013	0.79	NA	0.76	0.75	NA	0.71	**0.77**	NA	0.73
PETS 2006	0.86	0.71	0.83	0.72	0.74	0.68	**0.78**	0.72	0.75
CD night vision	0.54	0.48	0.51	0.63	0.75	0.60	**0.58**	**0.58**	0.55
CD PTZ	0.63	0.02	0.59	0.6	0.70	0.57	**0.61**	0.05	0.58
CD thermal	0.81	0.96	0.78	0.69	0.43	0.65	**0.75**	0.60	0.71

6. Conclusion

In this paper, a method is proposed for moving object segmentation and classification from surveillance videos. Moving object segmentation is based on statistical modeling combined with temporal domain. Suitability of the non-parametric statistical hypothesis testing tool for moving object segmentation has been explored in this work. Further, object classification is achieved by ensemble of geometric, shape and texture features with PNN classifier. A five class classification has been attempted in this work. Experimentation has been carried out by using the challenging benchmark IEEE PETS, IEEE CD and DARPA Neovision2 datasets which contain sequences captured from thermal, PTZ, aerial and night vision sensors. Qualitative results indicate the robust nature of proposed method in the case of cluttered and illumination variations environment. Further, comparative analysis with related works has also been presented for showcasing the stability of proposed method. In future, application of deep multi-layered networks shall be investigated for segmentation and classification.

Table 3. Quantitative results of proposed classification method

Dataset Sequence	Training Samples	Precision	Recall	F-measure
PETS 2013 S2.L1 View1	70%	0.91	0.9	0.90
	50%	0.87	0.9	0.88
	30%	0.9	0.89	0.89
PETS 2013 S2.L1 View7	70%	0.9	0.73	0.81
	50%	0.91	0.86	0.88
	30%	0.85	0.85	0.85
PETS 2001 Camera 1	70%	0.94	0.88	0.91
	50%	0.95	0.87	0.91
	30%	0.91	0.84	0.87
PETS 2001 Camera 2	70%	0.76	0.75	0.75
	50%	0.75	0.74	0.74
	30%	0.72	0.71	0.71
Neovision2 002	70%	0.76	0.89	0.82
	50%	0.75	0.87	0.81
	30%	0.7	0.94	0.8
Neovision2 012	70%	0.99	0.94	0.96
	50%	0.98	0.9	0.94
	30%	0.95	0.9	0.92

References

[1] Zappella, Luca, Xavier Lladó, and Joaquim Salvi, *Motion segmentation: a review*, in*Proc. Int.* conference on Artificial Intelligence Research and Development: Proceedings of the 11th International Conference of the Catalan Association for Artificial Intelligence, 2008, IOS Press.

[2] Olatunji I.E., Cheng CH, *Video Analytics for Visual Surveillance and Applications: An Overview and Survey*. In: Tsihrintzis G., Virvou M., Sakkopoulos E., Jain L. (eds) Machine Learning Paradigms. Learning and Analytics in Intelligent Systems, (2019), vol 1. Springer.

[3] Stauffer C and Grimson, *Adaptive background mixture models for real-time tracking*,

in*Proc.* of the IEEE Computer Society Conference on Computer Vision and Pattern Recognition (CVPR), 1999, vol. 2.

[4] Chen Y., Chen C., Huang C and Hung Y, *Efficient hierarchical method for background subtraction*. Pattern Recognition, 2007, vol. 40, no. 10, pp. 2706-2715.

[5] Armanfard N., Komeili M and Kabir E, Ted: A texture-edge descriptor for pedestrian detection in video sequences. Pattern Recognition, 2012, vol. 45, no. 3, pp. 983-992.

[6] Elgammal A., Duraiswami R., Harwood D and Davis L, Background and foreground modeling using nonparametric kernel density estimation for visual surveillance, inProc. of the IEEE, 2002, vol. 90, no. 7, pp. 1151-1163.

[7] Sadaf A., Ali J and Irfan , Human identification on the basis of gaits using time efficient feature extraction and temporal median background subtraction. International Journal of Image, Graphics and Signal Processing, 2014, vol. 6, no. 3, pp. 35-42.

[8] Subudhi B., Ghosh S and Ghosh, Change detection for moving object segmentation with robust background construction under Wronskian framework. Machine Vision and Applications, 2013, vol. 24, no. 4, pp. 795-809.

[9] Sergi Caelles, Kevis-Kokitsi Maninis, Jordi Pont-Tuset, Laura Leal-Taixe, Daniel Cremers, Luc Van Gool; *One-Shot Video Object Segmentation*, *inProc.* IEEE Conference on Computer Vision and Pattern Recognition (CVPR), 2017, pp. 221-230.

[10] Pavel Tokmakov, Karteek Alahari, Cordelia Schmid; *Learning Video Object Segmentation With Visual Memory*, in*Proc.* IEEE International Conference on Computer Vision (ICCV), 2017, pp. 4481-4490.

[11] Jeong, J.; Yoon, T.S.; Park, J.B.: *'MOSnet: moving object segmentation with convolutional networks'*, Electronics Letters, IET Digital Library , 2018, 54, (3), p. 136-138.

[12] Vecchio, S. Palazzo, D. Giordano, F. Rundo and C. Spampinato, *MASK-RL: Multiagent Video Object Segmentation Framework through Reinforcement Learning*, in*Proc.* IEEE Transactions on Neural Networks and Learning Systems. *2020.*

[13] Denman S., Fookes C and Sridharan S, *Improved simultaneous computation of motion detection and optical flow for object tracking*. In Proceedings of the International Conference on Digital Image Computing: Techniques and Applications (DICTA), 2009, pp. 175-182.

[14] Verri A., Uras S and De Micheli E, *Motion segmentation from optical flow*, in*Proc.* of Alvey Vision Conference, 1989, pp. 1-6.

[15] Klappstein J., Vaudrey T., Rabe C., Wedel A and Klette R, *Moving object segmentation using optical flow and depth information*, in*Proc.* Pacific-Rim Symposium on Image and Video Technology, 2009, pp. 611-623.

[16] Murali S and Girisha R, Segmentation of motion objects from surveillance video sequences using temporal differencing combined with multiple correlation, inProc. Sixth IEEE International Conference on Advanced Video and Signal Based Surveillance (AVSS), 2009, pp. 472-477.

[17] Lim T., Han B and Han J. H, *Modeling and segmentation of floating foreground and background in videos*. Pattern Recognition, 2012, vol. 45, no. 4, pp. 1696- 1706.

[18] Lee ChaoYang C. L., ShouJen Lin and Yang C, *An efficient continuous tracking system in real-time surveillance application*. Journal of Network and Computer Applications, Special Issue on Trusted Computing and Communications, 2012, vol. 35, no. 3, pp. 1067-1073.

[19] Ahn J., Choi C., Kwak S., Kim K and Byun H, *Human tracking and silhouette extraction for human robot interaction systems*. Pattern Analysis and Applications, 2009, vol. 12, no. 2, pp.167-177.

[20] Cheng F and Chen Y, Real time multiple objects tracking and identification based on

discrete wavelet transform. Pattern Recognition, 2006, vol. 39, no. 6, pp. 1126-1139.

[21] Liu C., Yuen P. C and Qiu G, *Object motion detection using information theoretic spatio temporal saliency*. Pattern Recognition, 2009, vol. 42, no. 11, pp. 2897-2906.

[22] Javed O and Shah M, *Tracking and object classification for automated surveillance*, in*Proc.* European Conference on Computer Vision, 2002, pp. 343-357.

[23] Liang Chung-Wei and Juang Chia-Feng, Moving object classification using local shape and HOG features in wavelet-transformed space with hierarchical SVM classifiers. Applied Soft Computing, 2015, vol. 28, pp. 483-497.

[24] Zang Q and Klette R, *Object classification and tracking in video surveillance*, in*Proc.* Tenth International Conference on Computer Analysis of Image and Patterns (CAIP), 2003, vol. 128, pp. 198-205.

[25] Asaidi H., Aarab A and Bellouki M, *Shadow elimination and vehicles classification approaches in traffic video surveillance context*. Journal of Visual Languages & Computing, 2014, vol. 25, Issue 4, pp. 333-345.

[26] Zhang Z., Huang K., Wang Y and Li M, View independent object classification by exploring scene consistency information for traffic scene surveillance. Neurocomputing, 2013, vol. 99, pp. 250-260.

[27] Zhang L., Li S. Z., Yuan X and Xiang S, *Real-time object classification in video surveillance based on appearance learning*, in*Proc.*the IEEE Conference on Computer Vision and Pattern Recognition (CVPR), 2007, pp. 1-8.

[28] Vijaya Lakshmi T.R., Venkata Krishna Reddy C. *Object Classification Using SIFT Algorithm and Transformation Techniques*. In: Mallick P., Balas V., Bhoi A., Zobaa A. (eds) Cognitive Informatics and Soft Computing. Advances in Intelligent Systems and Computing, Springer , 2019, vol 768.

[29] T. Mahalingam, M. Subramoniam, A robust single and multiple moving object detection, tracking and classification, Applied Computing and Informatics, 2018.

[30] Shapiro L. G. and Stockman G. C, *Computer vision*, (Prentice Hall, 2001).

[31] Dodge and Yadolah, *The concise encyclopedia of statistics*, (Springer Science & Business Media, 2008).

[32] Donald F. Specht, *Probabilistic neural networks*, Neural Networks, Volume 3, Issue 1, Science Direct, 1990, Pages 109-118.

[33] Hu M. K, *Visual pattern recognition by moment invariants*. IRE Transactions on Information Theory, 1962, vol. IT-8, pp.179-187.

[34] Goyette N., Jodoin P., Porikli F., Konrad J and Ishwar P, *Changedetection.net: A new change detection benchmark dataset*, in*Proc.* IEEE Computer Society Conference on Computer Vision and Pattern Recognition Workshops, 2012, pp. 1-8.

[35] Maddalena L and Petrosino A, *The SOBS algorithm: What are the limits?*, *inPorc.* IEEE Computer Society Conference Computer Vision and Pattern Recognition Workshops (CVPRW), 2012, pp. 21-26.

[36] Chandrajit M., Girisha R. and Vasudev T, *Motion Segmentation using Chi-Square Test Statistics*. In Proceedings of the 2nd International Conference on Emerging Research in Computing, Information, Communication and Applications (ERCICA 2014), Elsevier, 2014, vol. 2, pp. 365-372.

A robust approach for people counting in dense crowd images using deep neural networks

K Sri Harsha[1], C Shoba Bindu[2] and E Sudheer Kumar[3]

[1] PG Scholar, Department of CSE, JNTUACEA, Anantapuram.
[2] Professor, Department of CSE, JNTUA, Anantapuram
[3] Research Scholar, Department of CSE, JNTUA, Anantapuram
Email: Harsha070@Outlook.com

Abstract. People counting in dense crowd images with deep neural networks are proved to be effective. The models like RCNN are able to predict the crowd count by head detection, using the CNN and selective search algorithm, but these approaches are very slow, as they involve computing convolutional operation for 2k regional proposals, involving no shared computations, besides the selective search algorithm itself is slow. In this approach a Faster R-CNN for head detection which uses a Regional Proposal Network (RPN) has been used. The region proposal network is a Fully Convolutional Neural Networks that generates region proposals, these regional proposal were fed in to RoI pooling layer and subsequently classified and localized, Thus Faster-RCNN reduces computation cost of convolutional operations by passing a image only once, sharing the convolutional operations and also using RPN for regional Proposals. Using the Faster R-CNN MAE and MSE are reduced compared to R-CNN.

1. Introduction

As with the day by day progress in the world, people from different communities collaborate and participate in conferences, debates where a large gathering of crowd takes place, a new innovation in technologies would also attract people in mass gatherings, which represent the frequent gathering of crowds, apart from that people gather in huge numbers for religious events, parties and form a crowd. Hence providing security becomes very important at this places, keeping the surveillance on the number of people attending would help for several statistical purposes and also helps in disaster management and better safety to the crowd can be provided. Estimating the number of people present in the crowd from an image, using a feature extractor machine would automate the work and make easy for humans to maintain the statistics and log records, although detecting the number of people through the features of an image as shown in figure 1 would become difficult as they work like a blind rule, because the hand engineered features like texture features of background may resemble with the person texture, hence estimating the count of a person may be more or the method may consider a person as a background when occlusion were there and ignore them.

Considering the features that can be extracted from an image of low resolution, these methods will make a huge error on predicting the count, on the other hand end-end regression techniques will be used by the models like RCNN [1] and other detection models in order to predict the count, these models make use of the neural networks to learn and predict the count from an image, although these deep neural networks make use of the regression methods they take a large amount.

Figure 1. (a), (b) represents the images where hand crafted features extracted were not much helpful for head detection.

of time to predict the count of people from image with almost extracting nearly 2k proposals for an image and processing the regions of the image and making recursive computation in order to estimate the count, which will be highly computational cost in-efficient and these convolutional operations were not shared. To avoid this inefficiency this approach uses a Faster R-CNN [3] model that make use of regional proposal network in order to get the regional proposals from a image passed single time through the convolutional neural network which will be more faster than the selective search algorithm .The Faster RCNN method also uses the shared convolutional operations for region proposals and detections reducing the computational cost. They nearly produce 300 regions, during the test time it is very handy as they were lesser than that of models [2] with selective search algorithm with 2k proposals having repeated computations.

The rest of the content is organized as section 2 contains the literature survey for Faster RCNN, section 3 describes the dataset preparation, section 4 describes the experimental setup, section 5 describes experimental methodology, section 6 describes results, and draw some conclusions in section 7.

2. Literature survey

Since the approach is using region proposals in order to detect the head, there were lot of hand crafted feature and learning techniques for regional proposals. So studying them will be helpful in understanding the regional proposal algorithms.

2.1. Expert and learning based methods

Detecting the object requires the features to be extracted and classified. But these hand crafted features extracted are not sufficient to predict the count in dense crowd image. There are other detection methods that consider the movement of objects in order to count but these methods fails in a large gatherings as the movement of objects will be negligible. So to detect the crowd people came up with the regression methods where the image is divided in to patches and extract the low level features. How ever these methods were successful for a less number of people in an image. The learning methods make use of end-end regression analysis like cnn to predict the count. The R-CNN [1] model with regional proposal algorithm selective search [7] is an learning problem which predicts the relation between the input and output through the regression analysis. Context aware cnn for person head detection [4] is another approach which make us of three different models based on CNN for detecting the head by using endend regression approach. Let us the look at some of the object detection and region proposal algorithm and understand their methods for better analysis.

2.2. Region proposals

There were many popular region proposal algorithms like selective search [7], edge boxes [11] and some other methods for regional proposals, these algorithms make use of learning methods like neural networks sliding window approach and hand crafted features. The selective search algorithm is a region proposal algorithm that make use of HOG feature descriptor [12] and segmentation techniques like efficient graph based image segmentation which forms a basis for selective search algorithms to merge the similar features and propose the regions. These algorithms greedily merge the pixels of image at different scales and aspect ratios producing nearly 2k proposals, and most of proposals will overlap, which make recursive computation to be done for these region proposals which are not shared and also takes a lot of time.

2.3. Object detection
For object detection different algorithms exist in the literature that can be studied. DPM [8] a discriminatively trained part based model is an detection approach that make use of the individual parts of an object and their deformed shapes to detect an object. A region based convolutional neural networks uses an region proposal algorithm to get the regions and take these regions to classify the objects in an image and also bounding box regressor in order to localize the objects. Fast-RCNN [5] an other approach for reducing the time complexity of RCNN [1] by using the region of interest (ROI) extraction and ROI pooling methods to fast up the computation. SPP [10] which provides an architecture to speed up the R-CNN forms as a support to the R-CNN. The Fast RCNN uses the convolutional operation shared to speed up the detection process. There were some other methods like overfeat [9] which were used for object detection. Regarding the crowd counting there were methods like Multisource multiscale counting in extremely dense crowd images [6] which predicts the count based on head detection. Composition Loss for Counting, Density Map Estimation and Localization is other object detection algorithms that make use of the feature descriptors like HOG, SIFT for object detection.

3. Dataset preparation
The present approach make use of the Hollywood Head dataset [4] and Shanghai Tech [2] and UCF _CC_ 50 [2] crowd dataset, for training the proposed approach is going to use Hollywood Head Dataset and only few images from the shanghai tech Dataset, and la some of the images in the Shanghai tech are labelled using labelimg tool in order to get the bounding boxes and train the network. table 1 shows the dataset information used for training.

Figure 2. (a) represent image from Hollywood heads dataset with bounding box drawn to get coordinates shown.(b) image from Shanghai Tech dataset.

Table 1. Number of images in Dataset and its Labels.

S.no	Dataset	No of Images	Bounding Coordinates	Class Labels
1	Hollywood Heads	2,16,719	x, y, w, h	Head, rest
2	Shanghai Tech	300	x, y, w,h	Head, rest

4. Experimental setup

Hardware and Software: python 3.6 with tensor flow 1.13 has been used for object detection.
Data Preparation: The dataset has been divided in to the 80 % training and 10 % for validation and 10 % for the test sets, The Dataset is from Hollywood dataset and Some of images from shanghai tech.
Optimizers: For fine tuning of RPN stochastic gradient descent with back propagation and learning rate of 0.0001 were used, and tuning of RCNN uses tuning end–end initially and then only the final layers. the RCNN ROI pooling layers were fine tuned with Stochastic Gradient Descent with back propagation [5].

5. Experimental methodology

Deep Neural Networks gained a lot of attention rapidly as they were used for feature extraction and classification both. Now the same convolutional operations are used for generating regional proposals too. This paper uses the transfer learning approach which uses the.

1) Pre-trained network to extract the features.
2) And these features are used to generate the region proposals by using region proposal networks.
3) Obtained regional proposals are fed in to the Fast RCNN model RoI pooling to classify the regions and predict the bounding box.

In the further section there is detailed discussion of Faster RCNN architecture.

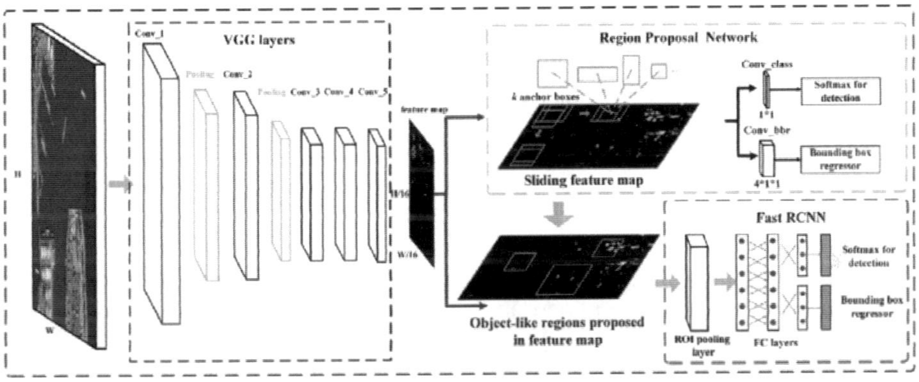

Figure 3. Faster R-CNN architecture [3].

5.1. Pre-trained cnn

The first part is going to use a pre-trained network that was trained on the image-net dataset. The network it uses is the VGG-16 network as in Simonyan and Zisserman model [14]. At first images are fed in to the vgg-16 to extract the features. Here the features are extracted from the middle layers. Since the approach is going to extract the features from middle layer of the network there is no restriction on the input size of the image. These extracted convolutional features are shared among the region proposal networks and the Fast R-CNN. During the training process convolutional features were trained corresponding to RPN and Fast R-CNN and they are fixed after that. The input to the network are images which are warped using the OpenCV library, these images propagate through 13x13 convolutional layers to produce the features.

5.2. Region proposal network

This method has been claimed as the faster region proposal method compared to the selective search. The RPN in Faster R-CNN [3] is based on Fully Convolutional Network [13]. With help of RPN the images are passed single time through a convolutional network and region proposal are generated from them which are used for detection, rather than processing multiple proposals through convolutional layers, which are not shareable as in the R-CNN. The regional proposal network takes the n x n spatial window of shared features extracted from the last layer that is convolutional features, that mean it applies a sliding window over shared convolutional features to get the region proposals along with the class object score. The regional proposal that were proposed are going to be of different scales and ratios which were defined as anchor boxes. The approach defines that a maximum number of regional proposals for each sliding windows position is going to be k that is (k anchor boxes), these anchor boxes are considered to have the center of object falls in to them. The number of anchor boxes k is defined on the different scales and aspect ratios that were considered. These sliding window are mapped to lower dimensional (512-d features). These lower dimensional features are passed in to two fully connected layer for classification and regression. The RPN is going to classify the proposals as foreground class or background class that is foreground means whether a object is there in the proposals or background that is proposal is not going to have the object. The proposal are classified as foreground or background based on the IoU over the Ground truths and the proposals. we set the proposals as foreground if IoU over ground truths and proposal is greater than 0.7 and if IoU is less than 0.3 then it is going to be considered a back ground class. The proposals that fall in between the thresholds are going to be ignored. The proposal anchor boxes that were generated may be more and overlap with each other for this we are going to use Non Max Suppression (NMS) where proposals with IoU greater than some threshold are going to be eliminated and finally you get the proposals. The bounding box regressor is going to adjust the proposals, during the training the bounding coordinates generated for background class are going to be ignored. The loss function that has been used for classification is log loss and for regression is robust loss function (smooth L1). For each scale and aspect ratio we are going to have k regressor boxes. The RPN is going to use the pre-trained image-net model and uses back propagation and stochastic gradient descent to reduce the loss. The general architecture that used for RPN is a n x n convolutional layer followed by two sibling 1x1 convolutional layers as shown in figure 3.

5.3. Fast R-CNN

The region proposals that were generated were fed to the Fast-RCNN network that uses the RoI pooling. The proposals were converted into the fixed size by using RoI pooling and are classified by the softmax layer and the network follows the bounding box regression to solve the localization errors. The fine tuning methods are applied at first end-end and then only to the unique layers of the RCNN. To understand the tuning of fast RCNN look at the Fast RCNN [5,3,10].

5.4. Training process

For the training Process Hollywood Head Dataset is going to be used. The input image is fed in to an convolutional network (vgg-16) and get the features extracted. The features along with the Ground Truth (Bounding Box Co-ordinates) are then passed to the region proposal network to get the anchor boxes and the bounding box coordinates. These anchor boxes along the with generated convolutional features are fed in to the Fast-RCNN model to get the classification of the object and generate the bounding boxes. Fine tuning of the network goes in an alternate way first fine tuning the RPN in an end-end manner is done and then the Fast R-CNN network. Later the Fast R-CNN is used as initialization of RPN network and then fine tune unique layers of RPN and Fast-RCNN, until predictions were correct.

5.5. Implementation details

The approach as said uses a VGG-16 to extract features, after passing through the convolutional layers. These backbone feature are given to the regional proposal network, a 2x2 sliding window is applied over the feature map in RPN, so that details of image were considered more during the testing, where the dense crowd images were passed. The scales the approach considers is 64,128 and 256 and with the aspect ratios of 1:1,1:2 and 2:1. The IoU for the non max suppression is going to be high, since the dense crowd have small heads where keeping less IoU would eliminate some region proposals. The network is trained for 40 epochs for RPN and the Fast R-CNN and later the convolutional layers were fixed and unique layer of RPN and Fast R-CNN were trained. During the testing the image is passed through the network and region proposal were given by RPN and head were detected and classified by Fast R-CNN. The figure 4 shows the block diagram during the testing where ground truth for bounding box were disabled.

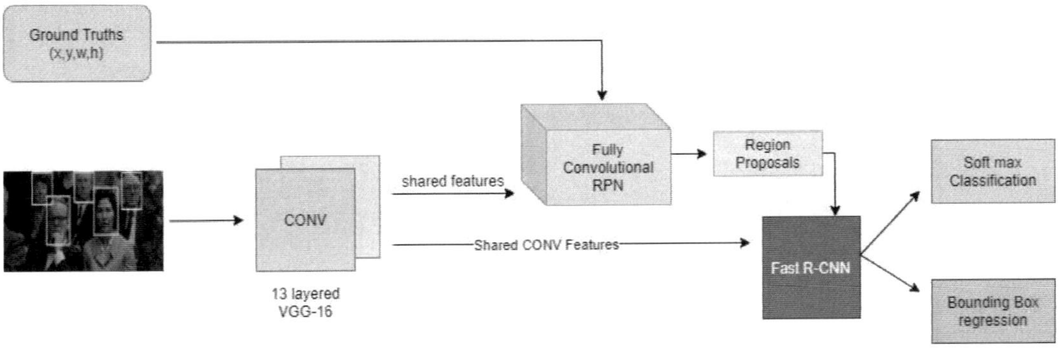

Figure 4. The Faster R-CNN [3] during the testing with ground truths disabled.

6. Results

The results for some of the images by Faster R-CNN model are shown in the figure 5.

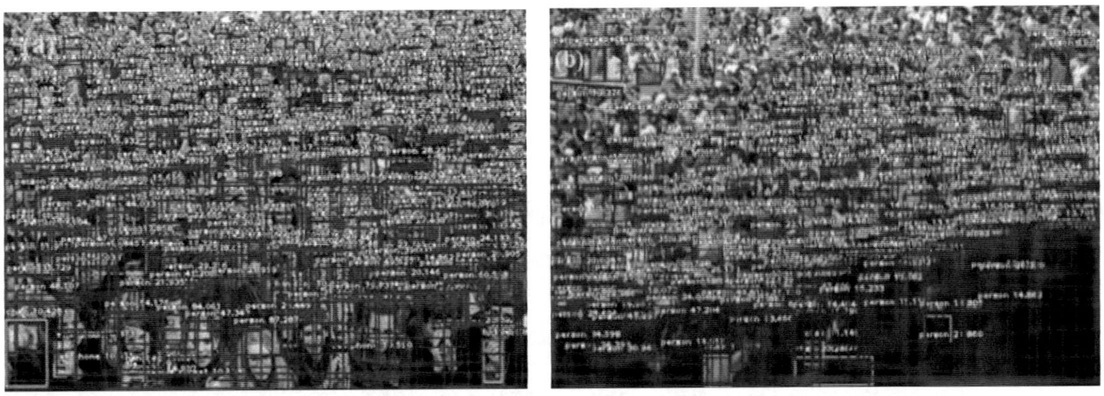

Figure 5. (a),(b) are results of head detection on some of the images of test set.

The approach have evaluated the Mean Absolute Error (MAE) and the Mean Squared Error (MSE) on Shanghai Tech and UCF CC 50 datsets. The methods corresponds to the robustness and accuracy of the algorithm you can see in table 2 and table 3 that the proposed model has performed better than the CNN and R-CNN model in the case of Shanghai Tech dataset and UCF cc 50 even though the errors were closed to the existing system the predictions of the count is much more faster and the model performed well over the R-CNN.

The below equations provide the calculations for MAE and MSE here y_i and y_o are the predicted and ground truth of the respective outputs, in this context of the paper y_i and y_o are the predicted count of the heads and ground truth count of the heads.

$$\text{MAE} = \sum_{i=0}^{n} (y_i - y_0) / n \quad (1)$$

$$\text{MSE} = \sum_{i=0}^{n} (y_i - y_0)^2 / n \quad (2)$$

Table 2. Comparisons on Shanghai tech dataset.

	Part A		Part B	
Method	MAE	MSE	MAE	MSE
Zhang et al., 2015 [15]	181.8	277.7	2.0	49.8
Zhang et al., 2016 [17]	110.2	173.2	26.4	41.3
Marsden et al., 2016 [18]	126.5	173.5	23.76	33.12
CNN	354	498	81	118
RCNN (Without Weighted Average)	254.32	395.5	60.4	79.8
RCNN (With Weighted Average)	241.6	369.6	43.2	61.4
Faster R-CNN	235.7	350.4	39.7	58.1

Table 3. Comparision on UCF CC 50 dataset.

Method	MAE	MSE
Rodriguez et al., 2011 [19]	655.7	697.8
Lempitsky and Zissermann, 2010 [21]	493.4	487.1
Idrees et al., 2013 [6]	419.5	541.6
Zhang et al., 2015 [15]	467.0	498.6
Hu et al., 2016 [16]	431.5	438.5
Zhang et al., 2016 [17]	377.6	509.1
Marsden et al., 2016 [18]	338.6	424.5
Chang et al., 2016 [20]	270.3	-
CNN	1242	1568
R-CNN (Without Weighted Average)	1161.9	1497.8
R-CNN (With Weighted Average)	1085.6	1427.1
Faster R-CNN	1069.7	1362.9

7. Conclusion

The object detection algorithms like R-CNN uses a Region proposals like selective search which are very slow and involving non shared convolutional operations, This paper has used a Faster RCNN which proved to have less error (MAE,MSE) in context of crowd counting and also faster than the existing approaches by sharing convolutional operations. How ever the Faster RCNN are not that much fast.we can go for some other methods which may be reliable in real time.

References

[1] Girshick R, Donahue J, Darrell T and Malik J 2014 *Rich feature hierarchies for accurate object detection and semantic segmentation* IEEE Conference on Computer Vision and Pattern Recognition.

[2] Shami M B, Maqbool S, Sajid H, Ayaz Y and Cheung S C S *2018 People counting in dense crowd images using sparse head detections* IEEE Transactions on Circuits and Systems for Video Technology **29(9)** pp 2627-36.

[3] Ren S, He K, Girshick R and Sun J 2016 *Faster r-cnn: Towards real-time object detection with region proposal networks* IEEE transactions on pattern analysis and machine intelligence **39(6)** pp 1137-49.

[4] Vu T H, Osokin A and Laptev I 2015 *Context-aware CNNs for person head detection* IEEE International Conference on Computer Vision pp 2893-901.

[5] Girshick R 2015 *Fast r-cnn* IEEE international conference on computer vision pp 1440-48.

[6] Idrees H, Saleemi I, Seibert C and Shah M 2013 *Multisource multi- scale counting in extremely dense crowd images* ICCV pp 2547–54.

[7] Uijlings J R, Van De Sande K E, Gevers T and Smeulders A W 2013 *Selective search for object recognition* International journal of computer vision **104(2)** pp 154-71.

[8] Felzenszwalb P F, Girshick R B, McAllester D and Ramanan D 2009 *Object detection with discriminatively trained part-based models* IEEE transactions on pattern analysis and machine intelligence **32(9)** pp 1627-45.

[9] Sermanet P, Eigen D, Zhang X, Mathieu M, Fergus R and LeCun Y 2013 *Overfeat: Integrated recognition, localization and detection using convolutional networks.* (*arXiv preprint arXiv:1312.6229*).

[10] He K, Zhang X, Ren S and Sun J 2015 *Spatial pyramid pooling in deep convolutional networks for visual recognition* IEEE transactions on pattern analysis and machine intelligence **37(9)** pp 1904-16.

[11] Zitnick C L and Dollár P 2014 *September Edge boxes: Locating object proposals from edges* In European conference on computer vision pp 391-405.

[12] Dalal N and Triggs B 2005 June *Histograms of oriented gradients for human detection* 2005 IEEE computer society conference on computer vision and pattern recognition Vol 1 pp 886-93.

[13] Long J, Shelhamer E and Darrell T 2015 *Fully convolutional networks for semantic segmentation* IEEE Conference on Computer Vision and Pattern Recognition.

[14] Simonyan K and Zisserman A 2015 *Very deep convolutional networks for large-scale image recognition* in International Conference on Learning Representations.

[15] Zhang C, Li H, Wang X and Yang X 2015 *Cross-scene crowd counting via deep convolutional neural networks* In Proceedings of the IEEE conference on computer vision and pattern recognition pp 833-41.

[16] Hu Y, Chang H, Nian F, Wang Y and Li T 2016 *Dense crowd counting from still images with convolutional neural networks.* Journal of Visual Communication and Image Representation **38** pp 530-9.

[17] Zhang Y, Zhou D, Chen S, Gao S and Ma Y 2016. *Single-image crowd counting via multi-column convolutional neural network.* In Proceedings of the IEEE conference on computer vision and pattern recognition pp. 589-97.

[18] Marsden M., McGuinness K, Little S and O'Connor N E 2016. *Fully convolutional crowd counting on highly congested scenes. arXiv (preprint arXiv:1612.00220).*

[19] Rodriguez M, Laptev I, Sivic J and Audibert J Y 2011 *Density-aware person detection and tracking in crowds* IEEE International Conference on Computer Vision pp 2423-30.

[20] Shang C, Ai H and Bai B 2016 *End-to-end crowd counting via joint learning local and global count* ICIP IEEE pp 1215–19.

[21] Lempitsky V and Zisserman A 2010 *Learning to count objects in images* NIPS pp 1324–32.

IOP Conf. Series: Materials Science and Engineering 1085 (2021) 012011 doi:10.1088/1757-899X/1085/1/012011

Machine Learning Techniques for Recommender Systems – A Comparative Case Analysis

Binu Thomas[1], **Amruth K John** [2]

[1]Associate Professor, Marian College Kuttikkanam (Autonomous)
binu.thomas@mariancollege.org
[2]Associate Professor, Marian College Kuttikkanam (Autonomous)
amruth.john@mariancollege.org

Abstract. Recommender System (RS) is one of the most popular applications of Artificial Intelligence which attracted researchers all around the world. Many machine learning algorithms are used to develop RSs. Choosing the best machine learning algorithm to provide users with a product or service is the most challenging task in the area of RSs. Now we are witnessing a paradigm shift in the purchase habits of people from in-shop to online resulting in the availability of online information exponentially growing every day. The ever-increasing online information and the number of online users create new avenues in RS. In an online shopping scenario, these systems must be able to recommend relevant items to the users. The RSs have to deal with the huge amount of information by filtering the relevant information based on the analysis made on the inputs made by the users during their online sessions. These systems can recommend appropriate items to users based on their interest and previous preference which can lead to increased sales. The three major techniques used to build a RS are content-based, collaborative based and hybrid-based. This paper presents the various applications of RSs and makes a detailed comparative study of different machine learning approaches used. The methodologies used for identifying research articles for analysis, the merits and demerits of different techniques in RSs and domain-specific applications of these techniques are well explained here with scientific review analysis.

1. Introduction

The Recommender Systems (RSs) are used in various domains like e-commerce, tourism, health, e-Learning, etc. These systems are widely used in the e-commerce field to recommend items to users of their interest. The items are recommended based on the buying history of customers. Based on the analysis, predictions can also be made on user preference. This technique can be considered as part of personalization because it helps each customer to buy products of his interest. This system can establish a good relationship with customers and can increase the sale rate if the recommendation is appropriate.

The RS in the e-commerce field has gained the attention of the business community. Amazon makes use of the RS to suggest the books frequently purchased by customers. Most of the RSs available today are based on the collected information of user's purchase history, explicit rating, and ownership data but no system is using all these simultaneously.

The development of RS can also contribute to the field of tourism. They guide tourists to manage huge information and allows them to take appropriate travel decisions. This system can suggest a suitable place that can match the appropriate activities to their profile. RSs are widely used as playlist generators for video and music services like Netflix, YouTube and Spotify. Amazon and social media platforms such as Facebook and Twitter make use of product recommenders. This system is one of the most successful applications of machine learning technologies in business. This helps the users to take quick decisions making in their online transactions and also improves the quality of their shopping experience.

These systems have also become significant in healthcare that can recommend healthcare information to patients as well as to health professionals depending on their requirements. The RS offers several opportunities in the medical field by using different machine learning techniques. The complex problems using unstructured data can be efficiently solved using different machine learning techniques.

Numerous works have been reported in the area of recommender systems for different applications like Movie recommendation, e-commerce, healthcare, etc. The recommendation in the e-Learning environment plays an important role in providing the accurate and right information to the learners. In the present scenario, e-Learning is completely transforming the learning habits of students and even adults. Intended learners are normally confused with the enormous number of courses available an online and keeping track of the relevance of these courses is also difficult. Since a huge amount of information is available on the internet, learners face the problem of searching for the right information and online recommendation system can solve the problem of information overload efficiently. This paper is an attempt to review the techniques used for RSs, the application domain areas of these techniques, merits and demerits of these techniques. In this paper, Section II introduces the methodology adopted in this study for the selection of source articles and reviews. Section III contains the literature review and a detailed discussion of merits and demerits. Evaluation matrices and domain areas are discussed in section IV. Finally, Section V concludes the paper.

2. Methodology

The review work of this paper accommodates the following phases for a systematic review of the literature. Fig.1 presents a specific set of steps followed in the management of this review paper.

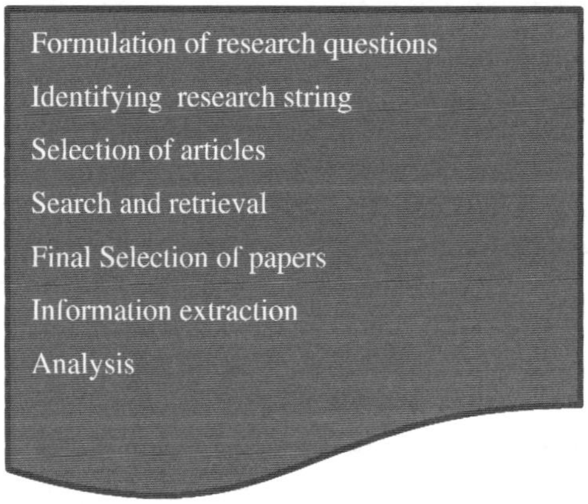

Fig 1. Structure of the review process

2.1 Research questions and Research String

The main purpose of this literature review is to comprehend what issues RSs could successfully address, how they are developed and evaluated, and in what ways or aspects they could be experimented with. To this end, we defined the following research questions:

Q1 What are the most relevant studies about RS?
Q2 Which data mining and machine learning techniques are used in RS?
Q3 What are the problems and challenges faced in RS?

Q4 How RS are evaluated?
Q5 Which are the specific domain areas where RS is used?
Q6 Which directions are most promising for future research?

We picked five digital libraries as the source of research papers under review (Fig.1). The strings *RS Applications, RSs Techniques, Review of RSs* are used for retrieving the research papers. A total of 27580 research articles were listed by the search from the digital libraries mentioned. We have included papers from three specific periods (2001-2005), (2009-2012), and (2016-2020) to review the initial stages, subsequent progress, and present situation in RSs. Only journal research papers and conference proceedings written in English are considered for the analysis. This resulted in the selection of 1450 research papers. In the next step, we used a fuzzy based technique for assigning scores to these papers based on the three parameters: Relevance, RS Techniques, and Domain Area. A fuzzy-based score between 0 and 1 is given to the papers in each category and research papers getting a score of 0.7 or above are retained for analysis (*expr. 1*).

Paper Score = $\sum_{i=1}^{3} S_i / 3$ *expr 1.*

Where S_i is the fuzzy score given for a paper in a category. Based on this selection criteria 42 research papers were selected for further analysis. The sources of these research papers are represented in Fig.2.

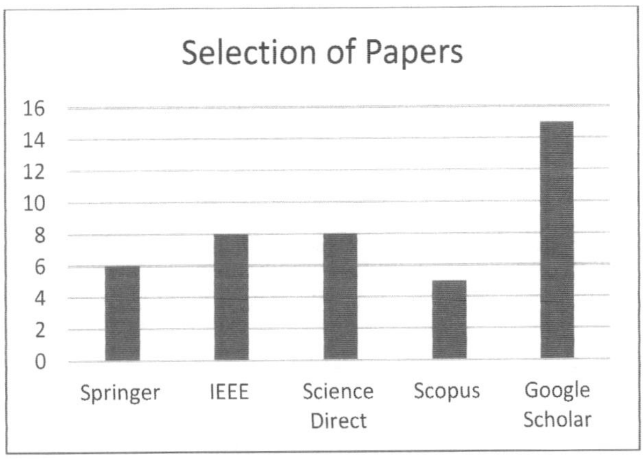

Fig 2. Sources of papers selected

The percentage of papers selected on the basis *RS techniques* as the selection criteria is represented in Fig.3.

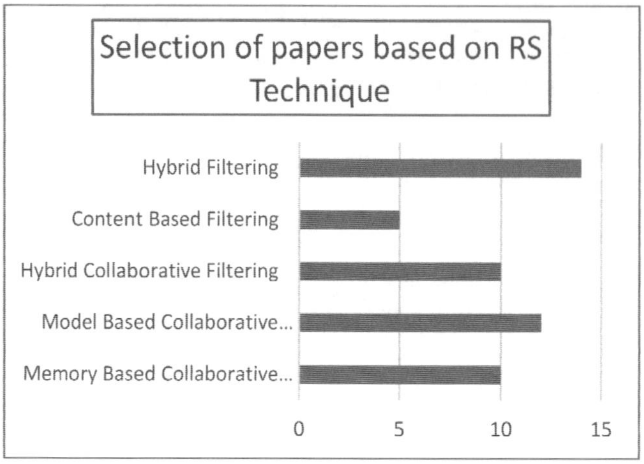

Fig 3. Paper selection based on the RS Technique

The number of papers retrieved based on the selection criteria *Domain area* are represented in Fig.4.

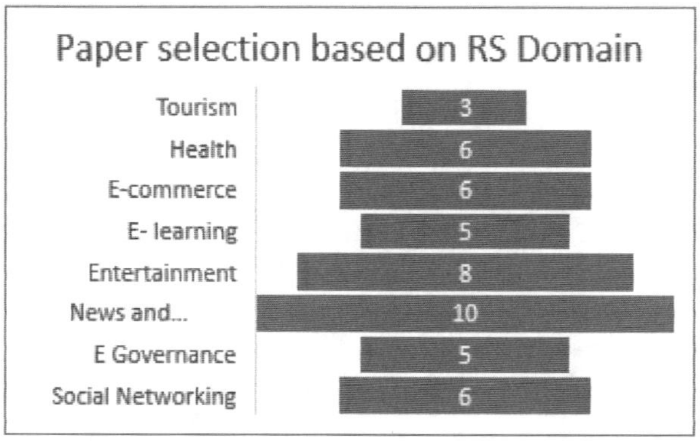

Fig 4. Paper selection based on Domain Area

3. Literature review

This paper presents a systematic literature review in the field of various RSs. The commonly used applications, recommendation techniques, and machine learning algorithms are discussed here. These works depict how RSs can provide personalized recommendations to online users so that they can reduce the time and effort in selecting the items or services according to their preferences. This section addresses the first Research Question and introduces the relevant studies in the area.

The study by Kumar [1] on the RS established that the recommendation system is an efficient tool to suggest recommendations on services to customers. Social media like Facebook, Google, Amazon, and Twitter proved that well-designed RS can provide accurate information to a large number of users. Item-based collaborative filtering algorithm is the commonly used technique in Amazon.com.

This technique can measure the number of users and the number of items independently. When using item-based collaborative filtering, we make a comparison between purchased and rated items of the user to related items. Then similar items are added to the recommendation list [2]. As per the study [3], it is observed that major problems in item-based RSs are the impact of context awareness, loss of neighbor transitivity, and sparsity.

The most commonly used classification of RSs is based on content-based, collaborative filtering, knowledge-based, and hybrid methods [6]. Content-based RS relies only on the past references of users to create their profile and to select the recommended items. But in collaborative filtering, it examines the behavior of similar users to identify the candidate items. The knowledge-based RS uses domain-specific knowledge for matching user requirements with items of possible interest. The hybrid recommender is the one that combines multiple approaches to produce the output. The purpose of combining different approaches is to improve the accuracy of the recommendation.

In most of the tourism RSs, collaborative filtering and hybrid approaches [4, 5] are used. Most of them use both explicit and implicit methods to identify user interest [7]. In this domain, the tourist details are recommended using a semantic network by combining collaborative filtering and content based filtering. The criteria include location, attraction type, price and time of travel, etc. A collaborative filtering approach is used for identifying user's interests and their ratings to attractions explicitly and the information of the user's social network is collected implicitly [9]. The location, time, and weather are used as the recommendation criteria. Another recommendation system in tourism based on a collaborative filtering approach and implicit criteria has been proposed by Yang and Marques. The two major issues of scalability and efficiency which need modern processing space and speed optimization to maintain user satisfaction are discussed here. In this work, collaborative filtering is used to build RS which can use the user ratings, information, and feedback collected from users to build recommendations [10].

The k Nearest Neighbours (kNN) recommendation algorithm is the commonly used algorithm in the recommendation processes using collaborative filtering. It is simple and can produce accurate results. But the low scalability [11] and sparsity [20] are the major drawbacks of this algorithm. The *k*NN algorithm focuses on similarity measures. The similarity calculation is performed between user to user, item to item, and user to item. By using similarity measures, similar users are assigned as neighbors to the user and items.

Major drawbacks found with the tourism RS are interactivity, adaptivity, and personalization. Even though it suggests user preferences, this system still needs user's help to plan their trip manually. The inability of automatic travel planning is the major problem to be addressed here. The social information and context of the user can be used to solve the problem. The major approaches like collaborative, content-based, and demographic approaches suffer from personalized recommendations. The hybrid techniques can be used to solve these problems of individuality.

Healthcare RS also makes use of a collaborative filtering approach to categorize diseases, based on symptoms and then suggest appropriate doctors to patients. While using a collaborative or a hybrid filtering approach, the RS must collect information about users to build recommendations [12]. One of the most popularly used machine learning algorithms in the healthcare industry is support vector machines. It works on a supervised learning model for classification, regression, and detection of outliers. This algorithm could predict the early symptoms and medicine for heart patients and help in saving the lives of millions of patients. This can also be used for protein classification, image segregation, and text categorization.

The following sections introduce some of the specific techniques, algorithms used in RSs. A detailed analysis of the performance evaluation of these methods is presented here and this section focuses on the second Research Question.

3.1 Singular Value Decomposition (SVD)

In data science, Singular Value Decomposition (SVD) is a standard technique for reducing dimensionality and SVD lays the foundation for RSs. This method is useful for the formation of latent factors (predictions) in the scope of RSs to resolve the problems faced by the approach of collaborative filtering [2]. The key attractions of this approach are scalability and predictive precision [3].

3.2 Convolutional Neural Network (CNN)

It is a type of neural network consisting of convolution layers capturing global and local computational characteristics to enhance efficiency and precision [2]. For modeling sequential data, this network is used.

3.3 Restricted Boltzmann Machine (RBM)

The RBM is an artificial neural generative stochastic network that discovers a distribution of probability over a set of inputs. It is a neural network of two layers that consists of a visible layer and a hidden layer [25,26]. The visible and secret layer does not interact with the intralayer. It utilizes approximation algorithms for gradient descent, such as contrastive divergence. This network is used for extraction operations in a supervised learning algorithm. These models are energy-based model, as shown in Fig 5, consisting of binary-valued hidden and visible units. The strength of the relation between the noticeable and hidden units Vi and Hj, respectively, is denoted by the weight matrix W. For these units, it incorporates bias weights. Increasing the number of hidden units contributes to the better performance of the model.

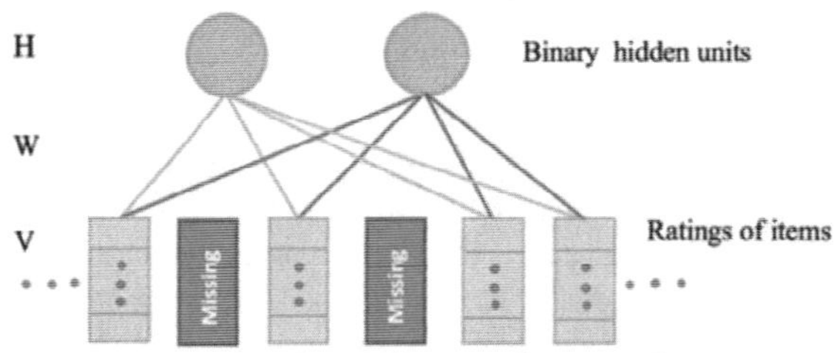

Fig 5. RBM Neural Network

Social network analysis (SNA) has been included in RSs as a result of the rapid growth of social networking resources in web-based systems [32, 33]. To enhance the user experience, RSs provide users with suggestions on their ability to communicate with other users in social interaction, such as online messaging, social comments, etc. These systems provide opportunities to make recommendations by using the social ties of users, especially for systems whose rating information is too scarce to perform Collaborative Filtering (CF). To present collaborative partnerships, Palau et al, [32] organized social networks and suggested steps to clarify how the Recommendation Process achieves collaboration.

Electronic government (e-government) [32] refers to the use of the Internet and other information and communication technologies to assist governments in providing better information. It is a conventional field for RSs in governance to propose web sites, documents, and news since such tools are growing rapidly. In most cases, the textual content is defined as a list of keywords that can be extracted from historical records, URLs and search engines, such as news, emails, documents and web pages, and many RSs are built based on the analysis of keywords and services for people and companies. An RS was proposed to help voters make choices in the e-election process [34], which uses fuzzy methods of clustering and offers information about candidates similar to the preferences of voters. To improve the trade exhibition RS for e-government, a hybrid fuzzy [36] logic-based recommendation system was developed.

Since these services expand rapidly, recommending websites, documents and news is a typical field for RSs. In most cases, a list of keywords that can be extracted from historical data is represented as textual content, such as news, emails, documents, and webpages. Based on keyword analysis, URLs, and RSs are constructed for search engines. In this field, probabilistic models such as the information retrieval techniques [37] are used. With each variable representing a keyword that takes into account frequency and location, contextual resources are converted into a vector (title or plain text).

e-Learning has become an important part of the education system [2] now. The user-based collaborative filtering algorithm is selected as the primary recommendation algorithm for online education [13]. Collaborative filtering based approaches give importance to the ratings given to items by users as the primary source of information for recommended learning. In many of the learning applications, the ratings are found very sparse and this causes collaborative based methods to reduce its performance in the recommendation.

The table below (Table 1) describes the various techniques used in different RSs. Based on the implementation techniques, the RSs can be divided into two, memory-based and model-based [4]. The memory-based method accesses the internal system database directly, whereas the model-based method uses the transaction data to create a model which can generate recommendation [14]. By accessing the database directly, the memory-based method is adaptive to data changes but requires large computation time depending on the data size [16]. The model-based method has a constant computing time regardless of the data size but not adaptive to data changes.

3.4 User-Based Collaborative Filtering (UBCF)

The UBCF method chooses clients with a comparable rating to a given thing as a client set [3]. It can foresee the client's appraising to another thing as indicated by others evaluating in a similar client set. The most testing undertaking of this calculation is to locate the best client set for the chose client. The principle center is to distinguish the neighbors with the most extreme similitudes with the chose clients [15]. In the wake of distinguishing the likeness of the client to other people, the comparable neighbors of clients as indicated by the closeness will be chosen and assembled into neighbor records [3]. The client appraisals to explicit things can be anticipated with the assistance of the rating history of neighbors to acquire the proposed suggestion results.

3.5 Item-Based Collaborative Filtering (IBCF).

To compare the comparability of different objects, the IBCF approach is used. This is used to predict a user's rating of a related item based on his current item rating [3]. Model-based methods can recommend items with the highest rank to a user, returned by the parametrized model [6]. The most popular algorithm in this category is the Slope one method. It is a linear regression model that learns through a set of simple predictors, one for each pair of two items with just one constant variable [17] so that this variable can represent the average difference between the ratings of two items. This method has fast computation and reasonable accuracy.

3.6 Matrix Factorization

These are the most successful latent models that can solve high sparsity problems. Matrix factorization became popular because of its enhanced scalability and predictive accuracy [18]. One of its significant contributions is its ability to uncover the hidden structure behind data. It is an effective technique used for information retrieval.

Using Matrix Factorization, the content-based filtering recommendation algorithms can be modified to a new algorithm called content-based filtering recommendation algorithm using HMM (Hidden Markov Model). The average precision of the HMM-based algorithm is higher, hence this algorithm is recommended for personalized systems [19].

3.7 Computational intelligence-based recommendation techniques (CI)

Computational intelligence approaches are commonly used in RSs to create recommendation models. Bayesian techniques, artificial neural networks, clustering techniques, genetic algorithms, and fuzzy set techniques are used in these CI approaches. For model-based RSs [32], Bayesian classifiers are popular and are often used to derive the Content-Based (CB) RSs model. Each node corresponds to an object when implementing a Bayesian network in RSs, and the states correspond to each possible vote value. For each object, there will be a collection of parent items within the network that represent its best predictors. As a basis for integrating both CB and CF methods, a hierarchical Bayesian network has also been introduced [36].

Table 1. Comparison of Recommendation System techniques

*T*echniques	Representative algorithm	Advantages	Disadvantages
Memory-Based Collaborative Filtering (CF)	User-Based CF Item Based CF	Simple to execute. Data addition is simple. Content should not be considered. Efficient scalability	Depends on the explicit suggestions. Cold start issue Trouble with Sparsity Scalability limited for huge dataset
Model-Based Collaborative Filtering	Slop-one CF	Enhances the efficiency of prediction. Improves issues with scalability and sparsity	Model is costly. Loss of information in a factorization matrix
Hybrid Collaborative Filtering	Combination of memory-based and model-based.	Overcome the sparsity limitations. Enhances the efficiency of prediction.	Complexity is increased. Challenging for implementation.
Content-Based Filtering	Content-Based filtering algorithm using Hidden Markov Model	No issue with scarcity and cold start. It ensures confidentiality.	Needs detailed description of items. Requires ordered user profile. Concern is material overspecialization.
Hybrid Filtering	Combination of Collaborative and content-based algorithm	Collaborative and content-based methods are complementary strengths and shortcomings.	Difficult to implement
Computational intelligence-based	Combination of Fuzzy Logic Neural Network Artificial Intelligence	It eases the overload of information, improves the processing of information and solves new problems.	Failure to predict correct results for varying situations. Poor in providing optimal shedding of load.

An Artificial Neural Network (ANN) is an interconnected node and weighted connection assembly that is inspired by the biological brain's architecture and can be used to create model-based RSs [35].

4. Discussion

4.1 RS Techniques and Challenges

RSs will provide users with personalized and targeted suggestions [38]. Therefore, they address the problem of data overload that plagues the users of the digital era. Different methods have recently been developed for building RSs, which can use either collaborative filtering, content-based filtering, or hybrid filtering. The technique of collaborative filtering has achieved a certain maturity and is one of the systems most frequently implemented. Content-based filtering techniques usually base their predictions on user data and, as in the case of collaborative techniques, they disregard contributions from other users. Although content-based filtering techniques have problems such as restricted content analysis, overspecialization is very important.
The challenges and limitations with content-based recommendations [23] are discussed in the following section and it addresses Research Question 3.

Over Specialization: The system does not suggest the items that are different from the item that the user has seen before. This may create a problem because the user wishes to try something new and the system may not allow this to work. So it is advised to make several options available to the user.

Limited content analysis problem – In this problem, we may represent two different items with similar attributes so that they cannot be distinguished.

New user problem: Since new users do not have enough ratings, the system may not be able to make accurate recommendations such a user.

It is advised to use Content-based filtering when there are limited users and fewer user items. The Collaborative Filtering method is best used when there is a large number of users and items [24]. The major drawbacks identified with Collaborative filtering techniques are listed below

Cold start – It is difficult to give suggestions to the new users because the system is not familiar with the new user's interests. This problem can be solved by surveying at the time of the users profile creation. Questions can be posed and based on the user's responses recommendations can be made.

Scalability – When there is a large number of users available, more resources are needed for information processing. The resources are spent to determine users with related purchasing patterns. This problem can be solved by using different types of filters.

Scarcity – A large number of users and items are available on online websites. Collaborative and other RSs can create a neighborhood of users using their profile using various recommendation techniques. If the user rates only a few items, it will be difficult to find out the user's interests and will be mapped to the wrong neighborhood. This is called a Data sparsity or scarcity problem.

Privacy – It is another major issue to be discussed in this system because the RS has to collect information like user location, bank data, etc. All online websites must make sure that the algorithms used can ensure data protection.

Another problem faced by the RS is Longtail items, the ones rated by fewer users. To solve this problem, it is important to consider other aspects like diversity along with accuracy.

This paper presented various algorithms and techniques used to build RSs. The algorithms and approaches discussed here have their advantages and disadvantages like user-based approaches are precise but not scalable, item-based approaches are scalable but not precise. Hybrid RS combines the best features of user-based and item-based algorithms but they are complex too. In this scenario, evaluation matrices are used to assess the effectiveness of RSs. This section focuses on Research Question 4.

4.2 Evaluation matrices

The most prominent evaluation metrics in the research literature measure the accuracy of the system's predictions. Accuracy is measured as the magnitude of error between the predicted value and the true expected value. Predictive accuracy is the RS's capacity to predict the expectations of a consumer for an

object. Mean absolute error is the traditional approach for calculating predictive precision (MAE). It is the absolute average difference between the estimated values and a user's real expected value.
The other possible evaluation metrics are listed below:

Novelty: A CF system can recommend items that the user was not aware of before. For many applications, novelty is one of the most valued characteristics of the CF system's recommendations. A system tuned for novelty will actively avoid recommending new stories that already a user has awareness.

Serendipity: Users are given recommendations for items that they would not have seen through their existing channels of discovery. A serendipitous scheme would imply news stories on subjects that have never been read about before. Researchers have researched how algorithms can be modified to facilitate serendipity and novelty [27], but it is difficult to quantify novelty because it involves live user trials where users indicate whether a suggestion is a novel.

Coverage: It is the proportion of the CF system's items known for which predictions can be made by the CF system. Variants such as the percentage of products that have the potential to be recommended to users can also be measured, as performance optimizations in recommendations can prevent certain items from ever being recommended [28].

Learning Rate: It tests how fast the CF system becomes an efficient taste indicator as information starts to arrive. These are calculated per-user, measuring the number of ratings a user has to provide before they receive personalized predictions of high-quality [29].

Confidence: It defines a CF system's ability to determine the likely quality of its predictions. Most CF systems produce rankings based on the expected rating that is most likely. A CF system that can accurately measure its confidence in a forecast can restrict recommendations to high trust, leading to a trade-off between fewer false predictions at the expense of reduced coverage and probably reduced novelty. Trust in predictions is measured in certain instances and shown to users to help them determine if the risk-return ratio is appropriate [30].

User satisfaction metrics: The metrics mentioned above are just a sample of potential metrics for assessment. There are several more metrics, in particular, that researchers may create. These metrics will provide users with a system, and calculate how the system is viewed by users.
It is typically achieved either by surveying the consumers or by calculating statistics for retention and usage.

4.3 Application Domain Areas

RSs and their applications reviewed in this paper are presented in Table 2. Here Research Question 5 is addressed and summarized. From the summary of RSs, the following important findings can be extracted: The major techniques used in the recommendation system are CB, CF, and Hybrid techniques, among these, Hybrid techniques are most popular. The CF techniques are widely used in e-resource RSs. The e-learning RSs have highly applied knowledge-based techniques. The social network-based RS and context awareness-based RS has also played a major role in recent application developments.

Computational intelligence methods, such as fuzzy logic, are used to deal with different uncertainties in all kinds of RSs. KB techniques, such as ontology and semantic tactics, are commonly incorporated with CF and CB recommendation methods into e-business RSs [31, 32]. The key explanation for this is that there is a high need for domain information in e-businesses [41] to help their recommendations.

Table 2 RS Techniques and application domain areas reviewed

Techniques / Domains	CB	CF	KB	Hybrid	Computational Intelligence	Social Network	Context-Aware	Group Aggregation
E-government	1	3	1	4	4			
E-business		1	3	3	4	1		
E-commerce/ E-shopping	3	1	4	1	4	2		
E-library	2	2		3	1			
E-learning	2		11		2			
E-tourism	5	9	9	9	3	2	11	
E-resource	9	16	6	15	8	1	1	
E-group Activity	9	5	2	5	1			2
Total	**31**	**37**	**36**	**40**	**27**	**6**	**12**	**2**

While RS applications have achieved great growth, there are still some problems that need more research with the advent of new applications for e-services. The hybrid recommendation approaches which combine CB, CF, and KB techniques are widely used in the e-library RSs. There are several cases, such as suggesting a TV program to a group of individuals who are unable to determine their users. In tourism scenarios, users need to negotiate online to engage in an activity together [39]. In these situations, for a whole community, individuals need online decision support. Traditional RSs make suggestions only for individual users, so it is suggested that community RSs (GRS) combine and match the individual preferences of group members to provide the group with satisfactory recommendations [42].

It is now possible to give customized and context-sensitive suggestions to mobile users with the increasing usage of Internet-accessing smartphones, and more mobile RSs would be needed. However, mobile data, which is heterogeneous, noisy, requires spatial and temporal auto-correlation, and has validation and generality issues [32,37] is typically more complex. Further Mobile-based analysis in the field of RS could make a significant contribution.

5. Conclusion

From the review of RSs, it is observed that there is an exponential growth of research in this field. Collaborative filtering is the most popular technique used for recommendation systems but the major drawback with this approach is its inability to solve the cold start problem. The commonly used technique next to CF is content-based filtering, but this technique also suffers from the same problem. Hence the hybrid approach, combining the other two techniques seem to produce a better result and it is becoming the most popular technique adopted in RSs. With the ever-increasing online usage in all aspects of human life due to the present pandemic situation, novel artificial intelligence-based systems are evolving in this area. In this paper, we have discussed the existing generation of RSs. There are possible extensions to these algorithms which can lead to more fruitful researches in this area.

References:

[1] F G. Adomavicius and A. Tuzhilin 2015 Context-aware RSs *Recomm Syst. Handbook, Second Ed.* 191–226

[2] Murad, D. F, Heryadi, Y, Wijanarko, B. D, Isa, S. M and Budiharto 2018 Recommendation system for smart LMS using machine learning: a literature review *Proc. Int. Conf. on Computing, Engineering, and Design (ICCED) IEEE* 113-118

[3] Venkatesan R and Sabari A 2020 Issues in various RS in E-commerce 2020 - A survey 2020 *Journal of Critical Reviews* **7(7)** 604-608

[4] Sahoo, A. K, Pradhan, C, Barik, R. K and Dubey H 2019 DeepReco: deep learning based health RS using collaborative filtering. *Computation* **7(2)** 25

[5] Sharma, M, Chauhan, N, Bansal, H and Stanciu L 2020 Digital Marketing and AnalysisTechniques: Transforming Internet Usage *New Age Analytics* 1

[6] Çano E and Morisio M.2017 Hybrid RSs: A systematic literature review. *Intelligent Data Analysis* ***21*(6)** 2017 Esmaeili, L, Mardani, S, Golpayegani, S. A. H and Madar, Z. Z 1487-1524

[7] Leila Esmaeili and Shahla Mardani et al 2020 A novel tourism RS in the context of social commerce *Expert Systems with Applications* Volume **149** ISSN 0957- 4174

[8] Alamdari, P. M, Navimipour, N. J, Hosseinzadeh, M, Safaei, A A and Darwesh A 2020 A Systematic Study on the RSs in the E-Commerce *IEEE Acces* **8** 115694-115716

[9] Singh, P. K, Dutta, P. K, Pramanik, A K D and Choudhury 2020 P RSs: An Overview, Research Trends, and Future Directions, *International Journal of Business and Systems Research*

[10] Li-Tung, Yue Xu and Yuefeng Li 2005 A framework for e-commerce oriented RSs, *Pro. Int. Conf. on Active Media Technology, Kagawa, Japan* 309-314

[11] Xue, G. R, Lin, C, Yang, Q, Xi, W, Zeng, H J, Yu,Y and Chen Z 2005 Scalable collaborative filtering using cluster-based smoothing *Pro. Int. Conf. on Research and development in information retrieval* 114-121

[12] Shih, Y. Y and Liu D. R 2005 Hybrid recommendation approaches: collaborative filtering via valuable content information *Proc. Int. Conf. on System Sciences 217b*

[13] Sikka, R, Dhankhar, A and Rana C 2012 A survey paper on e-learning RS. *International Journal of Computer Applications* **47(9)** 27-30

[14] Sarwar, B, Karypis, G, Konstan, J and Riedl J 2001 April Item-based collaborative filtering recommendation algorithms *Proc. Int. Conf. on World Wide Web* 285-295

[15] Dou, Y, Yang, H and Deng X 2016 A survey of collaborative filtering algorithms for social RSs *Proc. Int. Conf. on Semantics, Knowledge and Grids (SKG)* IEEE 40-46

[16] Thorat, P. B, Goudar, R. M and Barve 2015 S Survey on collaborative filtering, content-based filtering and hybrid recommendation system. *International Journal of Computer* Applications **110(4)** 31-36

[17] Dou, Y, Yang, H and Deng X 2016 A survey of collaborative filtering algorithms for social RSs *Proc. Int. Conf. on Semantics, Knowledge and Grids (SKG)* IEEE 40-46

[18] Ravi, L and Vairavasundaram S 2016 A collaborative location based travel recommendation system through enhanced rating prediction for the group of users *Computational intelligence and neuroscience*

[19] X Luo, Y Xia and Q Zhu 2012 Incremental Collaborative Filtering recommender based on Regularized Matrix Factorization *Knowledge-Based Systems* vol **27** 271–280

[20] J Bobadilla and F. Serradilla 2009 The effect of sparsity on collaborative filtering metrics in *Proc. Int. Conf. on Database (ADC '09)* 9–17

[21] Li, H, Cai, F and Liao Z 2012 Content-based filtering recommendation algorithm using HMM *Proc. Int. Conf. on on Computational and Information Sciences* IEEE 275-277

[22] Z Gao Z Li and K Niu 2016 Solutions for problems of existing E-Commerce recommendation system in Consumer Electronics- Taiwan(ICCE-TW) *Proc. Int. Conf. on IEEE* 1-2

[23] Soanpef Sree Lakshmi et al 2014 (IJCSIT) International Journal of Computer Science and Information Technologies *RSs: Issues and challenges* Vol **5(4)** 5771-72

[24] Siddhesh Jagtap, Yash Mane, Tushar Kadam and Trupti Dangge 2019 Survey paper on RSs Volume:**06** Issue:12 Dec 2019 *International Research Journal of Engineering and Technology (IRJET)*

[25] Yedder, H B Zakia, U Ahmed, A and Trajkovic, L 2017 Modeling prediction in RSs using restricted boltzmann machine *IEEE Proc. Int. Conf. on Systems Man and Cybernetics (SMC) Banff AB Canada* 5–8

[26] Belle, A,Thiagarajan, R. Soroushmehr, S M, Navidi, F Beard, D A and Najarian K 2015 Big data analytics in healthcare Biomed Res. Int. 370194

[27] Karypis G 2001 Evaluation of Item-Based Top-N Recommendation Algorithms *Proc. Int. Conf. on Information and Knowledge Management (CIKM)* **1** 247-254

[28] Sarwar, B Karypis, G, Konstan and J.A Riedl 2000 J Application of Dimensionality Reduction in RS- A Case Study ACM WebKDD Web Mining for E-Commerce Workshop Boston Massachusetts

[29] Schein, A I, Popescul, A, Ungar and L H 2001 Generative Models for Cold- Start Recommendations *Proc.Int. Conf. ACM SIGIR Workshop on RSs New Orleans Louisiana*

[30] Schafer J B, Frankowski D, Herlocker J and Sen S 2007 Collaborative Filtering RSs. In: Brusilovsky P, Kobsa A, Nejdl W (eds) The Adaptive Web *Lecture Notes in Computer Science* vol **4321** Springer Berlin Heidelberg

[31] D Ben-Shimon A, Tsikinovsky, L Rokach, A Meisles, G Shani and L Naamani 2007 RS from personal social networks *Advances in Intelligent Web Mastering* Springer 47-55

[32] Lu, J, Wu, D, Mao, M, Wang W and Zhang G 2015 Recommender system application developments : A survey *Decision Support Systems* **74** 12–32

[33] X Guo and J Lu 2007 Intelligent e-government services with personalized recommendation techniques *International Journal of Intelligent Systems* **22** 401-417

[34] L Terán and A Meier 2010 A fuzzy RS for eElections in: K. Andersen, E Francesconi Å Grönlund T van Engers (Eds.) *Electronic Government and the Information Systems Perspective* Springer Berlin Heidelberg 62-76

[35] P De Meo, G Quattrone and D Ursino 2008 A decision support system for designing new services tailored to citizen profiles in a complex and distributed e-government scenario *Data and Knowledge Engineering* **67** 161-184

[36] N Zheng and Q Li 2011 A RS based on tag and time information for social tagging systems *Expert Systems with Applications* **38** 4575-87

[37] R Jäschke, L Marinho, A Hotho, L Schmidt-Thieme and G Stumme 2007 Tag

Recommendations in folksonomies in: J Kok, J Koronacki, R Lopez de Mantaras, S Matwin, D Mladenič, A, Skowron (Eds.) *Knowledge Discovery in Databases: PKDD* Springer Berlin Heidelberg 506-514

[38] A Hotho, R Jäschke, C Schmitz and G Stumme 2006 Information retrieval in folksonomies: search and ranking in: Y Sure J Domingue (Eds.) *The Semantic Web: Research and Applications* Springer Berlin Heidelberg 411-426

[39] J Gemmell, T Schimoler, M Ramezani, L Christiansen and B Mobasher 2009 Improving folkrank with item-based collaborative filtering *Proc. of the ACM RecSys'09 Workshop on RSs and the Social Web ACM New York NY USA*

[40] X Amatriain, A Jaimes, N Oliver and J Pujol 2011 Data mining methods for RSs in: F Ricci, L Rokach, B. Shapira, P B Kantor (Eds.) *RSs Handbook* Springer US 39-71

[41] K Yu, V Tresp and S Yu 2004 A nonparametric hierarchical bayesian framework for information filtering *Proc. Int. Conf. on Research and Development in Information Retrieval ACM Sheffield United Kingdom* 353- 360

A recurrent neural network forecasting technique for daily $PM_{2.5}$ concentration level in Southern Kerala

Sherin Babu[1], Binu Thomas[2]

[1]Assistant Professor, Department of Computer Science, Assumption College Autonomous, Changanacherry, Kottayam, Kerala, India
[2]Associate Professor, Department of Computer Applications, Marian College Autonomous, Kuttikkanam, Idukki, Kerala, India

Abstract. The natural environment and public health are seriously impacted by air pollution. One of the most dominant contributors to air pollution around the world is particulate matter $PM_{2.5}$. Predicting air pollution in advance has considerable importance for the regulation of people's health and to implement pollution control strategies for air quality management. The study was conducted for Thiruvananthapuram district, the southernmost region of Kerala. The data for the period from 1st July 2017 to 31st December 2019 were collected from the Central Pollution Control Board (CPCB) website. To predict the daily $PM_{2.5}$ concentration, Recurrent Neural Network (RNN) based Long Short-Term Memory (LSTM) was used here. LSTM was built on the inputs of four meteorological parameters, namely average temperature, wind speed, wind direction and relative humidity and air pollutant parameter $PM_{2.5}$ values. Performance evaluation of the prediction model has conducted, and the results showed that the model attained considerable prediction accuracy.

1. Introduction

Air pollution means contamination of the air in the earth's atmosphere by harmful gaseous substances, particulates and biological molecules [1]–[3]. Air pollution is considered a severe threat to public health across the globe. As a result of civilization advancement, globalization and rapid industrial development, the impact of human activities on the natural environment is rising and, unfortunately, in most cases, unfavourable for the existence of the environment. Both temporal and spatial properties are dependent on air pollution. The key one is factors related to meteorology, which vary in spatiotemporal form. The temperature, humidity, amount of rainfall, wind speed, wind direction, etc. make air pollution differ from place to place. Traffic density and traffic congestion are other important causes of air pollution. The zone with a higher level of traffic has lower ambient air quality. Particulate matter is the sum of all solid and liquid particles suspended in the air such as dust, pollen, soot, smoke and liquid droplets[4]. $PM_{2.5}$ refers to the atmospheric particulate matter that has a diameter of fewer than 2.5 micrometres, which is about 3% of the diameter of human hair. $PM_{2.5}$ particles in the air reduce visibility and cause the air to appear hazy when its levels are elevated. $PM_{2.5}$ particles can penetrate deeply into the lungs, and hence exposure to high concentrations of $PM_{2.5}$ will cause respiratory and cardiovascular diseases. Several studies indicated that long-term exposure to $PM_{2.5}$ has resulted in elevated mortality rates. To control parameters within reasonable limits, prediction mechanisms for

IOP Conf. Series: Materials Science and Engineering 1085 (2021) 012012 doi:10.1088/1757-899X/1085/1/012012

pollutant concentration $PM_{2.5}$ in advance have become necessary [5]. We need to carry out the $PM_{2.5}$ prediction to provide an early air quality warning.

2. Related Work

Several $PM_{2.5}$ prediction methods are developed by researchers, based on statistic models and machine learning techniques. A deep learning model to forecast $PM_{2.5}$ concentrations based on an auto-encoder and bidirectional long-term short-term memory (Bi-LSTM) was explored to demonstrate the association between $PM_{2.5}$ and multiple climate variables by Zhang et al.[6]. Wen et al. proposed a spatiotemporal convolutional long-term short-term memory neural network extended (C-LSTME) model to predict $PM_{2.5}$ concentration in China [7]. Park et al., developed a convolutional neural network (CNN) model for estimating daily 24-h averaged ground-level $PM_{2.5}$ values in United States of America, by incorporating aerosol optical depth (AOD) data, meteorological fields, and land-use data [8]. Kaur and Mandal established that non-linear autoregressive network with exogenous input (NARX) is superior in predicting daily $PM_{2.5}$ concentrations of Delhi area, compared to other neural network models [9]. A hybrid deep learning method based on multiple attention-LSTM neural networks was presented by Yuan et al., to predict the $PM_{2.5}$ concentration in Hefei city [10]. Xing et al. created a deep learning model termed as temperature-based deep belief networks (TDBN) to predict the daily concentrations of $PM_{2.5}$, in Chaoyang Park in Beijing of China [11]. Geetha and Prasika developed a deep learning technique Long-Short Term Memory (LSTM) used for predicting various air pollutants, including $PM_{2.5}$ [12].

Lu et al. developed novel deep belief network (DBN) integrated with multilayer restricted Boltzmann machines and a single-layer back-propagation network model for the haze level prediction. This model used both air quality and the meteorology data, and then the haze prediction was carried out by using a competitive adaptive-reweighed method [13]. Another way aimed at $PM_{2.5}$ forecasting is developed by Zhou et al. Here Multivariate Bayesian Uncertainty Processor (MBUP) is combined with an artificial neural network (ANN) to perform the $PM_{2.5}$ prediction [14]. A combination of Convolution Neural Network and Long Short-Term Memory, called the ConvLSTM model, was proposed by Le et al., which automatically manipulates the spatial and temporal characteristics of the $PM_{2.5}$ air pollution data in Beijing city [15]. In [16], the authors claimed that the LSTM model outperformed all other neural network models in forecasting hourly $PM_{2.5}$ values in two big cities of South Korea. Ma et al. proposed a Geo-LSTM that have taken into account of the spatial-temporal correlation of air pollutant concentrations from different monitoring stations and then run the forecast model to predict the $PM_{2.5}$ values [17].

3. Methodology

This work aimed to develop an RNN model for predicting the daily $PM_{2.5}$ value for the southern part of Kerala. RNN is a type of deep learning based artificial neural network designed to process sequential data and recognize patterns in it. The Recurrent Neural Network is a type of feedforward neural network with an internal memory capable of processing input sequences. RNNs remember the inputs and the context as they have internal memory, enabling users to have more flexibility in the types of data that models or networks can process. RNNs are a powerful tool when data is sequential, and the next data point depends on the previous data point [18], [19]. Since they understand the context in a sequence, RNNs can produce better results/predictions. Several studies show that the current $PM_{2.5}$ concentration has a clear association with past results.

A type of recurrent neural networks is Long Short-Term Memory (LSTM) networks, which makes it easier to recall past data in memory. Hochreiter & Schmidhuber introduced LSTM to prevent the issue of capturing long-term dependencies by an RNN [20]. In each unit of the hidden layer, the LSTM network comprises a memory block containing three types of gate functions, namely the input gate, forget gate, and output gate. A block operates on an input sequence and each gate within a block uses the sigmoid activation function to control whether or not they are activated. By using back-propagation, it trains the model. As current pollutant concentrations are influenced by the past concentration of pollutants and past meteorological factors, LSTM was used to derive the time-series features. Also, LSTMs are popular and efficient for analyzing sequential data and time series, and it is used in this research study. Before feeding the data to the RNN model, data collection, analysis for feature correlation, and data preprocessing were carried out to get better prediction from the RNN model. All these data were collected from the Central Pollution Control Board (CPCB) website. The data collected contained daily $PM_{2.5}$ values and various meteorological parameters. The best meteorological parameters for the study were selected using correlation method. Then data preprocessing was carried out to handle missing data in the dataset. After that, by setting the best hyperparameters, the LSTM-RNN model was constructed and optimized. Then, the model was trained on the training dataset and then predictions were made on the test dataset. Finally, the model was evaluated using statistical measurement parameters to verify it's performance.

3.1. Study Area
This study is conducted for the southernmost district of Kerala named Thiruvananthapuram. For the analysis, the monitoring station located at Plammoodu (Latitude: 8.51N, Longitude: 76.94), Thiruvananthapuram, Kerala's capital, was selected. The Kerala Pollution Control Board owns and maintains this station. The rapidly rising urban areas and the development of housing, business and industrial units have caused the air to become increasingly polluted, thus raising the area's pollution rate

3.2. Data Set
The dataset consists of the daily values of meteorological parameters and daily values of $PM_{2.5}$ pollutant concentrations ($\mu g/m^3$). The meteorological parameters were selected after correlation based feature selection process. The parameters chosen are average temperature, wind speed, wind direction and relative humidity. The data of 914 days, from 1st July 2017 to 31st December, 2019 have been collected. The details of the data set are listed in Table 1.

Table 1. Details of the dataset.

Location	Plammoodu, Thiruvananthapuram
Unit	Daily
Dataset span	914 days
Meteorological Parameters	Average temperature, wind speed, wind direction and relative humidity
Air pollution parameter	$PM_{2.5}$

The actual distribution of $PM_{2.5}$ values over the above specified period is given in figure 1. The figure showed that the peak values of $PM_{2.5}$ are observed in the winter season, i.e., from November to January. It is also noted that the concentration of $PM_{2.5}$ is small during the rainy season, i.e., in the months of June, July and August.

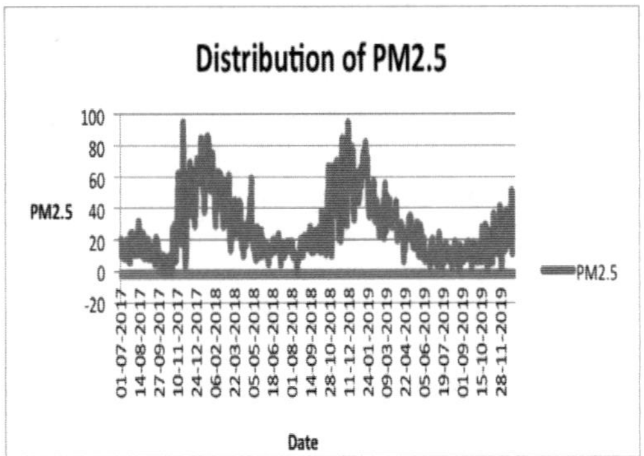

Figure 1. $PM_{2.5}$ distribution from July 2017 to Dec 2019

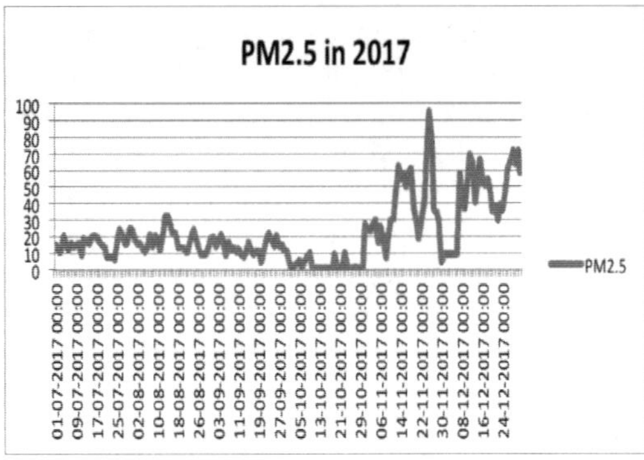

Figure 2. $PM_{2.5}$ distribution from July 2017 to Dec 2017

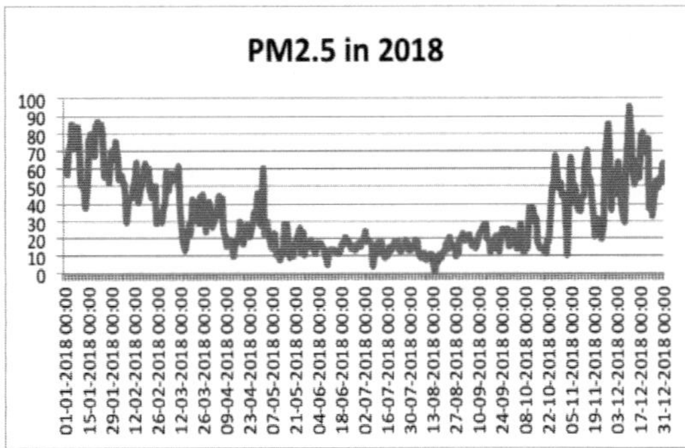

Figure 3. $PM_{2.5}$ distribution from Jan 2018 to Dec 2018

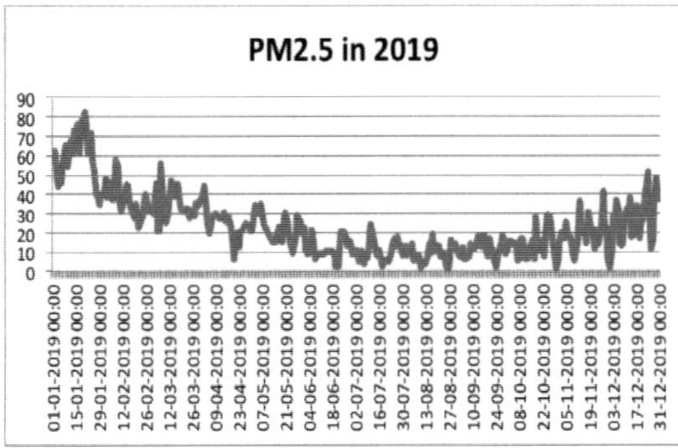

Figure 4. $PM_{2.5}$ distribution from Jan 2019 to Dec 2019

Clearly, it can be stated that the $PM_{2.5}$ distribution is not stationary, and the data values are time-dependent. Year-wise distribution of $PM_{2.5}$ in 2017, 2018 and 2019 are given in figure 2, figure 3 and figure 4 respectively. From the year wise distribution of $PM_{2.5}$ depicted in the plots, showed that there is slight decline in the daily $PM_{2.5}$ values over the years from 2017 to 2019. This is a positive indication that air pollution in the area is getting reduced.

3.3. Data Pre-processing

The data is pre-processed before it is given for RNN model training. The pre-processing steps used in this study are described here:

Missing value handling - The dataset contained missing values due to instrument malfunction, power failure or connection issues. Those values are imputed using iterative imputation model. Iterative imputation refers to a process where each feature is modelled as a function of the other features. Each function is sequentially imputed, one after the other, enabling previous imputed values to be used in the prediction of subsequent features as part of a model.

Feature scaling - It is a method used to normalize the range of independent variables or features of data.

The MinMaxScaler class is used to transform the data. On the training dataset values, the scaling coefficients called Min and Max values are determined and then applied to scale the test dataset. The default value range returned by MinMaxScaler is 0 to 1. Google Colab environment is used for developing the model. In the model, there are five layers in total, including an input layer, three intermediate layers and an output layer. Activation function used here is relu and the optimizer used is Adam. The output layer contained only one node that produces daily $PM_{2.5}$ values. The whole $PM_{2.5}$ dataset is divided in to training dataset and test dataset. Out of 914 total instances, 80% of data (731 cases) is contained in the training dataset, and 20% data (189 cases) is used for testing the model. The LSTM model is made to fit on all of the training data. Here the LSTM is made to run for 1500 epochs. Then at each step, a prediction is made based on the test data. The $PM_{2.5}$ concentration has a range from 1 to 96, and its histogram plot is as shown in figure 5. The histogram is right-skewed.

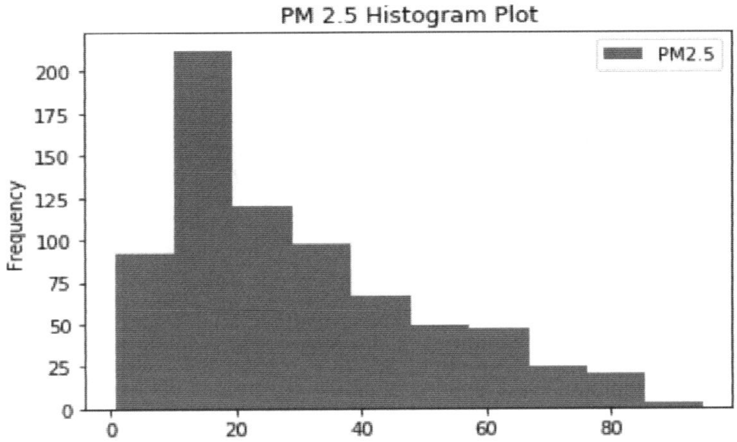

Figure 5. $PM_{2.5}$ values and frequency – histogram plot

4. Results and Discussion

After the training process, the model is used for making predictions. As the first stage, the model is trained on the training data, and then in the next stage, for each test record in the test data set, $PM_{2.5}$ value is predicted. The testing data are used to predict the $PM_{2.5}$ concentration values for a given day, and to test the performance of the proposed model. The root mean squared error (RMSE), mean absolute error (MAE), correlation coefficient (R^2), and mean absolute percentage error (MAPE) are employed to evaluate the performance of the model and the prediction accuracy. R^2 denotes the coefficient of determination, which is a statistical measure that denotes the amount of the variation for a dependent variable on an independent variable in a regression model. RMSE is a standard way to measure the error of a model in predicting quantitative data. The mean absolute error of a model is the mean of the absolute values of the individual prediction errors over all instances in the test data set. The models with lower RMSE and the higher R^2-values are said to be acceptable. Table 2 shows the values of the above said evaluation parameters on the LSTM model.

Table 2. Performance of the LSTM $PM_{2.5}$ prediction model.

Model	R^2	RMSE	MAE
LSTM	0.872	7.618	5.106

This is the initial machine learning-based research study conducted for the estimation of air pollutant $PM_{2.5}$, done in the Southern region of Kerala. Also, the study utilized the correlation of meteorological parameters in predicting the air pollutant concentrations

5. Conclusion

The findings demonstrated that the LSTM model could be used with proper consistency to predict the daily $PM_{2.5}$ concentration in the Thiruvananthapuram area. This forecasting model data can help the government officials to take the appropriate steps and provide the public with early notices to take safety precautions. As this is a preliminary research work carried out for the estimation of $PM_{2.5}$ using basic LSTM model, there exist opportunities for utilizing other advanced RNN models for $PM_{2.5}$ estimation. The model can be enhanced by the consideration of the multi-station data, different air pollutant concentrations and meteorological data.

6. References

[1] Brunekreef Bert and Stephen T Holgate 2002 *The lancet* **360** 9341 1233-42
[2] Yin P, Guo J, Wang L, Fan W, Lu F, Guo M, Moreno S B, Wang Y, Wang H, Zhou M and Dong Z 2020 *Environmental Science & Technology Letters* **7** 2 95-101
[3] Madrigano J, Kloog I, Goldberg R, Coull B A, Mittleman M A and Schwartz J 2013 *Environmental health perspectives* **121** 2 192-6
[4] Wu X, Wang Y, He S and Wu Z 2020 *Geoscientific Model Development* **13** 3 1499-151
[5] Masood A and Ahmad K 2020 *Procedia Computer Science* **167** 2101-10
[6] Zhang B, Zhang H, Zhao G and Lian J 2020 *Environmental Modelling & Software* **124** 104600
[7] Wen C, Liu S, Yao X, Peng L, Li X, Hu Y and Chi T 2019 *Science of the total environment* **654** 1091-9
[8] Park Y, Kwon B, Heo J, Hu X, Liu Y and Moon T 2020 *Environmental Pollution* **256** 113395
[9] Kaur M and Mandal A 2020 *Proc. Int. Conf. on Contemporary Computing and Applications* (Lucknow/IEEE) pp 40-3
[10] Yuan H, Xu G, Lv T, Ao X and Zhang Y 2020 *Analytical Letters* 1-2
[11] Xing H, Wang G, Liu C and Suo M 2020 *Neural Networks* **133** 157-65
[12] Geetha S and Prasika L 2019 *International Journal of Mechanical Engineering and Technology* **10** 1 1026-32
[13] Lu H, Song J, Di T, Kurdestany J M and Wang H 2018 *Tehnički vjesnik* **25** 2 519-27
[14] Zhou Y, Chang LC and Chang FJ 2020 *Science of The Total Environment* **711** 134792
[15] Le VD, Bui TC and Cha SK 2020 *Proc. Int. Conf. on Big Data and Smart Computing* (Pusan/IEEE) pp 55-62
[16] Qadeer K, Rehman WU, Sheri AM, Park I, Kim HK and Jeon M A 2020 *Applied Sciences* **10** 11 3984
[17] Ma J, Ding Y, Cheng JC, Jiang F and Wan Z 2019 *Journal of Cleaner Production* **237** 117729
[18] Mandic D and Chambers J 2001 *Recurrent neural networks for prediction: learning algorithms, architectures and stability* Wiley
[19] Rather AM, Agarwal A and Sastry VN 2015 *Expert Systems with Applications* **42** 6 3234-41
[20] Hochreiter S and Schmidhuber J 1997 *Neural computation* **9** 8 1735-80

Acknowledgement

We thank the Central Pollution Control Board (CPCB), Government of India for providing the dataset, needed to carry out this work.

Performance Analysis of Circular Shape Microstrip Antenna for Wireless Communication System

S.Kannadhasan[1], R.Nagarajan[2]
[1]Assistant Professor, Department of Electronics and Communication Engineering, Cheran College of Engineering, Tamilnadu, India, kannadhasan.ece@gmail.com
[2]Professor, Department of Electrical and Electronics Engineering, Gnanmani College of Technology, Tamilnadu, India, krnaga71@yahoo.com

Abstract:-In this paper proposes a broad band circular strip antenna based on numerical techniques. For the operating frequencies, a circular patch antenna is installed on FR-4 substrate material 4.2. The proposed antenna uses the finite element method to design an operating frequency from 2GHz to 10GHz. Substrate height is 1.6mm. The circular patch has a circumference of 8.82mm. The antenna displays good output below -16dB is the return loss value and the antenna gain is obtained in the range from 2.4dBi to 10.02dBi. The proposed circular patch antenna is designed and analysed the various parameters like VSWR, Gain, Return Loss, Directivity and Radiation pattern.

1. Introduction

Most of the microstrip antennas have different applications, particularly in the communications, defence and medical fields. In nature, the microstrip antennas are small in size and light in weight. The microstrip antenna is demand due to its more bandwidth available for wireless applications with the correct operating frequency. Most microstrip antennas are used to operate at lower lever frequencies to the higher frequency frequencies of microwave frequencies. The microstrip antenna consists of a patch above the dielectric substratum, and is grounded below the dielectric material [1]-[5]. In, a small rectangular double-band microstrip two separate slotted antennas are produced rectangular single-band microstrip antennas of L and plane slotted. It consists of a rectangular feed patch. The radiation pattern in the height is directional Azimuth plane and Omni-directional. The detailed analysis of patch antenna and how a probe is built and rendered of the IE3D circular microstrip patch antenna [6]-[8].

Microstrip square patch antenna was proposed are medium weight, low profile, low development expense, durability and manufacturing ease. Wireless technology convergence allows it appealing for applications such as high efficiency for aircraft, spacecraft, satellite applications, rocket applications and embedded applications. They have poor radiation quality, however, low power, high Q and very small bandwidth frequency. Microstrip patch antennas are named according to their shape of the patch of radiation. There are various radiation patch types available, including rectangle, rectangular, circular, elliptical, triangular, circle and ring, square, rectangular and circular antennas are simple to use to build and evaluate all antenna parameters. Patch antenna is better to build and can easily be monitored for radiation [9]-[14]. Furthermore the physical size circular patch antenna at the same design frequency is 16 percent smaller than the rectangular antenna. Microstrip antenna has a radiation patch on one part of the dielectric substrate and has a floor plane. The other hand of the substratum and its radiation from the fringing areas between the parchment periphery and plane of the field.

2. Mathematical Model Design of Proposed Antenna

The antenna have modelled with the aid of electromagnetic simulator on circular microstrip patch antenna with acceptable operating frequencies from 2GHz to 10GHz using finite element model theory The proposed antenna is designed to operate the different frequencies on a FR4 substratum with a dielectric constant 4.2 and substrate height 1.6mm. The patch antenna is a material used for copper conductors that operates on the finite element model theory. The surface mesh is used for triangular shaped mesh, and for tetrahedron shaped with two and three dimensional regions, respectively. The design parameter is shown in Figure 1 only for the patch radius with the help of dominant TM_{10} mode.

Figure.1. Circular Patch Antenna Structure

The radius of an antenna with circular patch is 8.22mm and the circular patch antenna length (L) and width (W) are measured by using the formula at 8.4mm and 4.2mm, respectively. The feed location is used for impedance of the microstrip line at 50ohm. Many of the microstrip patch antenna nowadays have revolutionary ideas of miniaturization design and production. The greater part of the microstrip patch antenna is used as quick and most of the microstrip patch antenna are used as simple and also presented with the EBG structure.

The proposed antenna has the operating frequencies that are typical for GPS applications with the slot radiation. The circular patch antenna is used in microwave applications with wireless links and other sources. The proposed antenna has been designed to miniaturize the antenna size for wide bandwidth GPS applications. Narrow bandwidth is the greatest drawback of the microstrip patch antenna. The proposed antenna is used to build the emphasis to increase the bandwidth and the limit is also low return loss with antenna performance. Table 1 shows design parameters of circular patch antenna.

. **Table 1.** Design Parameters of Circular Patch Antenna

Sl.No	Parameters	Range
1.	Operating Frequencies	2 GHz to 10 GHz
2.	Radius of the Patch	8.82 mm
3.	Substrate dielectric material	FR4
4.	Dielectric Constant	4.2
5.	Thickness of Substrate	1.6mm
6.	Feeding techniques	Line fed
7.	Ground Plane	30mm X 30mm
8.	L*W	8.4mmX4.2mm
9.	Impedance	50Ω
10.	Bandwidth	180MHz

Most satellite applications require the antenna which is a gain feature to enhance the unidirectional, and the multidirectional is used to analyze the system. The patch antenna width increases, and efficiency, as well as bandwidth decreases. The antenna proposed for FEM model can be designed to be resonant in transverse magnetic modes. The various applications are used in everyday life for antenna design with the necessary wireless requirements, small in size and light weight. The microstrip patch antenna is used for different shape antenna designed with suitable configurations and is modified using different with rectangular, circular and triangular patches. Based on the electromagnetic spectrum the various parameters are analysed with the help of different slots in TM_{10} is shown in the Figure 2.

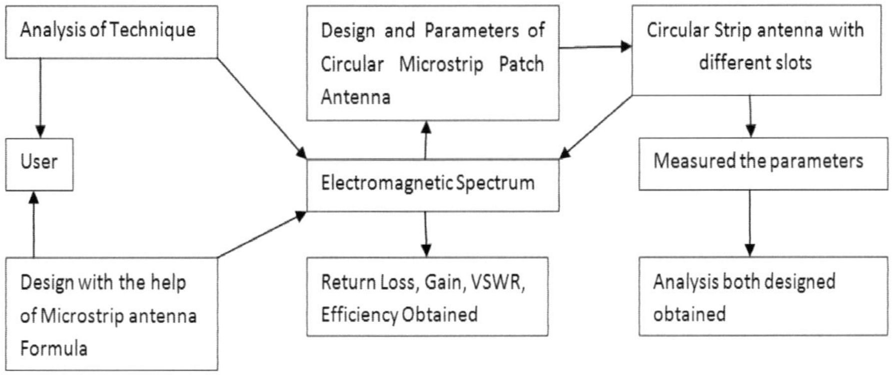

Figure 2. Block Diagram of the Proposed Circular Patch Antenna

The radius patch of the antenna is measured by using the formula. The radius patch of the antenna is measured by using the formula

$$a = \frac{F}{\left(1+\frac{2h}{\pi\varepsilon_r F}\left(\ln\left(\frac{\pi F}{2h}\right)\right)+1.7726\right)}\frac{1}{2} \quad \text{----[1]}$$

Where

$$F = \frac{8.791X10^9}{f_r\sqrt{\varepsilon_r}} \text{---------[2]}$$

DGS is carried out by applying a defected form to a floor plane such that the shielded current delivery can be disrupted based on the shape and the dimension of the defect. The interruption of the shielded current distribution will impact antenna input impedance and current flow. The agitation and electromagnetic waves may also be regulated by the layer of the substrate. The simple TM11 resonant mode for a non-slotted circular patch microstrip antenna may be divided into two independently resonant modes similar to one another and the antenna has characteristics for a dual band antenna by choosing different sizes of the circular slots in the patch.

3. Results and Discussion

The proposed antenna shows 2GHz to 10GHz at resonant frequency and obtained a return loss value of less than -16dB is shown in Figure 3. The proposed antenna of VSWR is achieved in the order of 2 as shown in figure 4. The results indicate that the patch and feed lines match perfectly with impedance. Figure 5 shows the antenna gain of the proposed antenna being achieved for 2.2dBi to 10.2dBi. The various parameters analysis of the proposed circular patch antenna is shown in table 2. The antenna is an essential element of the transmission chain since it affects the signal propagation and receiving in free space depends on the form of protected area network and pollution standards to select a circular antenna. Table 1 analyzes the various design parameters and shows Design Parameters of Circular Patch Antenna strong results of VSWR, Gain, Return Loss, Directivity and Radiation pattern.

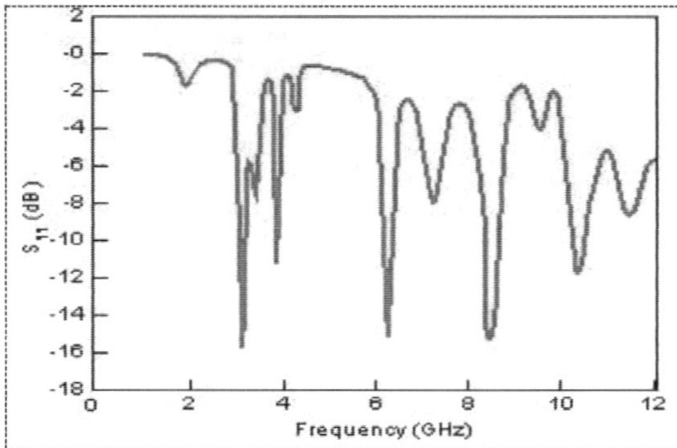

Figure 3. S_{11} of the Circular Patch Antenna

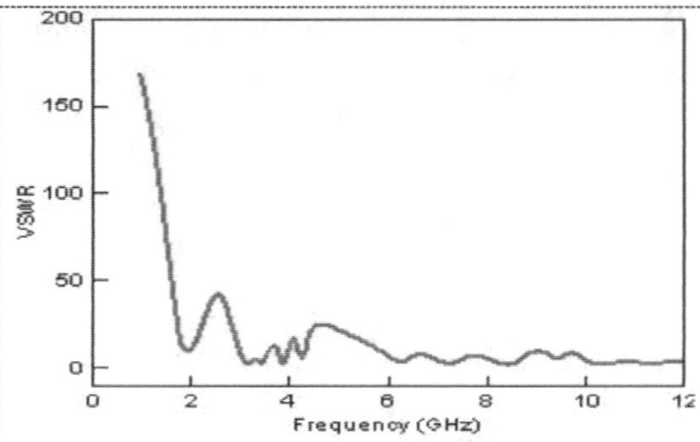

Figure 4. VSWR of the Circular Patch Antenna

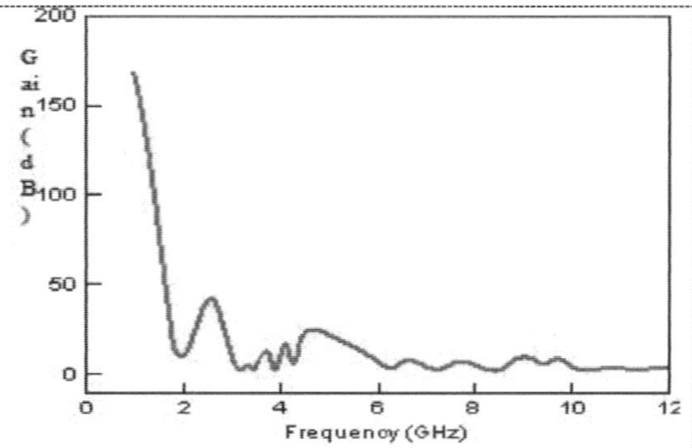

Figure 5. Gain of the Circular Patch Antenna

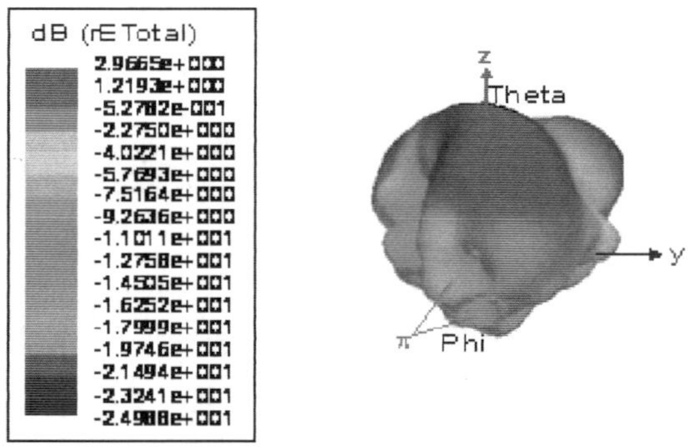

Figure 6. 3D Gain Plot of Circular Patch Antenna

Table 2. Analysis of the Proposed Circular Patch Antenna

Operating Frequency (GHz)	Gain (dB)	S11 (dB)	VSWR	Efficiency (%)
2.5	1.2	-5	1.18	60
3.5	1.4	-8	1.28	65
4.5	2.6	-9.5	1.3	72
5.5	3.2	-10.25	1.35	78
6.5	5.6	-12.5	1.56	80
7.5	7.8	-14.68	1.68	82
8.5	8.2	-15.25	1.89	85
9.5	10.2	--16	1.98	88

An antenna is associated with the source through a function of impedance transfer line 50Ohms to ensure the optimal transmitting of power between power and antenna, so it is important to ensure matching impedance the adaptation makes the cancellation of the reflective coefficient S11 the antenna input. Figure 7 shows the radiation efficiency of the proposed antenna at 89.1 percent for E-field. Figure 8 shows the radiation efficiency of the proposed antenna at 88.1 percent for H-field. The antenna proposed is to design the number of circular slots to be displayed and the parameters obtained for the various features for the wireless. During the coupling period the isolation was achieved with the help of electromagnetic signal to reduce the fading of the signal and gain is also reached to the maximum level. The return loss for the different circular slots is estimated in the order of less than -16dB with a bandwidth of 180MHz.

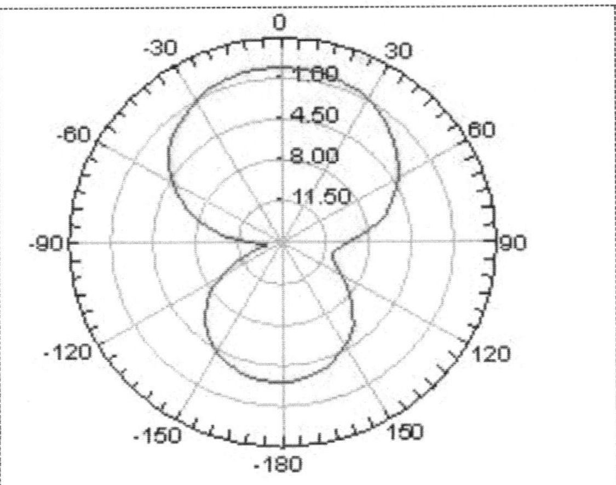

Figure 7.a. Radiation Pattern of Circular Patch Antenna for both Upper Resonant Frequency and Lower Resonant Frequency at E-Plane

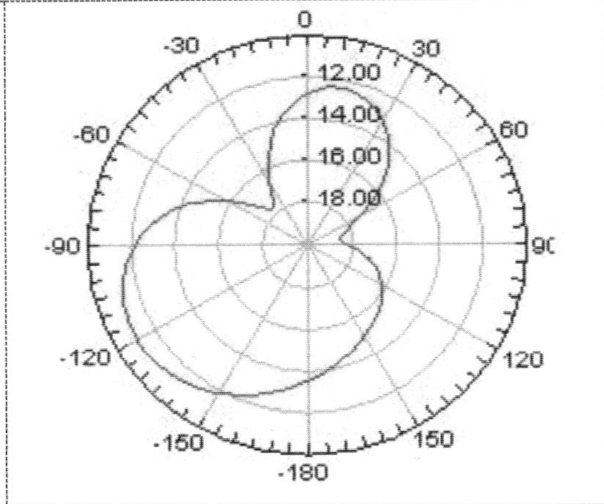

Figure 7.b. Radiation Pattern of Circular Patch Antenna for both Upper Resonant Frequency and Lower Resonant Frequency at E-Plane

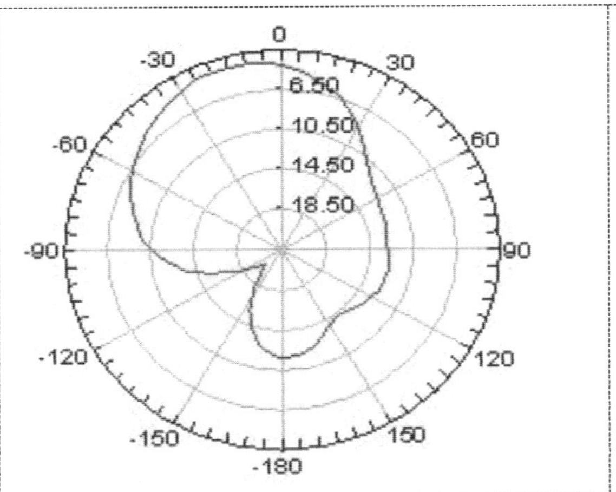

Figure 8.a. Radiation Pattern of Circular Patch Antenna for both Upper Resonant Frequency and Lower Resonant Frequency at H-Plane

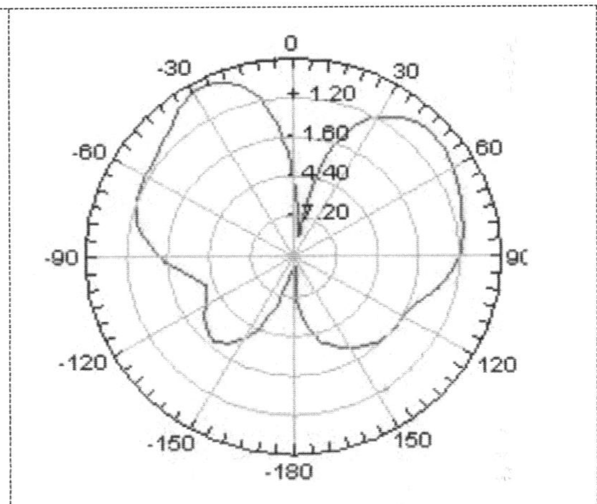

Figure 8.b. Radiation Pattern of Circular Patch Antenna for both Upper Resonant Frequency and Lower Resonant Frequency at H-Plane

4. Conclusion

This paper shows strong results for taking advantage of the low profile antenna in various antenna parameters. The antenna patch and feed position measurements are used to realize matching with the antenna output and impedance. The antenna displays good output below -16dB is the return loss value and the antenna gain is obtained for 2.2dBi to 10.02dBi. VSWR is achieved in the range of 2. The results of the designed antenna are also used to measure the performance of the proposed antenna for Wireless applications.

References

1. Vasujadevi M, Dr.P Siddaiah, 2019, "Design and simulation of array DGS using HFSS", *International Journal of Innovative Technology and Exploring Engineering*, pp. 47-49,Volume 8, Issue 3,
2. Togcuoglu, N. B., Albayrak, Y., Saylik, M. N., Daye, M. A., Bal, M., Imeci, M., & Imeci, T, 2017, "Circular patch antenna with circular and rectangular slots", *25th Signal Processing and Communications Applications Conference (SIU)*

3. Ayyappan, Manoj. B, Jagadish Chadran, March 2016, "Design and Analysis of Circular Microstrip Antenna at 5.8GHz with Fr-4 Substrate," *International Journal of Advanced Research in Electrical, Electronics and Instrumentation Engineering*, Vol. 5, Special Issue 4.
4. K.Soundarya, S.Saravanan, R. Srividhya, May-2016, "Design of multi-band reconfigurable square spiral antenna for WLAN & GPS applications," *International Journal of Advanced Computational Engineering and Networking* "Volume -4, Issues-5.
5. S Nagakishore Bhavanam, 2016, "Design of a Novel Coaxial Feed Triple Frequency Patch Antenna with Slots and Shorting Pin", *ELSEVIER Journal "Procedia Computer Science*", ISSN: 1877-0509, Vol 85, pp.345–351.
6. Ken Paramayudha, AriefBudi Santiko, Yuyu Wahyu, Folin Oktafiani, " Design and Realization of Circular Patch Antenna for S-Band Coastal Radar", *2016, International Conference on Radar, Antenna, Microwave, Electronics, and Telecommunications (ICRAMET)* , DOI: 10.1109/ICRAMET.2016.7849595
7. Keshav Gupta, Kiran Jain, Pratibha Singh, 2015, "Analysis and Design of Circular Microstrip Patch Antenna at 5.8 GHz," *International Journal of Computer Science and Information Technologies*, Vol. 5 (3), 2015, 3895-3898
8. N. Boskovic, B. Jokanovic, F. Oliveri, D. Tarchi, 2015, "High Gain Printed Antenna Array for FMCW Radar at 17 GHz" *12th International Conference on Telecommunication in Modern Satellite, Cable and Broadcasting Services (TELSIKS)*, DOI: 10.1109/TELSKS.2015.7357760
9. V. P. Sarin, V. Deepu, C. K. Aanandan, P. Mohanan, and K.Vasudevan, 2009, "Wideband printed microstrip antenna for wireless communications, "*IEEE Antennas Wireless Propag. Lett.*, vol. 8, pp. 779–781,.
10. C. H. Chang and K. L. Wong, 2009, "Printed -PIFA for penta-band WWAN operation in the mobile phone," *IEEE Trans. Antennas Propag*., vol. 57, pp. 1373–1381.
11. Nasimuddin and Z. N. Chen, 2009, "Wideband multilayered microstrip antennas fed by coplanar waveguide-loop with and without via combinations," *IET Microw. Antennas Propag.*, vol. 3, pp. 85–91,.
12. S. Kannadhasan and R.Nagarajan, September 2020, "Performance Design and Development of Environmentally Safe W-Shaped Structure Antenna for Wireless Applications," *Journal of Green Engineering (JGE)*, Volume-10, Issue-9, pp-4558-4565,
13. M. John and M. J. Ammann, 2007, "Wideband printed monopole design using a genetic algorithm," *IEEE Antennas Wireless Propag. Lett.*, vol.6, pp. 447–449,.
14. S. K. Oh, H. S. Yoon, and S. O. Park, 2007, "A PIFA-type varactor-tunable slim antenna with a PIL patch feed for multiband applications," *IEEE Antennas Wireless Propag. Lett.*, vol. 6, pp. 103–105,.

IOP Conf. Series: Materials Science and Engineering 1085 (2021) 012014 doi:10.1088/1757-899X/1085/1/012014

Segmentational color map analysis to identify retinal detachment

L. Poongothai[1], Dr. K. Sharmila[2]

[1]Research Scholar, Department of Computer Science, VISTAS & Assistant Professor, Department of Computer Science Dr.MGR Janaki College for Women, Chennai, India.
[2]Associate Professor, Department of Computer Science, VISTAS, Chennai, India.

Abstract: The recent perilous times entails the importance of good health for humans. This can be pertaining the maintenance of good health in various organs of the human body. With the increase of digital platforms, and the necessity to stay longer hours with gadgets and electronic devices, the human eye takes a toll in terms of absorbing radiations that maybe harmful. An imperative check of the eye is essential from time to time, as the luminous membrane that aids in vision needs to hold firm to the various components within itself. Although there are various treatments associated with the detection of eye detriments, several gremlins tend to loom at large. These eye diseases can arise from numerous reasons ranging from genetic disorders to accidents and vision loss due to stimulated injuries, infections and dysfunctional development of tissues/cells/nerves within the eye. Retinal detachment (RD) is one of the profound problems that persist amongst many individuals scaling over a diverse range of ages. The risk in relevance to untreated detachment of the eye leads to purblind and sometimes complete loss of vision. The previous paper concerning retinal detachment of human eye has focused on various algorithmic and post-processing approaches, but failed to identify the origination and symptoms that could aid in meticulous diagnosis. The indagation in this paper pivots on the unsheathing of attributes that explicates the disassociation of the retina from the underlying layers of the eye using the fundus image with cost-effectiveness and swifter analysis. An important characteristic that is emphasized in this study is the segmentation of eye floaters in the vitreous fluid of the eye. These eye floaters are color mapped to identify the presence of detached tissues in the eye. The histogram analysis using the grayscale thresholding value further aids in providing a detailed distinction of the normal eye from the detached eye. The simulations are carried out in MATLAB, and the results have been obtained successfully.

1. Introduction

The human eye is a cardinal organ of the human body. This organ of the human body is compared and holds close relevance to a camera in the way that it processes the image through light and signal passing. The tri-layered structure of the human eye is segregated into chambers on the inner portion. Each layer holds significant importance in enabling vision. The sclera is the outer layer membrane which is opaque, and consists of the cornea with the mild bulge in front. The intermediate layer is the choroid, it is a vascular portion holding the tethering tissues which connect the membranes. This choroid coating lies between the sclera and the retina [15], and can have varying depths of thickness. Pro-centered in the choroidal layer is the iris, with a small aperture called as the pupil [8]. The inner-most structure is the retina which comprises of two sub-layers for visual, sensory processing and for signal generation [9]. The anterior, posterior and the vitreous chambers of the eye fulfill the functionalities of vision through the lens aperture and capturing the images using the macula in the retinal lobe. These chambers regulate the functioning of the retinal detailing, pupil contraction, vitreous fluid generation and maintaining the elasticity of the membranes [15]. The correlation between the retina, the fluids, lens and the cornea are depicted in the figure below.

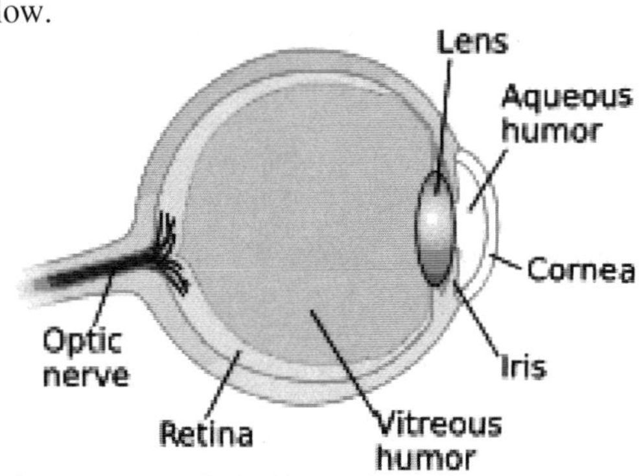

Fig 1. Anatomy of the Eye with relevance to Retina

The interrelation that the retinal blood vessels have with the concomitant parts helps in sending the right signal to the brain to effectuate normal vision. However, all these functionalities can be in jeopardy if there are any pre-existing diseases or genetically transmittable inherent detriments that may have an impact on the retina [1]. If the retina is not conjoined at the earlier phases, permanent vision loss may subsequently follow [10]. Depending upon the degree of detachment in the ocular region, RD can be stratified into rhegmatogenous, tractional and exudative [2], [7].

Retinal detachment in the most traditional form of clinical analysis is evaluated using the ultrasound-hinged mode of diagnosis. Ophthalmologists consider this tool of evaluation as irrefutably conventional mode of testing the defective retinal detached eye of an individual [5], [6]. Nonetheless, the challenges pertaining this ultrasound mode of evaluation lie in the obscure and non-transparent nature of the vitreous fluid that flows through the retinal membrane. This opaqueness can be caused due to the existence of bleeding, cataract and aberrations in the vitreous fluid, thereby making it an onerous task for light to seep through. Bio-medical depictions and representations in the form of Magnetic Resonance Imaging (MRI) [3], Optical Coherence Tomography (OCT) [4], color fundus photography [2] and ultrasound imaging modes of operation have been extensively explored for multifarious studies adopted for enhancing the diagnosis of the visual lurgies, malaises and detriments.

This paper analyzes to determine the attributes that constitute in giving rise to the detachment of the retinal tissues in the human eye. This distinct differentiation cognized between a normal human eye with no detachments, and one that holds disengaged tissues is analyzed through the histogram analysis of thresholding, and through the eye floaters which exceed in number in the vitreous gel. This agnizing of the symptoms usually take expensive scanning through clinical procedures. The proposed indagation identifies these symptoms through mitigated costs and faster time of processing by utilizing the image processing and segmentation techniques. The paper is regimented with section II expounding the existing methodology, the proposed study is explicated in section III. Section IV delineates on the results, and the final section provides a compendium of the work with the future implementation to be done.

2. EXISTING METHODOLOGY

The existing research pertaining to retinal detachment have been explored in various domains and clinical practices. The ultrasonographic analysis have been used in various image pre-processing analysis [12],[13],[5] and detailed by various researchers in the field. The ROI detection of the detached section with adequate computation of the disassociated distance has augmented in identifying the tear more precisely [5], [6]. Further exploration on the existing methodologies also entailed the detachment identification by evaluating the thickness of the choroid gel through the percentage area calculation of the detached retina (PADR) [14].

3. PROPOSED INDAGATION

Retinal detachments are identified through a list of parameters that enable in analyzing the symptomatic changes in the human eye. These attributes may include the following and are not necessarily restricted to the below symptoms [16]:

- Increase number of eye flashes otherwise called as photopsia.
- Quantitative increase in the number of eye floaters in the vitreous fluid.
- Loss of vision.
- Mitigated peripheral vision.
- A lingering shadow over the visual cortex.
- Red clusters leading to dilated pupil
- Tunnelled vision
- Scarred retina leading to fibrosis.

The clinical practices and existing methodologies have had extensive research in identifying and precisely locating the tear of retina in the human eye. Nevertheless, the cause of origination is a lesser trodden path. The indagation presented follows a series of computational and pre-processing methods to agnize this germination of detachment for an individual through the segmentation of eye floaters observed in the vitreous humor. Figure 3 presents a graphical explanation of the sequence of evaluations done for identifying the number of floaters in the eye.

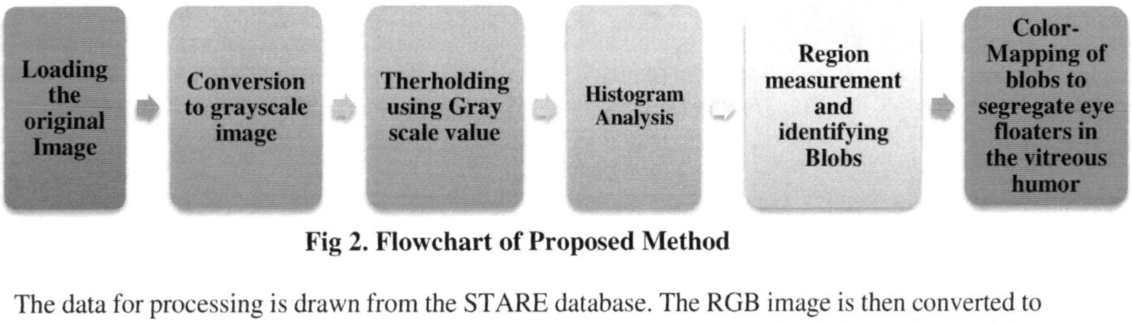

Fig 2. Flowchart of Proposed Method

The data for processing is drawn from the STARE database. The RGB image is then converted to a grayscale image for initial preprocessing and noise reduction. Preprocessing bolsters the precision and interpretability of an image. The aim of pre-processing is to generate an improvised image that suppresses unwanted distortions, and parallelly enhances some image features which are vital for further processing and characterization. An image with good quality helps in rendering the right pixel accuracy of the targeted features.

The reason for converting the RGB image to a grayscale image is to compress the image to lesser number of bits as compared to the former. The grayscale image can also help in swifter processing of the subsequent methods when compared to the RGB image. The ensuing step after this image transformation is the method of thresholding. The process of segmenting the image based on the pixel ranges is performed through this method [21]. This study utilizes the grayscale value for thresholding and converts it into a binary image comprising pixels which can either be 0 or 1. The image thresholding method is computed through the below formula [22]:

$$f(m,n) = \begin{cases} 1 \; if \; g(m,n) \geq Thresh \\ 0 \; Otherwise \end{cases}$$

Where f(m,n) indicates the image procured from the threshold grayscale image pixel analysis, and the g(m,n) refers to the coordinate points m and n in the chosen image.

The next process of execution entails the histogram analysis which signifies the intensity distribution of the grayscale pixels in the image. This graphing method can also indicate the frequency of occurrence of a pixel value in the chosen image [17]. The histogram analysis for this study is depicted in the following section for a normal and a detached eye. Since this study focusses on identifying the parameters which best indicate the detachment of retina in the eye, the parameter of increasing eye floaters is taken into consideration. Eye floaters [23] are those that curtail the vision of the eye, and holds the property of floating in the vitreous fluid. Figure 4 depicts these structures more explicitly.

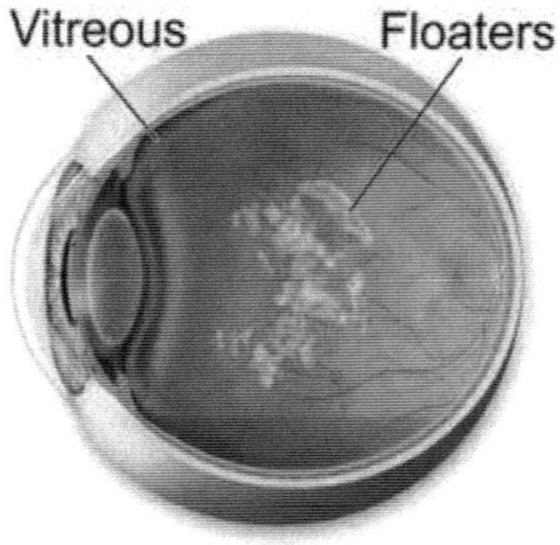

Fig 3. Eye floaters in the vitreous gel

In order to segment the number of floaters, it is necessary that the image holds different pixel intensity for the floaters observes in the vitreous gel. This motivated the incorporation of color-mapped pixel analysis [21], which helps in agnizing the structures that exist in the image. This is therefore implemented by using the label2rgb function that helps in perceiving the labelled compartments in different pixel intensities. The label matrix choses the intensity of the pixels based on the color map, pixel coordinate position and the quantity of structures that are to be labelled [20]. The outcome of this process helps in identifying the eye floaters which increase when there is detachment of retina. The region property [19] delivers the various attributes such as the area, centroid, eccentricity, orientation, pixel values, mean intensity, minimum and maximum intensities associated for the chosen image. These properties are a great aid in differentiating the retinal detached eye to that of the normal eye in a significant manner. The result of the cluster of processes is shown in the subsequent section.

4. SIMULATION RESULTS

The simulation of this study is executed in MATLAB, the initial process entails the original fundus image obtained from the STARE database to be given as input. The flow of execution is designed in a manner that any image when loaded, is converted to a grayscale image for further processing. Figure 4 illustrates the initial steps of implementation for the fundus image procured for retinal detachment.

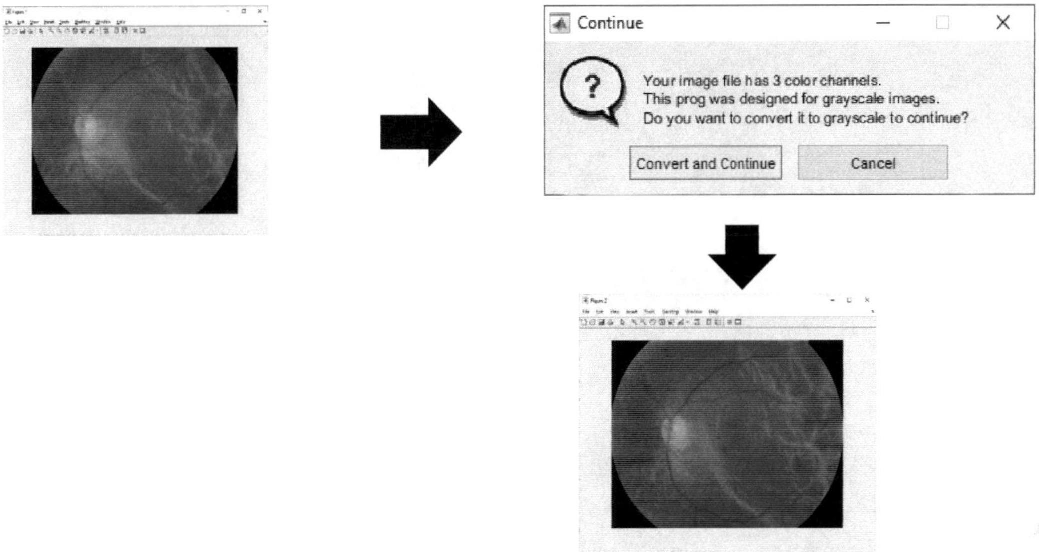

Fig 4. Image Loading and Conversion to grayscale

The next outcome of the process is indicative of setting the grayscale value to perform the process of thresholding and histogram analysis. This grayscale value is fine-tuned according to the user's perspective of the desired outcome from the image chosen. The value of grayscale chosen for this study is equated to 112 for the detached and normal eye. The ensuing result depicts the thresholding segmentation of the grayscale to a binary image along with the histogram result which identifies the pixel intensities observed for the same.

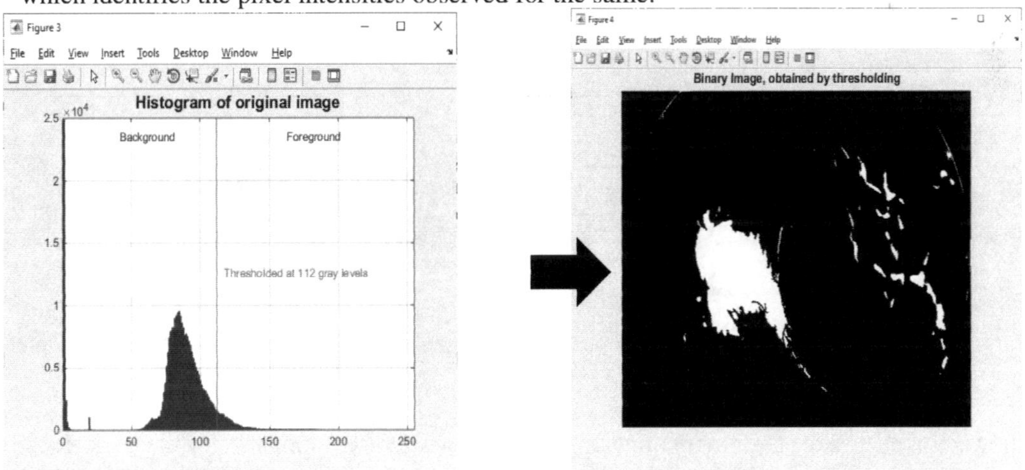

Fig 5. Histogram analysis and thresholding to obtain Binary Image

The final result is crucial in understanding the increase in the number of eye floaters observed when there is retinal detachment. The process of color-mapping illustrates the eye floaters more distinctively than what is observed in a normal eye. Figure 6 presents the color-mapped segmentation of the eye floaters which helps in the facile analysis of the detachment.

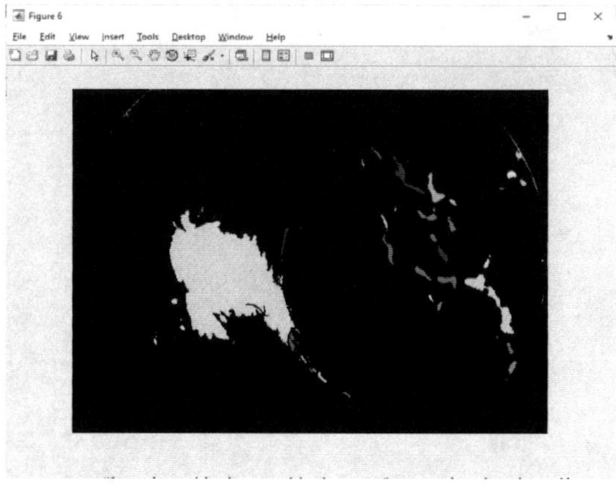

Fig 6. Segmentation using color-mapped intensity matrix

The following figure illustrates a comparison drawn between a normal eye and a detached eye, where the histogram variations and the absence of floaters in seen. This helps in understanding the analysis in a better manner.

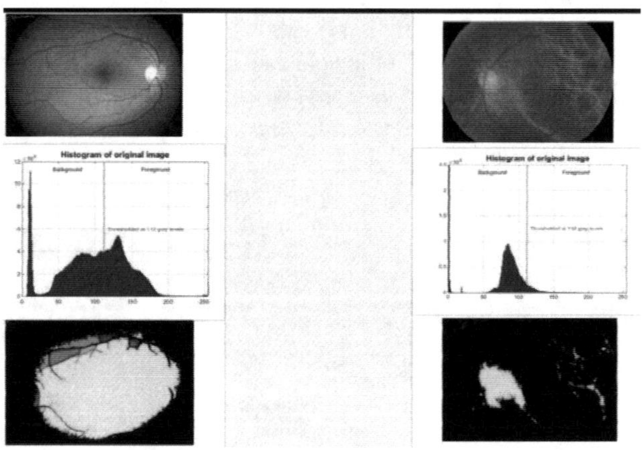

Fig 7. Comparison drawn from a Normal eye (Left) and the Retinal Detached eye in the right

The result depicts the uniformity of peaks in the histogram for a normal eye, while the sudden peak in the detached eye. The normal eye also indicates that there are no floaters in the vitreous gel, while the floaters seen in the detached eye is higher in number. The region of area for the detached eye when compared to the normal eye also significantly increased due to the detached layers, this parameter also has been helpful in classifying the retinal detachment from the database of fundus images chosen as input.

5. CONCLUSION AND FUTURE WORK

Eye floaters have been a cause of concern for effective vision in a human eye. However, the emphasis of this study is to efficaciously identify, detect and process the detachment in the human eye as compared to a normal eye. The indagation has successfully analyzed a parameter that best depicts the origination of retinal detachment from its underlying layers. This comparison done between the normal eye and the detached eye can help in increasing automated approaches instead of physical presence of patients to identify the floaters in the vitreous humor. It also proffers better cost-efficacy along with mitigated time consumption. The future work entails analyzing more parameters relevant to the detachment of retina through computational procedures.

REFERENCES

[1] Poongothai .L and Dr.K.Sharmila April 2020 Augmenting idiosyncrasy of retinal images by feature detachement using pixel thresholding (Alochana Chakra Journal) 2231-3990 Volume IX, Issue IV.

[2] Subhadra Jalali Smt. Kumari Santhamma 2003 Retinal detachment (Community Eye Health) Vol 16 No. 46.

[3] Damianidis, C. H., et al. 2010 Magnetic Resonance Imaging and Ultrasonographic Evaluation of Retinal Detachment in Orbital Uveal Melanomas. (The neuroradiology journal 23.3 () 329- 338).

[4] Jaffe, Glenn J., and Joseph Caprioli. 2004 Optical coherence tomography to detect and manage retinal disease and glaucoma American journal of ophthalmology 137.1 (): 156-169.

[5] Rajeev Gupta 2018 Automated Detection of Retinal Detachment from Ocular Ultrasound using Image Thresholding International Conference on Signal Processing and Integrated Networks

[6] Yuen Aoi Chee January 2019 Image Processing Analysis for Ultrasound Retinal Detachment Images Advances in Environment, Biotechnology and Biomedicine.

[7] LozaBekalo January 2019 Automated 3D retinal layer segmentation from SD-OCT images with neurosensory retinal detachment IEEE access.

[8] Coleman, D. Jackson, and Robert L. Jack. B-scan ultrasonography in diagnosis and management of retinal detachments Archives of ophthalmology 90.1 (1973): 29-34.

[9] Malhotra, A., F. J. Minja, et al. (2011) *Ocular Anatomy and Cross-Sectional Imaging of the Eye* Seminars in Ultrasound, CT, and MRI 32(1): 2-13.

[10] Sconfienza, L. M., F. Lacelli, et al. (2010). *"High-resolution, three-dimensional, and contrast-enhanced ultrasonographic findings in diseases of the eye"*. Journal of Ultrasound 13(4): 143-149.

[11] Chang HJ, Lynm C, Golub RM. *"Retinal Detachment"*. JAMA. [12] Shinar, Zachary, 2012;307(13):1447.doi:10.1001/jama.2012.320

[12] Linda Chan, and Michael Orlinsky. "Use of ocular ultrasound for the evaluation of retinal detachment." The Journal of emergency medicine 40.1 (2011): 53-57.

[13] Coleman, D. Jackson, and Robert L. Jack. "B-scan ultrasonograph in diagnosis and management of retinal detachments." Archives of ophthalmology 90.1 (1973): 29-34.

[14] Rajeev Gupta 2018 A novel method for automatic retinal detachment detection and estimation using ocular ultrasound image" Multimedia Tools and applications.

[15] Eye Anatomy and Function, University of Michigan, https://www.uofmhealth.org/health-library/hw121946

[16] Subhadra Jalali, Smt Kannuri Santhamma, 2003 Retinal Detachment (Community Eye Health) Vol.16, No.46.

[17] Komal Vij, Yaduvir Singh, "Enhancement of Images Using Histogram Processing Techniques", Int. J. Comp. Tech. Appl., Vol 2 (2), 309-313, ISSN: 2229-6093

[18] Mazen A.Hamdan , Prof. Ziad A.Alqadi , Bassam M.Subaih, November 2016 A Methodology to Analyze Objects in Digital Image using Matlab", IJCSMC, Vol. 5, Issue. 11, pg.21 – 28, ISSN 2320–088X.

[19] Nilima Kulkarni, February 2012 Color Thresholding Method for Image Segmentation of Natural Images International Journal of Image Graphics and Signal Processing, DOI: 10.5815/ijigsp.2012.01.04 .
[20] J. R. Parker September 1991 Gray level thresholding in badly illuminated images IEEE Transactions on Pattern Analysis and Machine Intelligence,.
[21] Rebecca Milston, Moptoma , Michele C. Madigan, Phdb,C , J. Sebag, Frcophth, 2016 Vitreous Floaters: Etiology, Diagnostics, And Management, http://dx.doi.org/10.1016/j.survophthal.2015.11.008,(Elsevier)

Implementation of Multivariate Logistic Regression Model for Cerebral Palsy Identification using Prenatal, Perinatal Risk Factors

K. Muthureka[1], U. Srinivasulu Reddy[2*], Mrs. B .Janet[3]

[1]Research Scholar, Department of Computer Applications, National Institute of Technology, Tiruchirappalli, India .
[2*,3] Department of Computer Applications, National Institute of Technology, Tiruchirappalli, India.
*Corresponding Author Eail ID:usreddy@nitt.edu

Abstract. Cerebral Palsy (CP), a static, neuro and motor disorder caused by brain injury in the time period of prenatal, perinatal and postnatal, is the major developmental disability affecting children's function. Children with CP in children cannot be curable but quality of life can be improve with the help of treatment such as surgery and therapy. Early identification is important to the CP children for starting the treatment. There are numerous Machine Learning (ML) algorithms used in health care for prediction and classification. One of the ML algorithms called Logistic Regression which is used for binary classification using univariate and multivariate. This study, is of interest to enable early identification of CP using prenatal and perinatal risk factors with help of Multivariate Logistic Regression.

1. Introduction

CP is a neurodevelopmental and motor control disorder, caused by brain injury at the time of birth or at the early childhood [1]. Children with CP face a lot of problems [2-5] such as mobility, communicating with others, intellectual disability, retinal issues, epilepsy and most of the children suffer body pain. Now-a-days, there are many ways to identify the CP in childhood through Birth History, Physical Examinations and Neuro Imaging. Diagnosing CP using Neuro Imaging is costly and developing countries provide limited facilities. Whereas Physical Examinations method need well-trained physicians. Finally, using Birth History is the easiest method of diagnose the CP.

Risk factors associated with CP can be grouped into three types [6,7] and these classification types are based on time period of child's birth, that is before birth called as prenatal, during birth called as perinatal and after birth called as postnatal.Figure 1 shows the classification of risk factor based on time peried.

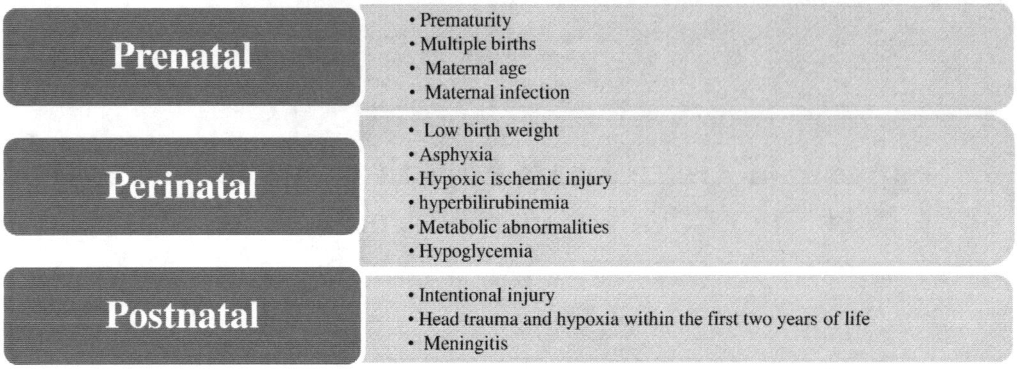

Figure. 1. Risk Factors of Cerebral Palsy

Perinatal asphyxia [8] is a major cause for CP. Birth asphyxia is a genuine global clinical issue and contributes enormously to morbidity and neonatal mortality [9].Reasons for perinatal birth asphyxia might be maternal or fetal. The individuals who endure asphyxia during childbirth may have opportunity to create neurological entanglements including epilepsy, cerebral paralysis and formative postponement [10]. Another risk factor of perinatal period is birth weight. Most of preterm babies are born under weight. They may fall under two categories suggested by World Health Organization (WHO) as born with Low Birth Weight (LBW) and Very Low Birth Weight (VLBW). Babies weighing less than 2500 grams come under Low Birth Weight (LBW) category and weighing less than 1500 grams comes under Very Low Birth Weight (VLBW) category, some babies may extremely low weight which is less than 1000 grams as per WHO guidelines. During child birth, Apgar score calculated is prominent As it is directly related to birth weight and gestational age [11]. Van et al. [12] and his team had been identified in the literature work that gestational age, birth weight, gender, and medical diagnosis for the mother are the risk factor of CP.

Gestational Age (GA) is prenatal risk factor of CP. Children born in preterm are highly affected by CP disorder. The risk increases with decrease in gestational age [13-15]. The risk factor of cerebral palsy below 33 weeks gestation is 30 times greater than those who are born at term and is approximately equals to 70 in 1000 deliveries. Figure 2 presents classification of birth term based on gestation week under WHO advise. Early prediction of CP is highly essential inorder to provide better treatment to the make their living easier.

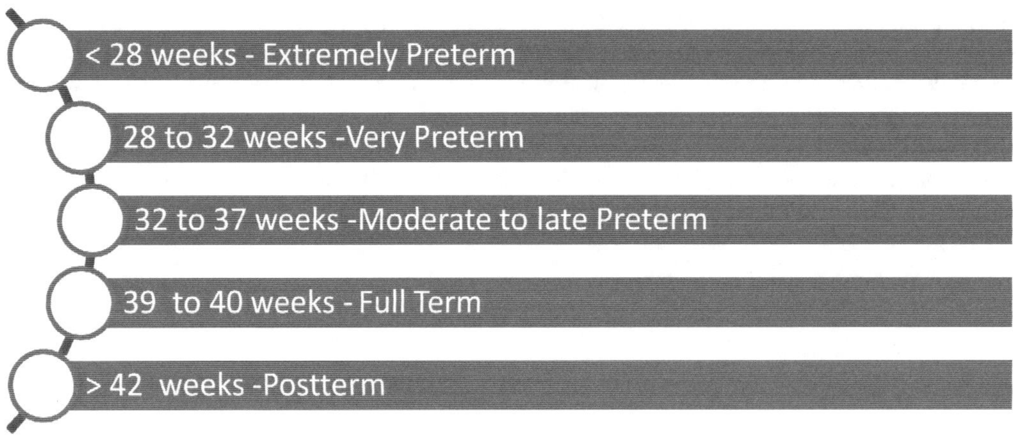

Figure. 2. Birth Term Classification

Machine Learning (ML) helps to make correct decisions based on observations and predictions. ML examines the areas of algorithms which makes high-end predictions on data accurately. ML automatically improves through its experience. The training data is used to build the model and the testing data is used to validate the built model. Regression is one of the Supervised learning Algorithm of ML to find relationship between independent and dependent variables. The most popular Regression algorithm is based on regression value. The Logistic Regression (LR) model outcome variable is binary which makes it distinguishable from other models and hence, in the current work LR was adopted to early identification of CP using prenatal and perinatal risk factors. Mostly patients' data are used to identify the influential attribute in predicting the given outcome [16].

2. Literature Review

Sukhov et al. [17] discussed about CP risk factors such as gestational age, age of mother, maternal race/ethnicity, twin, placenta abruption, mild to severe birth asphyxia, fetal distress, birth defects and birth trauma. This study found gestational age is risk factor of CP and also prematurity has the greatest impact upon the CP people's future development.

Farin et al. [8] explained that the mother's age is cause for the perinatal asphyxia (CP risk factor). In this study they used 159 cases and split into two group. Case group contains 53 participants and control group contains 106participants. They found 6.6% of mothers under the age of 18 years at pregnancy, 90.8% of mothers in the 18–35 year age group, and 2.6% of mothers over the age of 35 years. Accuracy prediction of this study is 87.8%.

Linsell et al. [18] explained that children with very low birth weight or children born in very preterm (less than 32wks) have poor neuro development. This study suggests that very low birth weight and very preterm factors prognostic values of CP.

Luanda et al. [19] did research on 56 spastic CP children. In his study, he tried both Univariate and Multivariate Logistic Regression to identify the clinical and neurophysiologic characteristics associated with active transcranial direct current stimulation (tDCS) in CP children. This study took 7 weeks to complete and result was 30% increase on the 6 Minute Walk Test (MWT).

Deepta et al.[20] found that perinatal insult of children with CP causes Optic Nerve Head (ONH) morphology which can serve as an indicator for severe neurologic damage . In this study, authors have used 54 consecutive patients with CP and Multivariate Logistic Regression for research. The result of this study predicted 70% of all CP patients.

Tosun et al. [21] Proposed to one of the machine learning algorithm, multivariate regression analysis to evaluate the relationship between BMD and possible risk factors. In this research, author's team worked with various categories of children and the categories are 30 CP Children,38 CP children with epilepsy,54 children with Epilepsy only and 30 Healthy children. Team found Low BMD rate in children with Epilepsy only have3.7%, only CP children have 50% and CP children with epilepsy have 39.5%.

Mann et al. [22] examined using multivariate regression analyses with 128 ambulatory CP children. In this study inspected the relationship between physical activity and walking performance to quality of life (QOL). Result of this study in the following manner physical QOL is 52.15, psychosocial QOL is 60.94 and total QOL is 56.55.

3. Materials and Methods

3. 1 Data collection

Based on literature review we prepared questionnaire for data collection. We collected data from mothers of CP children and normal children. The questions were used to get information about child's age, gender, mother's medical history, gestational age, birth weight, asphyxia etc. We met parents of CP children at rehabilitation center and took a face to face interview. Table 1 shows the characteristics of participants.

Table 1. Participant's characteristics

Age	**Min -3 Years**	**Max - 23 Years**
Gender	Female – 56%	Male - 44 %
Birth weight	Min – 1100g	Max – 3750g
Gestation age (weeks)	Min -30	Max – 40
Asphyxia	Yes – 36% No – 64%	

3. 2 Experimental setup

The current study to find the result using Multivariate analysis was conducted using LR. In the Multivariate analysis, risk factors were significantly associated with CP were taken as birth weight, gestational age and birth asphyxia. In order to build our proposed predictive model, we have exploited the capabilities of LR. The general formula of LR based on Sigmoid function can be expressed as follows.

$$P = \frac{1}{1+exp^{-z}} \quad (1)$$

where P represents the probability of an event occurrence (dependent variable); z represents the independent variables linear combination function and z can also be expressed as follows.

$$Z=\beta_0+\beta_1 x_1 +\beta_2 x_2 \ldots \beta_n x_n \quad (2)$$

where β_0 denotes the expected mean value of when x=0-; x_n refers to the independent variables; β_n denotes to the regression coefficients of each independent variable and n represents the number of the independent variables.

By applying the linear combination function in equation (1) for the used tests as independent variables to determine an outcome (Cerebral Palsy Identification (CPI)); the formula of the predictive model that computes the probability of occurrence of CPI depending on the predictability of used tests can be written as below:

$$P = \frac{1}{1+exp^{-(\beta 0+\beta 1*T1+\beta 2*T2+\beta 3*T3)}} \quad (3)$$

Where, P represents the probability of occurrence of an outcome (Cerebral Palsy Identification) based on the selected independent variables. β indicates to the regression coefficients of each included blood test. T1, T2, T3 refers to the included test parameters ("birth weight, gestational age and birth asphyxia") respectively. We implement this work using python in jupyter notebook. The following section tells about implementing process

3.3 Data analysis.

- First we have to preprocess the data. In preprocessing we encode the categorical values and we check the null value and visualize the location of missing values using seabornheatmap.
- Assign features where X contains birth weight, gestational age and birth asphyxia and Y contains affect and then split data into training and testing.
- After defining the model, we evaluate the performance using predict proba() , which returns the matrix of probabilities that the predicted result is equal to one or zero.

```
[0.01918849 0.98081151] [0.3353625  0.6646375 ]
[0.08370583 0.91629417] [0.37049067 0.62950933]
[0.0507244  0.9492756 ] [0.04963239 0.95036761]
[0.03174891 0.96825109] [0.11093587 0.88906413]
[0.36928274 0.63071726] [0.4675367  0.5324633 ]
[0.13281487 0.86718513] [0.16790579 0.83209421]
[0.70203577 0.29796423] [0.40454557 0.59545443]
[0.26975981 0.73024019] [0.25609295 0.74390705]
[0.02649558 0.97350442] [0.06930902 0.93069098]]
```

- Calculate the accuracy.

Accuracy of logistic regression classifier on test set: 0.83

- Receiver Operating Characteristic (ROC) curve is a useful tool for predicting the probability of a binary outcome. ROC curve used to plot the false positive rate in x-axis and true positive rate in y-axis based on number of different candidate threshold values between 0.0 and 1.0. Figure 3 indicates false positive rate and true positive rate.

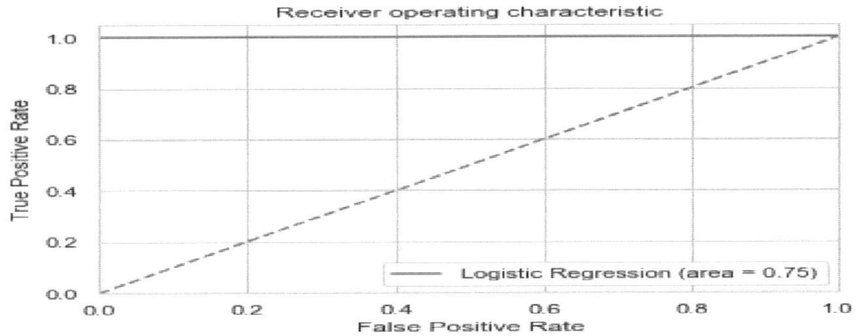

Figure. 3. ROC Curve

4. Conclusion

An early identification is more important to the CP. We can easily find CP using causative risk factor. This study tried with minimum number of CP children's birth history data which are collected directly from the mother. To the best of our knowledge, this is the first report of CP identification using birth weight; Gestational age and birth asphyxia with machine learning algorithm multivariate logistic regression and the accuracy is 83%.

References

[1] Zhang. B , Zhang Y, 2008, Classification of cerebral palsy gait by Kernel Fisher Discriminant Analysis , *International Journal of Hybrid Intelligent Systems*, **5(4)**, pp 209–218.

[2] Yesim Garip, Sumru Ozel, Ozgul Bozkurt Tuncer, Gizem Kilinc, Feryal Seckin , Tansu Arasil, 2016, Fatigue in the mothers of children with cerebral palsy, *Disability and Rehabilitation*, pp 757- 756.

[3] Nimbalkar S, Raithatha S, Shah R, 2014, A qualitative study of psychosocial problems among parents of children with cerebral palsy attending two tertiary care hospitals in Western India, ISRN Family Medicine, 2014:769619.

[4] Steven J. Korzeniewski, Jaime Slaughter, Madeleine Lenski, Peterson Haak, Nigel Paneth, September 2018, The complex aetiology of cerebral palsy, *Nature Review Neurology,* **14**, pp 528-543.

[5] Zeynep Alpay Savasan , Ali Yilmaz , Zafer Ugur, Buket Aydas, Ray O. Bahado-Singh Stewart F. Graham, 2 February 2019, Metabolomic Profiling of Cerebral Palsy Brain Tissue Reveals Novel Central Biomarkers and Biochemical Pathways Associated with the Disease: A Pilot Study ,*Metabolites*.

[6] Mahendra Rana, JyotiUpadhyay, Amita Rana, Sumit Durgapal, Arvind Jantwal, 2017 , A Systematic Review on Etiology, Epidemiology, and Treatment of Cerebral Palsy, *International Journal of Nutrition, Pharmacology, Neurological Diseases ,Wolters Kluwer Medknow*, **7(4)**, pp 76-83.

[7] Moshe Stavsky, Omer Mor, Salvatore Andrea Mastrolia,Shirley Greenbaum,Nandor Gabor Than, Offer Erez, 2017, Cerebral Palsy Trends in epidemiology and Recent Development in Prenatal Mechanisms of Disease, Treatment, and Prevention , *Frontiers in Pediatrics*.

[8] Soleimani F, Vameghi R, Biglarian A, 2013, Antenatal and Intrapartum Risk Factors for Cerebral Palsy in Term and Near-term Newborns, *Archives of Iranian Medicine*,

16(4), pp 213 – 216.

[9] Chayasak Pitsawong ,Prisana Panichkul, October 2011, Risk Factors Associated with Birth Asphyxia in Phramongkutklao Hospital , *Thai Journal of Obstetrics and Gynaecology* **19**, pp 165-171.

[10] Haider BA, Bhutta ZA, 2006, Birth asphyxia in developing countries: current status and public health implications, *Current Problems in Pediatric and Adolescent Health Care*, **36(5)**, pp 178-188.

[11] Tove S.Rosen,DavidBateman, 17 march 2010, Chapter 1 - The Effects of Gender in Neonatal Medicine , *Principles of Gender-Specific Medicine* (Second Edition) , pp 3-17.

[12] Van Lieshout P., Candundo, H, Martino. R., Shin,S, Barakat-Haddad, 2017, Onset factors in cerebral palsy: A systematic review, *NeuroToxicology*, **61**, pp 47–53.

[13] Hafstrom M, Kallen K, Serenius F, 2018, Cerebral Palsy in Extremely Preterm Infants, *Pediatrics*, **141(1)**, e20171433.

[14] Himmelmann K, Uvebrant P, 2014, The panorama of cerebral palsy in Sweden. XI. Changing patterns in the birth year period 2003-2006, *Acta Paediatrica* , **103(6)**, pp 618–624.

[15] Oskoui M, Coutinho F, Dykeman J,Jetté N, Pringsheim T, 2013, An update on the prevalence of cerebral palsy: a systematic review and meta analysis, *Developmental Medicine & Child Neurology*, **55(6),** pp 509– 519.

[16] Evangelia Christodoulou, Jie Ma, Gary S. Collinsb,, Ewout W. Steyerberg, Jan Y. Verbakel, Ben Van Calster, 2019, A systematic review shows no performance benefit of machine learning over logistic regression for clinical prediction models, *Journal of Clinical Epidemiology*, **110**, pp 12-22.

[17] Sukhov, A., Wu, Y., Xing, G., Smith, L. H., Gilbert, W. M, 2011, Risk factors associated with cerebral palsy in preterm infants, *The Journal of Maternal-Fetal & Neonatal Medicine*, **25**, pp 53-57.

[18] Linsell, L., Malouf, R., Morris, J., Kurinczuk, J. J., Marlow, N, 2016, Prognostic factors For cerebral palsy and motor impairment in children born very preterm or very low birth weight : a systematic review, *Developmental Medicine & Child Neurology*, **58(6)**, pp 554-569.

[19] Luanda A. Collange Grecco, Claudia Santos Oliveira, Manuela Galli, Camila Cosmo, Natália de Almeida Carvalho Duarte , Nelci Zanon , Dylan J. Edwards Felipe Fregni ,July 2016, Spared primary motor cortex and the presence of MEP in cerebral palsy dictate the responsiveness to tDCS during gait training, *Frontiers in Human, Neuroscience*, **10**, Article 361.

[20] Deepta Ghate, Vettaikorumakankav Vedanarayanan, Abdulbaset Kamour, James J. Corbett , Sachin Kedar, 2016, Optic nerve morphology as marker for disease severity in cerebral palsy of perinatal origin, Elsevier, pp 25-31.

[21] Ayse Tosun, Seda Erisen Karaca, Tolga Unuvar, Yakup Yurekli, Cigdem Yenisey, Imran Kurt Omurlu, 2016, Bone mineral density and vitamin D status in children with epilepsy, cerebral palsy, and cerebral palsy with epilepsy, *Childs Nerves System,* **33**, pp 153–158.

[22] Kilby Mann, Elaine Tsao , Kristie F. Bjornson, 2016, Physical activity and walking Performance : influence on quality of life in ambulatory children with cerebral palsy (CP), *Journal of Pediatric Rehabilitation Medicine: An Interdisciplinary Approach*, **9**, pp 279–286.

ECG based Decision Support System for Clinical Management using Machine Learning Techniques

Sireesha Moturi [1], Dr. Srikanth Vemuru [2], Dr. S. N. Tirumala Rao [3]

[1]Research Scholar, KLEF, Vaddeswaram, India, Assoc. Prof., Narasaraopeta Engineering College, Narasaraopet, India
[2]Professor, KLEF, vaddeswaram,India
[3]Professor& HOD, Narasaraopeta Engineering College, Narasaraopet, India,
sireeshamoturi@gmail.com

vsrikanth@kluniversity.in

nagatirumalarao@gmail.com

Abstract. Heart disease prediction system using ECG is to predict heart disease using ECG signals. Heart is the next major organ comparing to brain, which has more priority in human body. Heart disease diagnosis is a complex task which requires much experience and knowledge. The huge amount of data generated for prediction of heart disease is too complex and voluminous to be processed by traditional methods. By using traditional methods doctors took lot of time to diagnosis the disease. So, an entropy based feature selection technique is used with classification algorithms in order to reduce the search space. The proposed model was tested on the real time dataset of NRI Hospital medical data. Using this system it is easier to predict the disease. It will also helpful for the doctors to take quick decisions.

1. Introduction

Heart attacks are one of the major reasons behind several deaths happening worldwide [1]. An electrocardiogram is a graphical record of the extent and course of the electrical movement that is created by contraction and relaxation of the ventricular and atria electrocardiogram is used to check the heart rate by placing several electrodes ECG because of their age, weight, high cholesterol level, obesity etc. Any change in the rhythm of the ECG signal is due of heart disease. According to the survey in 2004 1.1 million people died from heart disease , a total of 72 million of these deaths are due to heart disease and 5.7 million were due to stroke A recent survey has found out that about 23.6 million people will die of cardiovascular disease by 2030. Chronic disease requires constant treatment to enhance the quality life of patients. Nowadays, it is estimated that 12% of natural deaths occur accidentally, 88% of which are cardiac. The identification of early heart beat rhythms plays a vital role in preventing heart disease. Now a day's data is generating more and more [14] so, Content extraction becomes a more challenging task in the today's world [7]. Information on the web is also of different types which contains structured, semi structured and unstructured kind of data and current websites present a larger wide variety of complexities than traditional ones[13]. Data mining is about describing the past and projecting the future for analysis. It also helps to derive information from large datasets [3] [4]. This involves data planning, evolution, data interpretation, modelling and deployment. To deal this complex data, there is a need of dominant tool like Machine Learning (ML) [2].

2. Literature Survey

The healthcare system produces massive quantities of data every day. Much of it isn't used successfully though. Animesh Hazra et al.[5] suggested that any of the latest work on heart disease prediction using data mining techniques analyses the various combinations of mining algorithms used and conclude which techniques are successful and efficient. A non-stationary signal, the electrocardiogram (ECG) is commonly used to measure the rate and frequency of heartbeats. A comparison of the overall pattern and shape of the ECG waveform helps doctors to identify possible illnesses. Currently, a Computer based diagnosis is conducted using some signal processing to diagnose an ECG-based patient. The feature extraction scheme for subsequent analysis specifies the amplitudes and intervals in the ECG signal or any other features there. Recently various research techniques for analysing the ECG signal have been developed [6].

3. Proposed Model

3.1. Experimental setup: This experiment was conducted on the Intel® Core™ i7 Processors with 64 bit Windows 10Pro machine. Anaconda 5.1.0 jupyter notebook Python distribution is used in this experiment. This dataset is gathered from real time data. We have collected the real time information in NRI hospital Vijayawada. We gathered the ECG records of the patients and note down the values in an excel sheet. The dataset consists of 9 attributes such as PR interval [12], QRS duration, QTC interval, QT interval, vent rate, P wave, T wave, QRS wave and problem. Out of these attributes, class variable is problem attribute and remaining attributes are used as the predictor variables. Figure 1 shows the process of the proposed system of Heart Disease prediction using ECG signals.

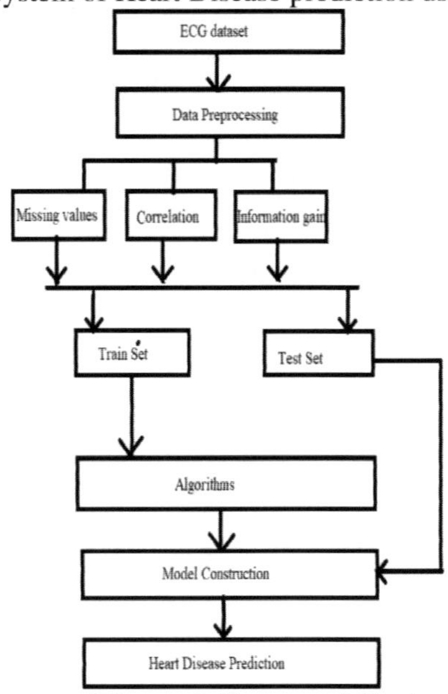

Figure 1 Proposed System

Figure 2 shows the Electrocardiogram of a healthy heart. Figure 3 shows normal ECG. Figure 4 shows abnormal ECG. Figure 5 shows Borderline ECG. Figure 6 Normal ECG except with rate. Table 1 displays attributes, range and description about the attributes.

s.no	Attributes	Standard range	Description
1.	Vent_rate	60-100bpm	It normally refers to the ventricular contraction rate, Nothing but heart beat.[13]
2.	PR_interval	120z`-200ms	The PR interval is the time from the beginning of the p wave to the startof the QRS complex.

IOP Conf. Series: Materials Science and Engineering 1085 (2021) 012016 doi:10.1088/1757-899X/1085/1/012016

3.	QT_interval	300-440 ms	The QT interval is the time from the start of the QRS series, representing ventricular depolarization to the end of the ventricular repolarization of the T wave.
4.	QTC_interval	400-440 ms	It is nothing but corrected QT interval.
5.	P_axes	110 ms	The p wave shows the first positive deflection on the ECG and atrial depolarization.
6.	T_axes	160 ms	T wave reflects ventricular repolarisation.
7.	QRS duration	80-120 ms.	The“QRS complex” is a variation of the Q wave, R wave and S. This reflects depolarization to the ventricle
8.	QRS_axes	<100 ms	QRS axes applies to frontal plane ventricular depolarization.
9.	Problem	-	Describes whether it is normal ECG or abnormal ECG or borderline ECG or normal except with rate.

Table 1 Attribute Description

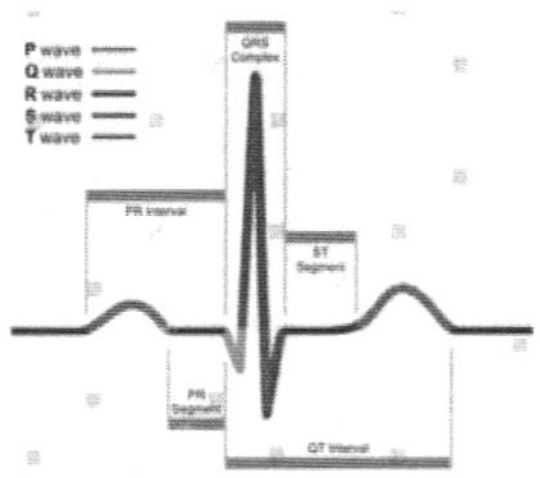

Figure 2. Electrocardiogram of healthy heart

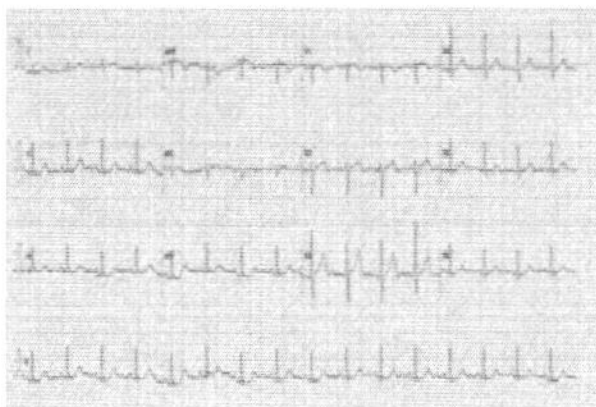

Figure 3. Normal ECG

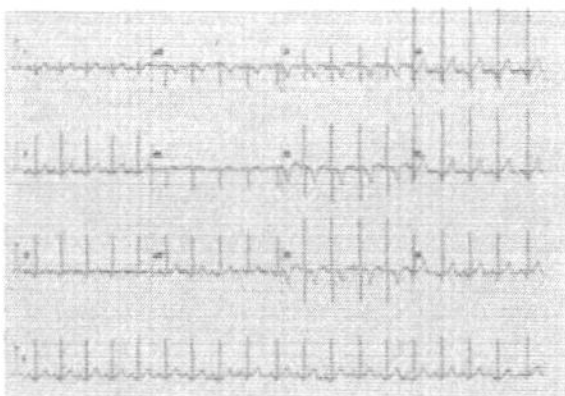

Figure 4. Abnormal ECG

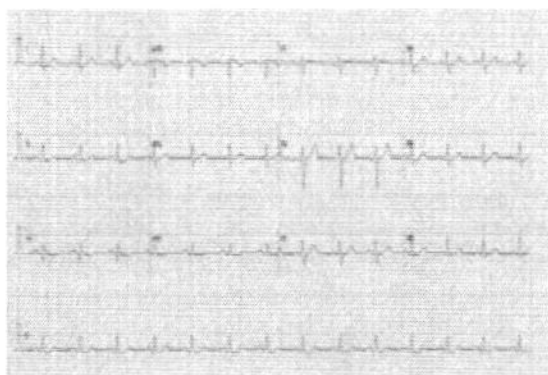

Figure 5. Borderline ECG

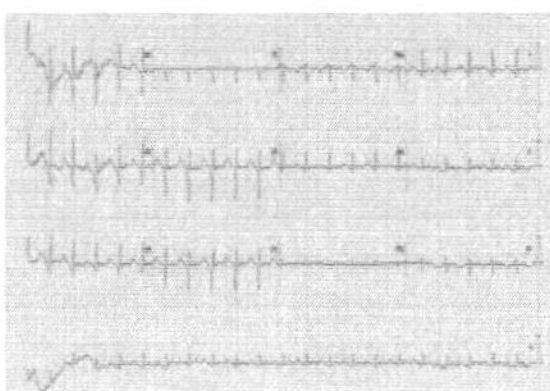

Figure 6. Normal ECG except with rate

3.2. Data Preprocessing

Data preprocessing is a data mining technique used to convert the raw data into an effective and usable format.

3.2.1Handling Missing Data

Sometime data may contain insignificant and missing values. Data cleaning is required to handle that portion. It includes managing data which are missing, noisy data, etc. It can be done by manually filling in the missing values, by mean attribute, median or by dropping the missing values. But, dropping missing values is not so good technique to opt because we miss the valuable data. From figure7 we can observe that the dataset is having missing values. In Figure 8 we replaced missing values with mean so that we don't have any missing values.

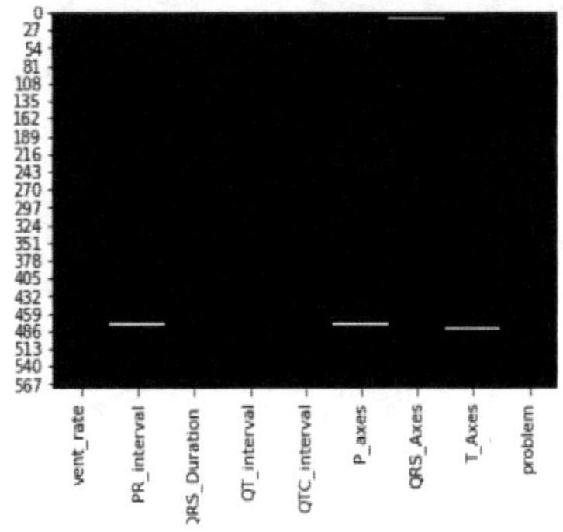

Figure 7. Heat map before filling missing values

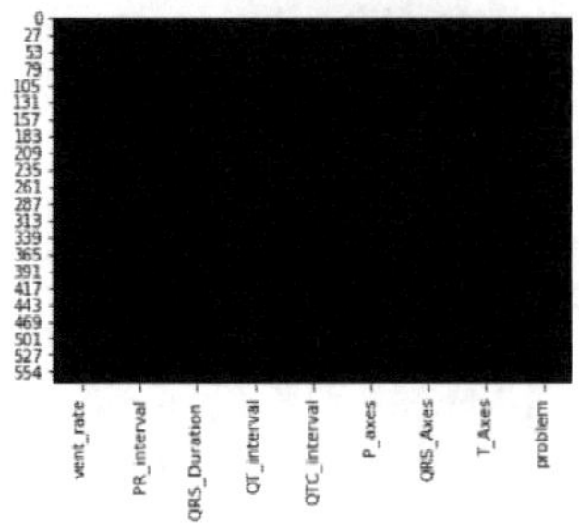

Figure 8. Heat map after filling missing values

3.2.2 Correlation

Correlation is a statistical measure that demonstrates how often two or more variables fluctuate together. A positive relationship shows degree to which those factors positively correlated to one other. A negative relationship shows degree to which those factors negatively correlated to one other. From the Figure 9 it is observed that the vent_rate and QT_interval are negatively related to one other. So, the attribute vent_rate got dropped.

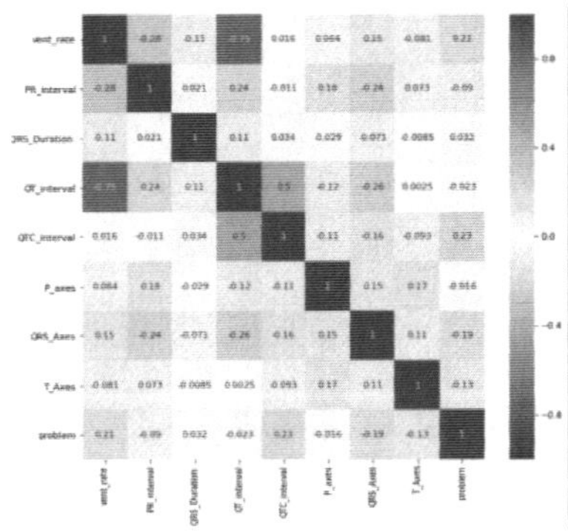

Figure 9. Heat map before filling missing values

```
vent_rate 0.36
PR_interval 0.19
QRS_Duration  0.21
QT_interval 0.29
QTC_interval 0.31
P_axes 0.25
QRS_Axes 0.4
T_Axes 0.49
```

Figure 10. Heat map after filling missing values

3.2.3 Information Gain

This is also one of the pre-processing techniques, which is used to measure the reduction in entropy. It is widely used to build a model like decision tree from a training dataset, to calculate the information benefit of a variable and to choose a variable that maximizes the benefit of information, thereby dividing the dataset into groups for successful classification. From Figure 10 it is observed that PR_interval has the lowest information gain. Information gain can be calculated using the following formula.

$$I(s_1, s_2, \ldots, s_m) = -\sum_{i=1}^{m} p_i log_2(p_i)$$

4. Implementation

4.1 Splitting the Dataset

IOP Conf. Series: Materials Science and Engineering 1085 (2021) 012016 doi:10.1088/1757-899X/1085/1/012016

Dataset is splitted into two parts such as the train set and test set. It means 80% of data into the train set and the remaining 20% of data into the test set as shown in Figure 11.

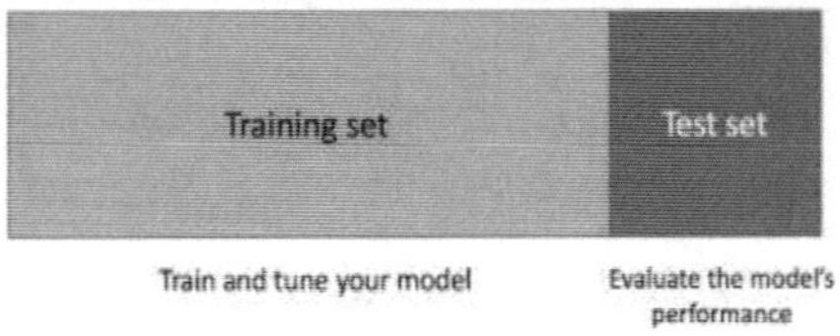

Figure 11. Splitting of dataset

4.2 K-Fold Cross Validation

It is a kind of resampling technique used to evaluate the Machine Learning Model. In every iteration one fold is used for validation or testing and the remaining folds are used for training purpose as shown in Figure 12.

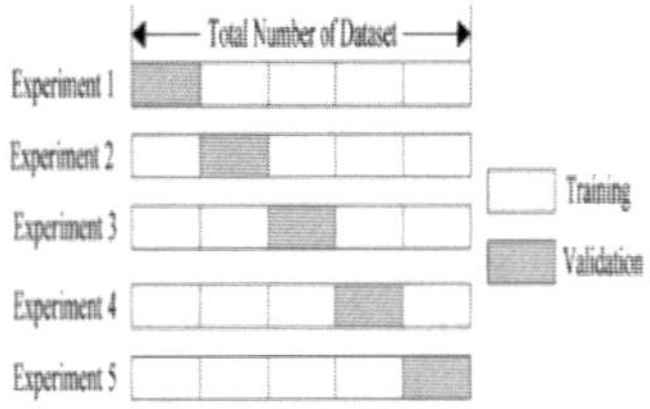

Figure 12. K-Fold Cross Validation

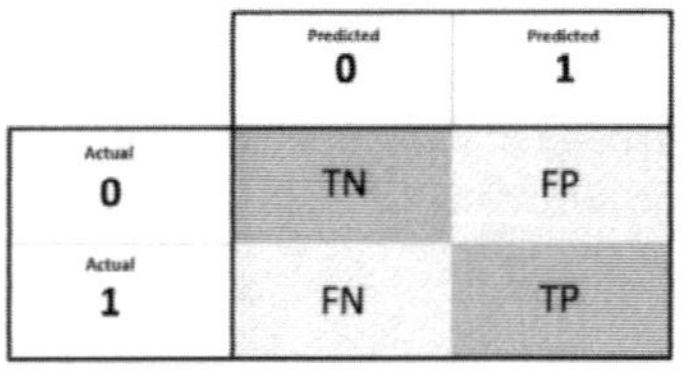

Figure 13. Confusion Matrix

4.3 Algorithms Used

There are many algorithms for classification in Machine Learning. The present paper gives a comparison between the performances of four classifiers: Decision Tree, Random Forest, Support Vector Classifier (SVC), and Gaussian Naive Bayes. The main aim of researchers is to evaluate the efficiency of these algorithms in terms of accuracy. For Performance Evaluation Confusion matrix is used to evaluate the performance of any classification algorithm as shown in Figure 13.

5. Result Analysis

Figure 14 shows the accuracy comparison of three classifiers (Decision Tree, Gaussian Naïve Bayes, SVC). From Figure 14 it is observed that the highest accuracy is 97% for Decision Tree and Gaussian Naïve Bayes. Figure 15 shows the accuracy comparison of three classifiers (Decision Tree, Gaussian Naïve Bayes, SVC) when K-fold cross validation is applied. From fig14 it is observed that the highest accuracy is 98.2% for Decision Tree. It is also observed that all the three classifiers achieves highest performance after applying K-fold cross validation.

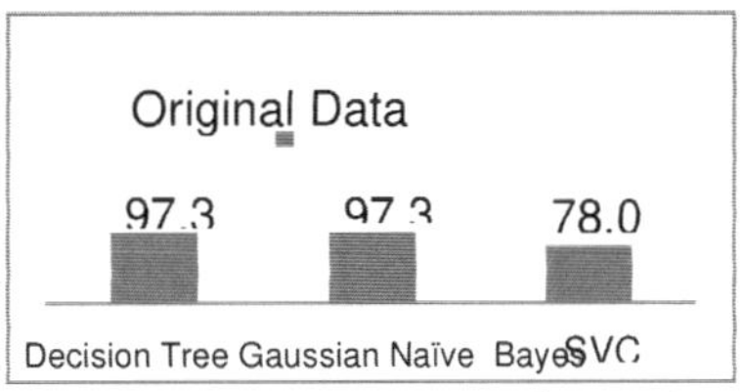

Figure 14. Comparison of Accuracy

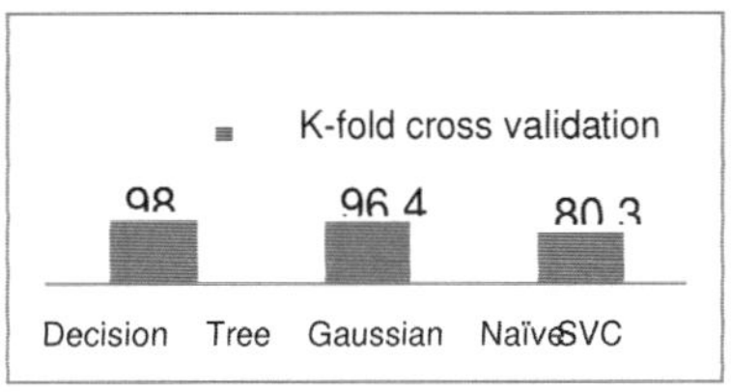

Figure 15. Comparison of Accuracy

Figure 16 shows the Comparison accuracy of three classifiers (Decision Tree, Gaussian Naïve Bayes, SVC) before and after applying Information Gain. It is observed that after applying Information Gain the accuracy of decision tree has increased from 97.3 to 98.2.

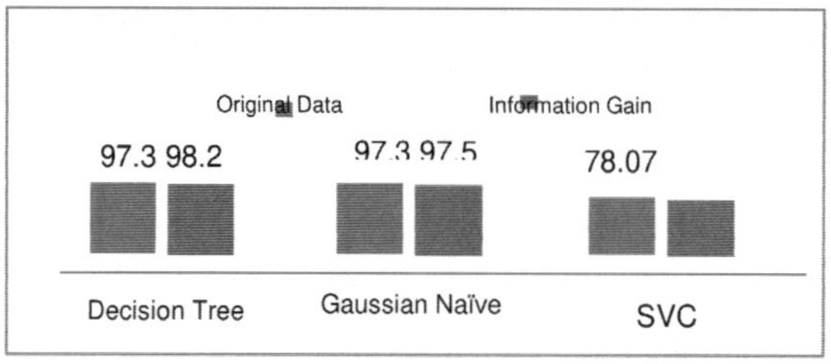

Figure 16. Before and after applying Information Gain

6. CONCLUSION

In this paper various pre-processing techniques were applied on the data with three Machine Learning algorithms like Decision Trees, Gaussian Naive Bayes and SVC in- order to predict presence or absence of heart disease. The accuracy varies for different algorithms. The highest accuracy 98.2% was achieved by Decision tree with Information Gain and K-Fold cross validation methods. By using this system we can reduce medical errors, enhance patient safety and improve patient outcomes. It is easier to predict the disease and it is also helpful for the doctors to make quick decisions.

7. References

[1] M.Sireesha, S.N.TirumalaRao and Srikanth V 2019 optimized Feature Extraction and Hyrid Classification Model for Heart Disease and Breast cancer Prediction *International journal of Recent Technology and Engineering* vol-**7**, No **6** pp 1754-1722.

[2] Prakash K.B and Dorai Rangaswamy M.A. 2016 Content extraction studies using neural network and attribute generation. *Indian Journal of Science and Technology* pp 1-10.

[3] M.Sireesha, Srikanth vemuru and S.N.Tirumalarao 2018 Coalesce based binary table: an enhanced algorithm for mining frequent patterns *International journal of Engineering & Technology*, vol. **7**, no.**1.5,** pp 51-55.

[4] M.Sireesha, S.N.Tirumalarao and Srikanth V 2018 Frequent itemset Mining Algorithms: A survey *Journal of Theoretical and Applied Information Technology* vol **96**, No.**3** pp 744-755.

[5] Animesh H, Subrata Kumar M, Amit G, Arkomita M and Asmita M 2017 Heart Disease Diagnosis and Prediction Using Machine Learning and Data Mining Techniques: A Review *Advances in Computational Sciences and Technology* Vol **10**, No **7** pp 2137-2159.

[6] Naveen Ku. Dewangan and S. P. Shukla 2015 A Survey on ECG Signal Feature Extraction and Analysis Techniques *International Journal Of Innovative Research In Electrical, Electronics, Instrumentation And Control Engineering* Vol. **3** Issue **6** pp 12-19.

[7] Prakash K.B and Rangaswamy M.A.D 2016 Content extraction of biological datasets using soft computing techniques *Journal of Medical Imaging and Health Informatics* pp 932-936

[8] M.Sireesha, S. N. TirumalaRao and Srikanth V 2020 Predictive Analysis of Imbalanced Cardiovascular Disease Using SMOTE *International Journal of Advanced Science and Technology* Vol **29** No **5** pp 6301 – 6311.

[9] A.Dhanasekar and Dr.R.Mala 2016 Analysis of Association rule for heart disease prediction from large datasets *International Journal of Innovative Research in Science, Engineering and Technology* Vol **5**, Issue **10** pp 18059-18063

[10] M.Sireesha, Srikanth V and S.N.Tirumala Rao 2020 Classification Model for Prediction Of Heart Disease Using Correlation Coefficient Technique *International Journal of Advanced Trends in Computer Science and Engineering* Vol **9**, No **2** pp 2116 – 2123.

[11] A. Calderon, A. Pérez and J. Valente 2019 ECG Feature Extraction and Ventricular Fibrillation (VF) Prediction using Data Mining Techniques *IEEE 32nd International Symposium on Computer-Based Medical Systems (CBMS)*, Cordoba, Spain, 2019, pp. 14-19

[12] https://www.healio.com/cardiology/learn-the-heart/ecgreview/ecginterpretation-tutorial

[13] Prakash K.B., Dorai R M.A., and Raman A.R. 2010 *Proc. Int. Conf. on Trendz in Information*

Sciences and Computing pp 28-31.
[14] M.Sireesha, Srikanth Vemuru and S.N.Tirumala Rao 2020 Int. Conf. On Machine Intelligence And Soft Computing.

IOP Conf. Series: Materials Science and Engineering 1085 (2021) 012017 doi:10.1088/1757-899X/1085/1/012017

An evaluation of deep learning approaches for detection of voice disorders.

Chinchu M S[1], Ms. Dr. B. Kirubagari [2]

[1] Assistant Professor at SJCET Palai & Research Scholar, Computer science and Engineering Department,India
[2]AssociateProfessor& Research Guide,Computer science and Engineering Department,India.
chinchums18@gmail.com, kirubacdm@gmail.com

Abstract. The human voice manufacturing system is a complicated natural device capable of modulating pitch and loudness Human sound frequency particularly. The part in which the folded is the primary source underlying internal and/or external factors often destroys vocal folds justification. Some changes consequences are reflected in functioning emotional state soul. Therefore it essential to identify variations at an early stage gives the patient an endangerment overcome any impact modernize their quality of life will-less detection disorders using depth study methods plays an important role as has been proven to facilitate the process. Many researchers have explored technologies for streamlined that can help clinics diagnose noise paper we present the conducted research activities.

1. Introduction

Voice order is a health issue that is neglected often by the people as the people neglected the superintendency of the voice or not desired to meet the professional doctor. Previous researches has developed voice disorder detection approaches using the sensory features and machine learning from the voice recordings rather than using the professional medical devices for the detection. There are two classes of voice recordings, like Pathological and Normal. These classes are explored by the increasing popularity of full-length learning, deep learning and transfer learning. The voice disorders are recognized well using the unstipulated viability of full-length learning and deep learning. Moreover, in the existing datasets, the voice disorder detection like the lack of generality and size of the dataset are the major drawbacks.

The speech is defined as the fundamental instinct of the human and voice in its subsystem. The study of vocalization is termed as, vocology [1]. The interaction of sound product of the pulmonary air pulses interacting with the trachea is the Normal sound. The Normal sound produced the periodic and/or aperiodic sounds by snatching the original vocal folds. Some of the noise disorders include the dysphonia and aphonia caused by multiple wiseacre vocalizations. The dysphonia is the partial sound loss, whereas the aphonia is the complete sound loss. The disorder in the voice occurs due to the gender, age and social group that has variations in the volume, quality, and tone [2]. The impact of the disorders in the malignant speech is not life-threatening. However, the speech disorders that are professional, social, and personal aspects of the liaison are not affected by the impact of the untreated speech disorders [3]. The increased public sensation of the voice care needs is caused due to the changes in the professional and lifestyle needs. Now-a-days, the citizen consults the voice therapists and laryngologists for maintaining the voice health. However, the examinations, like stroboscopy, laryngoscopy or videos are expensive and difficult. Speech processing is a developing research zone that focuses on aspects of the speech processing from full-length extraction to visualization support systems.

2. Voice Disorder

The disorder in communication science are categorized into five types, such as language disorders, speech disorders, cognitive and social communication disorders and swallowing disorders [5]. According to the Classification Manual for Voice Disorders, the hypo-functional and hyper-functional vocal fold pathologies are classified as,

- Injury or Trauma-based,
- Inflammatory conditions, such as acute laryngitis,
- Structural lesions, such as vocal polyps and vocal nodules,
- No laryngeal aero digestive disorders, consisting of bronchitis and reflux disorder among others.
- Neurological conditions, like abductor palsy, adductor palsy and spasmodic dysphonia,
- Systemic conditions, such as, hypothyroidism and hyperthyroidism,
- Psychological disorders and psychiatric disorders that are functional and mutism
- Undiagnosed and otherwise not specified.
- Other disorders, such as muscle tension dysphonia [5]

Mass-based pathologies of various vocal folds are very popular. The vocal polyp and vocal nodule occurred due to the external influence and persistent tissue inflammation [6]. In these circumstances, the voice production is conceptually rough and not economical, whereas the vocal fold closure is incomplete. Even though the no vocal folds in the nonphotromatic voice disorders, the quality of the voice is decreased along with increased laryngeal tension and noise fatigue. The nonphotromatic voice disorder includes functional voice disorder and muscle tension dysphonia. For evaluating the voice, the Multi-parametric assessment protocol is taken into consideration [2], [7]. The laryngeal imaging from the laryngoscopy and stroboscopic, patient inters- view, standardized psychoacoustic methods for perceptual analysis, evaluation of subjective voice, assessment of aerodynamic and acoustical analysis is essential for the structured approach. In order to distinguish the normal voice from the dysphonia and/or aphonia, the recent technologies are developed by the scientists for developing tools for acoustical analysis. The tools should be developed to solutions in financial that are reliable in dynamic clinical settings and sensitive in detecting the changes in the sound. If enough research is done on acoustic analysis, the result is an automatic discovery.

Recently, numerous solutions are developed in the pathological speech detection field [21], [22]. For instance, Al-Nasheri et al.[23] designed a extraction method for the classification and detection of voice pathology using entropy and autocorrelation by determining the variegated frequency bands. According to the study, the vocal polyps, vocal cysts and vocal paralysis are caused due to the impairment of voice. For the classification and detection, the frequency band is considered between 1000 Hz and 8000 Hz. The support vector machine is used as a classifier and each voice sample consists of sustained vowel /a/. The accuracy obtained for the detection of specimen are, 92.79%, 99.69%, and 99.79% for Saarbrucken Voice Database (SVD), Massachusetts Eye and Ear Infirmary (MEEI) and Arabic Voice Pathology Database (AVPD), respectively.

3. Voice Disorder Database

The dataset consists of the recordings of pathological and normal sounds. The audio recording has continuous conversation and regular vowel pronunciation. From the survey, it is observed that most of the researchers use the standard database, like SVD, MEEI and AVPD. The information from the three databases described in Nasheri et al.[8] are considered. Other researchers used the dataset from the private dataset in association with the local hospitals are considered. In the dataset, the audio recording of pathological and normal sounds is taken into consideration. The recordings in the dataset comprises of continuous conversation or regular vowel pronunciation. The standard databases, like Saarbrucken Voice Database (SVD), Massachusetts Eye and Ear Infirmary (MEEI), and Arabic Voice Pathology Database (AVPD). The public misogynist Saarbrucken Voice Database is also used by most researchers [20]. In public misogynist Saarbrucken Voice Database, the voice recordings comprises of the samples of the subject with the recordings of vowels, /a/, /i/, and /u/ that are produced at high, normal, low and low-high-low pitches. The recording length ranges from 1 to 2 seconds for sustained vowels.

3.1. Feature Extraction

The feature extraction is the process of extracting the relevant feature from the test set. The feature extraction process is carried on for reducing the feature vector size with respect to the dimension of the feature extraction. The redundant features are removed in the feature extraction process. This process reduced the computation time and improved the accuracy for helping the learning algorithm. The complexities can be further determined using other machine learning methods. In the NLP, repetitive neural networks are applied frequently when compared with the other models. The data is targeted for particular time range in the RNN. Similar features are extracted using multiple extraction methods. For the voice disorder detection system, the first step is the feature extraction. The frequently used methods for the feature extraction and acoustic analysis are discussed below.

3.1.1. Mel-Frequency Cepstral Coefficient (MFCC).

The MFCC is a standardized process for extracting features that utilize knowledge of the human auditory system. A popular audio feature extraction method is the Mel-frequency septic coefficient (MFCC) with 39 features. The number of features is small enough to motivate us to understand the audio information. 12 parameters are related to the magnitude of the frequencies. It provides us enough frequency channels to analyze the audio. The process of extracting the MFCC features from the single frames [12], [13] are given as,

a) Computing the coefficients of discrete Fourier transform.
b) Using Mel spaced triangular filter for filtering.
c) Computing the energies of sub-band.
d) Computing the coefficients of discrete cosine transform.

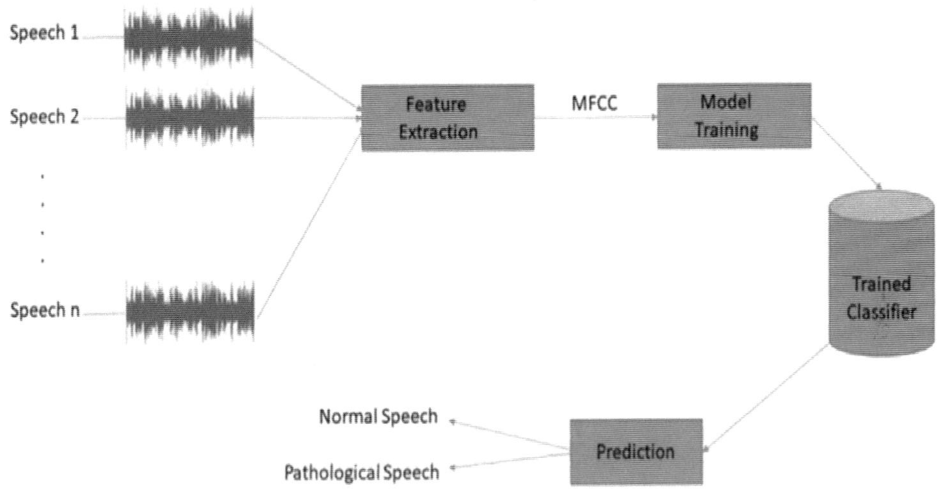

Figure 1.Voice disorder Identification Using MFCC

3.1.2. Neural Networks.

The neural network is modelled on the basis of human brain for recognizing the patterns. The sensory data is interpreted with the form of labelling, machine perception, or raw input clustering. All the real-world data, like sound, images, time series or text are translated by the patterns that are contained in vectors and numeric. The features from the other algorithms are extracted for the classification and clustering process. The neural networks are used as a layer helps in categorizing and clustering the features that are stored and managed. For instance, the inputs are used for grouping the unlabelled data depending upon the similarities and when there is a labelled dataset for training, they classify the data. The correlation between the person name and image pixel are established using the deep learning, which is known as a static forecast. The exposure to the data is done adequately with the help of same token for establishing the relationship within the future events and current events in the deep learning study. There is no need of worrying for the study time of the deep learning. Then, the number is predicted and string of numbers is read for the given time range in the deep learning study. The unstructured data are applied with the labelling data from the trained in-depth learning network for providing access to the input for improving the performance. The accuracy of this method depends on the number of data that can be trained by the network. In the in-depth learning network, the raw data is provided in the form of image and the individual is represented with 90 percent chance from the input data.

Deep Neural Networks, The Deep learning network uses the single-hidden-layer neural networks for separating from their depth. The data is passed through a multistep process for the recognition of pattern. The layers in in-depth learning networks are trained with specific features depending on the output of the preceding layer. As you progress further into the neural network, the features that your nodes can distinguish are more complex because they combine and reassemble features from the previous layer. The automatic feature extraction is performed in the deep-learning without the intervention of humans [7, 8]. In the deep-learning network, more access is provided to the input by training the labeled data and applying to the unstructured data. The accuracy of the deep-learning network depends on the data for training the net. When compared to the previous algorithms, the ability of the deep learning to learn from the unlabeled data provides a huge advantage. The likelihood is assigned to the specified label or outcome by the deep learning networks.

3.1.3. Convolutional Neural Network (CNN)

CNNs are the neural networks used for object identification in scenes, similar cluster images and sorting the images based on scene. For instance, CNN are used for identifying the street signs, faces, individuals, platypus, tumours and other aspects of data. The images are considered as volumes in the conversion chains. The conversional chains are three-dimensional objects rather than the flat canvases that measured the height and width. All the colours are combined together to form the colour spectrum in the digital colour images with red-blue-green (RGB) encoding. The images of different colour levels are stacked together on the top of another using the conventional network. Hence, the standard colour image is considered as a rectangular box in the convolutional network in which the height and width are measured with the help of the number of pixels in the dimension, whereas the depth is indicated by the three layers that represents the letter in RGB. The channels are considered as the depth layer.

Huiyiet al. [8] developed a voice detection approach using the CNN. In this approach, the pathological speech recordings are given as the network input. The weight in the CNN is trained with the help of Convulsive Deep Confidence Network (CDBN). The input data structure is determined using the statistical methods in the generative model. The weights are improved by training the CNN using the supervised back-propagation learning algorithm. Better results are achieved during the classification through this approach.

Musaed alhussein and Ghulam Muhammad et al. [9] designed a Mobile Healthcare Framework for voice detection using deep learning. The sounds in the detection system are recorded with the smart mobile devices. The voice signals are pre-processed and provided to the CNN. The CNN models are used along with the transfer learning method, especially the VGG-16. The detection accuracy is up to 97.5% in the voice pathology detection process.

Alice Rueda and Sridhar Krishnan et al. [10] designed a voice detection method using the Fourier-based synchrosqueezing transform in the CNN. One of the major challenges in the study of dysphonia voice is the small dataset. The in-depth learning methods that does not have fitting are more difficult to apply. Large amount of training data is required by the CNN and the sound is limited for the data enhancement methods. The data size is enhanced using the data augmentation technique, such as Fourier-based Synchronous Caching Transformation (FSST). Besides increasing the size of the data, FSST method provided better study with the CNN than the Short-Time Fourier Transform (STFT). Different vowels are constructed in the CNN using the dynamic stopping algorithm. For the first selected small dataset, the results are trained by improving the CNN in advance. The fitting is bypassed using larger dataset by improving the performance using the increased number of convex layers. More reliable and meaningful results are generated using the larger database.

Huiyi Wu and John Soraghanet al. [11] developed a Voice Detection approach using CNN. The sound is classified as, healthy or pathological by extracting the sound features through acoustic analysis with the help of signal processing tools. The automatic feature are extracted and classified into normal and random sounds by the spectrogram specific voice recording in the database. The accuracy obtained for the dataset containing the 482 organic dysphonia speech files and 482 normal dysphonia speech files are set as, 88.5% for training, 66.2% for evaluation, and 77.0% for test data.

3.1.4. Recurrent Neural Network (RNN)

RNN are the type of NN algorithms used for continuous processing of the data like time range data, voice or written natural language. A version of RNN is used by the Deep Mind for playing the video games along with the autonomous agents. At each subsequent step, the RNN returned the output to the RNN from $n-1$ step without affecting the n step with the help of feedback loop. When compared to the traditional computers, the RNN allows continuous and parallel calculations. The RNN is in the shape of human brain that consists of large network of neurons connected together to translate the input stream into a series of output motor [14]. One input and output is received and generated at a given time using the Feed forwarded networks. Same type of obstruction is not faced by the RNN. When submitted to the RNN, the image data is appeared as a sequence and they are not continuous. For example, if the image is a handwritten word then, the cursive image is transformed into a letter. The word ending is predicted by figuring out the beginning of the word. Thus, the image portion is considered as a series of letters. The problems in the traditional RNNs are overcome by the feedback network, known as Long Short-Term Memory (LSTM). The LSTM solved the problems occurred in the previous tasks.

- In noisy input sequences, the extended patterns are temporally recognized.
- The events in the noisy input streams are separated widely by recognizing the temporal order.
- The temporal distance between the events are conveyed by extracting the information.
- The non-smooth and smooth periodic trajectories, precisely timed rhythms are generated in stable.
- The high-precision real numbers are stored robustly across the extended intervals of time.

3.15. LSTM Networks

The long-term dependency is dealt with the help of RNN. The Long-term memory networks are a type of RNN that is commonly known as, LSTM. The long-term dependence is avoided by designing the LSTM clearly. One of the default behaviors of the LSTM is the storage of information over long intervals of time. There are more repetitive modules in the neural networks.

Vibhuti Guptaet al. [16] developed a voice disorder model using LSTM. The performance of the voice disorder method is determined by considering the original 400 testing samples without labels. The LSTM model is used for the classification by extracting the features using different feature extraction methods.

Danish Raza Rizvi1et al. [17] designed a Deep Learning Model for detecting the voice disorder. The Parkinson's disease (PD) is predicted by determining the LSTM and DNN. In this method, the accuracy obtained is 97.12% and 99.03% for varying sizes. Junli Gaoet al. [18] designed a Effective LSTM Repeat Network for voice disorder method. The training effect is enhanced using the LSTM network along with FL for eliminating the impact of the standard ECG beat data on training the model. This method provided accuracy, recall, accuracy, specificity, and F1 score of 99.26%, 99.26%, 99.30%, 99.14%, and 99.27%, respectively.

4. Conclusion and Future Works.

In this research, we have analysed the major research activities related to detection approach of voice disorders system. This paper is based on the organization of the contents, sound distortions, feature extraction, and DL techniques used. The theoretical aspects of voice disorders, feature extraction methods, and DL methods, and performance of the significant research activities in this field are reviewed. From the standard database, the subjects of certain diseases are selected for performing the experimental activities. In most papers, sounds are classified as normal and pathological. However, PD and Alzheimer's disease are specifically targeted in few studies. The research activities in this field have great potential as they apply to the community. With this in mind, we need to develop more standard databases and discover new features. The standard database conducts future research activity experiments by selecting a few randomly selected voice disorders from SVD. It was decided to classify sounds as normal and pathological using DL techniques.

References

[1] *A Survey on Machine Learning Approaches for Automatic Detection of Voice Disorders. Journal of Voice, Vol. &&, No. &&, 2018.(Placeholder1).*
[2] *Development of the Arabic Voice Pathology Database and Its Evaluation by Using Speech Features and Machine Learning Algorithms.*
[3] *Towards Robust Voice Pathology Detection-Investigation of supervised deep learning, gradient boosting, and anomaly detection approaches across four databases.*
[4] *Fundamentals of Speech Recognition. Englewood Cliffs: PTR Prentice Hall; 1993*
[5] *High-Resolution Motor State Detection in Parkinson's disease using convolutional neural network-2020 Franz M. J. Pfister,*
[6] *Sunghwan Yoo, Isha Gujrathi, Masoom A. Haider & Farzad Khalvati,Prostate cancer Detection using deep convolutional neural networks -2019.*
[7] *Pavol Harar∗, Jesus B. Alonso-Hernandez†, Jiri Mekysk Voice Pathology Detection Using Deep Learning on Mobile Healthcare Framework, volume 6, 2018.*
[8] *Mazin Abed Mohammed 1,*OrcID,Karrar Hameed Abdulkareem 2OrcID,Salama A. Mostafa Augmenting dysphonia voice using Fourier-based synchro squeezing transform for a CNN classifier, 978-1-5386-4658-8/18/$31.00 ©2019 IEEE.*
[9] *Huiyi Wu, John Soraghan Senior Member, IEEE, Anja Lowit, Gaetano Di Caterina Convolutional Neural Networks for Pathological Voice Detection-2018*
[10] *Convolutional Neural Networks for Raw Speech Recognition, Convolutional Neural Networks for Raw Speech Recognition http://dx.doi.org/10.5772/intechopen.80026.*
[11] *Convolutional Neural Networks for Pathological Voice Detection, Conference: 2018 40th Annual International Conference of the IEEE Engineering in Medicine and Biology Society (EMBC).*
[12] *Speech Recognition Using Convolutional Neural Networks. International Journal of Engineering & Technology, 7 (4.6) (2018) 133-137.*
[13] *Mahmoud Keshavarzi 1, Tobias Goehring 2, Justin Zakis, Use of a Deep Recurrent Neural Network to Reduce Wind Noise: Effects on Judged Speech Intelligibility and Sound Quality-2018.*
[14] *Vibhuti Gupta ,Voice Disorder Detection Using Long Short Term Memory (LSTM) Model-2018.*
[15] *Danish Raza Rizvi1, Iqra Nissar2, Sarfaraz Masood*3, Mumtaz Ahmed4, Faiyaz Ahmad5 An LSTM based Deep learning model for voice-based detection of Parkinson's disease-2020.*
[16] *An Effective LSTM Recurrent Network to Detect Arrhythmia on Imbalanced ECG Dataset-2019*
[17] *Development of Output Correction Methodology for Long Short Term Memory-Based Speech Recognition-2019*

[18] *S.Girirajan, R.Sangeetha, T.Preethi, A.Chinnappa ,Automatic Speech Recognition with Stuttering Speech Removal using Long Short-Term Memory (LSTM) -2020*

[19] *Barry, B. Saarbruecken Voice Database. Institute of Phonetics; Saarland University. Available online:http://stimmdb.coli.uni-saarland.de/ (accessed on 23 February 2017).*

[20] *Gupta, R.; Chaspari, T.; Kim, J.; Kumar, N.; Bone, D.; Narayanan, S. Pathological speech processing:*
   a. *State-of-the-art, current challenges, and future directions. In Proceedings of the 2016 IEEE International*
   b. *Conference on Acoustics, Speech and Signal Processing (ICASSP), Shanghai, China, 20–25 March*

[21] *Danubianu, M.; Pentiuc, S.G.; Schipor, O.A.; Tobolcea, I. Advanced Information Technology-support of improved personalized therapy of speech disorders. Int. J. Comput. Commun. Control 2010, 5, 684–692.2016;pp. 6470–6474*

[22] *Al-Nasheri, A.; Muhammad, G.; Alsulaiman, M.; Ali, Z.; Malki, K.H.; Mesallam, T.A.; Ibrahim, M.F. Voice Pathology Detection and Classification using Auto-correlation and entropy features in Different Frequency Regions. IEEE Access 2017, 6, 6961–6974*

Answer Book Valuation Using Semantic Similarity

P. Sunilkumar
M.Tech Scholar, Department of CSE, Government Engineering College Idukki
E-mail: sunilkumar.p.kochi@gmail.com

K.S. Sunil
Assistant Professor, Department of CSE, Government Engineering College Idukki
E-mail: sunilks@gecidukki.ac.in

Abstract. Conventional examination systems are a tedious task for the universities and the faculties. It takes a lot of time and human effort for the valuation of the answer books, updation of the marks and the declaration of the results. Automation of descriptive answer book assessment would be useful for universities and academic institutions to simplify the valuation system to a large extend. It ensures a uniform valuation and also helps to publish the results without much delay after the examination. We design a system for automatic assessment of descriptive answer books of technical subjects. Semantic similarity is a metric used to assess the similarity between documents and it gives the degree of similarity as a numeric value. It can be used to value the students' answer books by checking the similarity with original answers given in the answer key and then award appropriate marks.

1. Introduction

In conventional examination systems, expertized faculties are valuing the answer books by comparing with the answers given in the answer key. It requires large amount of time and efforts to complete the examination process and declare results. Since the valuation is a manual process there may be chances to occur human errors during valuation. Automated valuation of descriptive answer book is an innovative idea and it has several advantages over the conventional method of valuation. Time and efforts required for the valuation can be reduced to a large extend and the human errors can be eliminated by using this method and hence it ensures a uniform valuation of the answer books. It is possible to declare the results without much delay after the examination and it is one of the great advantages of this system from the students point of view.

The proposed system will scan answer books and the answer keys and convert them in to images. These images will be converted to text format using handwritten word image classification in machine learning. Using semantic similarity methods, the students' answers are compared with the answer keys and assigning appropriate marks.

Daniel Keysers, et al. [3] describes Googles online handwriting recognition system that can support 22 scripts and 97 languages. The important objective of the system is to provide a fast and highly accurate text entry for smart phones as well as touch enabled devices. D. Thenmozhi, et al. [2] proposes a path to find text similarity between a pair of texts by considering various factors. It uses a method to determine set of clauses present in the texts by resolving conjunctions in complex sentences that identify hidden triples from the text. An approach for finding out the semantic similarity between concepts in Knowledge Graphs such as WordNet and DBpedia is presented in Ganggao Zhu, et al. [4]. It comes up with a semantic similarity method, namely wpath, to combine KG based and corpus based approaches. Ming Liu, et al. [6] proposes an innovative semantic matching method for large documents in the academic and educational clusters. Lexical and semantic similarity measure approach for long answer evaluation is used in Riya Goswami, et al. [8]. This paper aims to introduce a model which programmatically evaluates the long answers from the examinee and hence reduce the time and effort of human intervention as well as make the evaluation procedure impartial to the entire user's. An innovative approach that trains a Fully Convolutional Network (FCN) to predict text line structure in document images is presented in Riya Goswami, et al. [8]. Jija Das Gupta, et al. [5] presents a novel ensemble classifier-based off-line handwritten word recognition system following a holistic approach. Xiangping Wu, et al. [9] proposes a new method for unconstrained offline handwritten word recognition by combining position embeddings with residual networks (ResNets) and bidirectional long short-term memory (BiLSTM) networks. Avani Sakhapara, et al. [1], have designed and implemented a machine learning-based subjective answer grader system (SAGS) using two algorithms, namely latent semantic analysis (LSA) and information gain (IG) for the generation of grades. A hybrid approach for sentence paraphrase identification is discussed in Muhidin Mohamed, et al. [7].

Our work: We proposed a neural network based classifier composed of Convolutional Neural Network (CNN), Recurrent Neural Networks (RNN) and Connectionist Temporal Classification (CTC) layers to recognize text from the handwritten words, which is a unique method for word classification which can give good performance over the above discussed models. We used Universal Sentence Encoder (USE) and Bidirectional Encoder Representations from Transformers (BERT) for obtaining semantic similarity of texts. These two are the state of the art methods being used in the semantic similarity based applications and research. Moreover these two methods are futuristic and will ensure more improvements in future.

2. Preliminaries

Artificial Neural Network is a mathematical model that tries to simulate the structure and functionalities of biological neural networks. A simple mathematical model (function) usually known as an artificial neuron forms the basic building block of every artificial neural network. *A Convolutional Neural Network (CNN)* is comprised of one or more convolutional and then followed by one or more fully connected layers as in a standard multilayer neural network. The architecture of a CNN is designed to take advantage of the 2D structure of an input image or other 2D input such as a speech signal. *Recurrent Neural Networks (RNN)* are similar to feed-forward neural network with no limitations regarding backloops. Here the information is transmitted in backward direction also. This creates an internal state of the network which allows it to exhibit dynamic temporal behavior. *Long Short Term Memory (LSTM)* is one of the recurrent neural networks topologies. In contrast with basic recurrent neural networks, it can learn from its experience to process, classify and predict time series with very long time lags of unknown size between important events. *Connectionist Temporal Classification (CTC)* is a type of neural network output and associated scoring function, for training RNNs such as LSTM networks to tackle sequence problems where the timing is variable. It can be used for tasks like handwriting recognition or recognizing phonemes in speech audio. *The Universal Sentence Encoder (USE)* encodes text into high-dimensional vectors. These vectors can be used for natural language tasks like text classification, semantic similarity, clustering etc. *Bidirectional Encoder Representations from Transformers (BERT)* has been a popular technique in Natural Language Processing (NLP) since Google open sourced it in 2018. BERT is the first deeply bidirectional, unsupervised language representation, pre-trained using only a plain text corpus.

3. System Design

The system which we propose is an automated answer book valuation, which value the scanned answer books and assign appropriate marks using machine learning methodologies for valuing answer books and assigning the scores of the students'. This system calculates the students' score by checking semantic similarity between students' answers and original answers. Here the answer books are the scanned images of students' handwritten answer sheets. Each pages in the answer book will be recorded as separate image files. Filename of the image file will be the combination students' register number and page number in the answer book. These answers are to be converted into text format for further processing.

The proposed system consists of two modules. Conversion of handwritten images to text format and the valuation of the answer books. In the first module, the answer books are scanned and converted into text format and in the second module we apply semantic similarity methods to find the degree of similarity between the students' answers and the original answers to assign students' scores. Figure 1 shows the different steps in the system design.

3.1. Conversion of Answers to text format

Conversion of handwritten images to text is a challenging task because of the huge variations in individual writing styles. The rise of artificial intelligence, along with machine learning and deep learning, are opening up almost limitless possibilities in this area.

In this module, the scanned handwritten answer books are converted to corresponding text format. Machine learning and image processing methodologies are being used for this image to text conversion. Several steps are involved in the conversion process. Figure 2 shows the functional block diagram of handwritten images to text conversion.

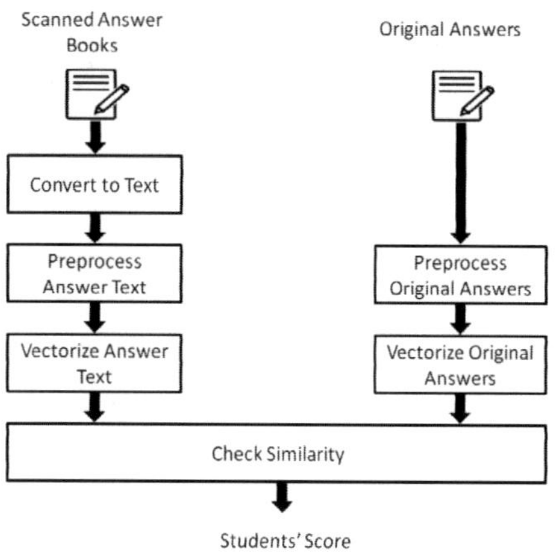

Figure 1. Answer book valuation

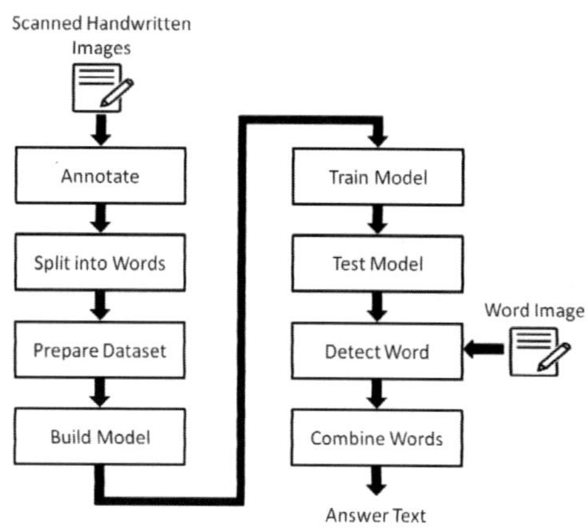

Figure 2. Images to text conversion

- **Annotation:** The co-ordinates of all the words available in the scanned handwritten images will be extracted. It helps to record individual word images from the answer books.
- **Split into Words:** The words in the scanned handwritten images will be copied into separate image files for further processing. It uses the coordinates obtained in the annotation step to identify the bounding boxes of individual words.
- **Prepare Dataset:** IAM handwriting dataset is used for train and test word classification model. It is a popular dataset used for several NLP applications. It contains 115,320 isolated and labeled words.

- **Build Model:** A neural network classifier will be built to process the dataset and to perform the word classification. The details of the model are described in Section 4.
- **Train and Test Model:** Training and testing will be done on the model with dataset so as to detect the text format for word images. The details of the step will be described in Section 4.
- **Detect Word:** Here the system detects the word in text format by inputting the word image into the classification model.
- **Combine Words:** All the detected words will be combined to form the answer text. Semantic similarity will be done on this answers with the answer key.

3.2. Answer Book Valuation
In this module, the converted answer texts are compared with the original answers using semantic similarity methods and assign the scores of each answers based on the similarity. The original answers in the answer key are also converted to text format using the steps discussed in Section 3.1. The detailed working of this module is given below.

- **Preprocessing of the answers:** In this step we preprocess the dataset used for semantic similarity. Preprocessing helps to improve the accuracy and efficiency of the similarity checking. The preprocessing include tokenization, removing stop-words, etc. Preprocessing will be done on students' answers as well as on the answer key.
- **Vectorize the answers:** Here we convert the answer text to vector format. Pretrained models are used for vectorizing. USE and BERT are used for for vectorizing the text.
- **Check similarity:** Here we check the semantic similarity between students' answers and original answers in the answer key. The scores will be calculated with respect to the degree of similarity. Scores of all questions attended by the students are combined to obtain the final score of the student.

4. Implementation
The system is implemented in Python as given in the design. Google Colab and Jupyter Notebook are used as the development environments. This system is implemented using TensorFlow and Keras libraries and the visualizations are done using TensorBoard.

4.1. Handwritten Text Recognition (HTR)
Our system uses a neural network based classifier for recognizing words. We identify the handwritten text by using CNN for feature extraction and RNN with CTC for sequence labeling. This design achieves good accuracy rate and demonstrates the potential for future improvement.

- **Preprocessing:** The images from the dataset may not be in same size, therefore we resized word images to 128x32 without distortion. We normalized the gray-values of the image which simplifies the task of the neural network. Data augmentation is done by copying the image to random positions or by randomly resizing the images.
- **Model Creation:** A classification model is built with 5 CNN layers, 2 RNN layers and one CTC layer. CNN layers are used to extract features from the word images. The output feature map has a size of 32x256. The popular LSTM implementation of RNNs is used in this classifier. It has more robust training-characteristics than vanilla RNN.
- **Training & Testing:** 95% of the dataset is used for training and 5% of the dataset is used for testing the model. 128x32 sized input is supplied to the model. CNN will extract features from the word images. The feature sequence contains 256 features per time-step. The RNN propagates relevant information through this sequence. The output sequence of

the RNN is mapped to a matrix of size 32x80 which is fed into the CTC layer. For loss calculation, we feed both the ground truth text and this matrix in to the operation. The ground truth text is encoded as a sparse tensor. The length of the input sequences must be passed to both CTC operations. We now have all the input data to create the loss operation and the decoding operation. Finally we get the output from the model.

- **Word Recognition:** 128x32 sized input image is supplied to the model which predict the word in text format. Combine these words to form the answers in text format. The detail of classifier model is: *Tensorflow version 1.4.0, Input size: 128 x 32, Training and Testing ratio: 95:5, Batch Size: 50, Epochs: 38, Optimizer: RMSPropOptimizer.*

4.2. Check Semantic Similarity
Semantic similarity is computed with students answers and the answer key. Students' score is computed from the degree of similarity obtained. We use two popular methods USE and BERT for checking the semantic similarity.

- **Convert to text:** We convert the handwritten answer books in to text format.
- **Preprocessing of students' answers & original answers:** Preprocessing is performed on both students' answer books and on the answer key.
- **Vectorize students answers and original answers:** Here we convert the answer text in to vector embedding. USE has the vector size of 512 and BERT has the vector size of 768. Vector size is fixed regardless of the size of text.
- **Check similarity:** Here we check the semantic similarity on vectors to find the degree of similarity between students answers and the original answers. Cosine similarity measure is used to find the similarity of vectors. The result of the similarity will be between 0 and 1. A value close to 1 will be treated as most similar answers or correct answers and value close to 0 will be treated as dissimilar answers or incorrect answers. The score will be calculated with respect to the degree of similarity.

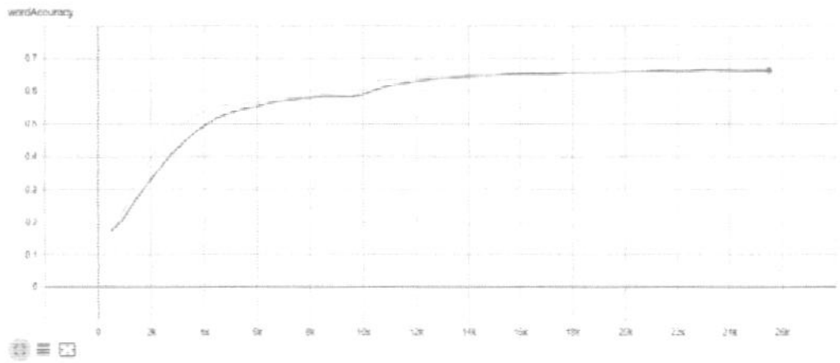

Figure 3. Accuracy - Testing

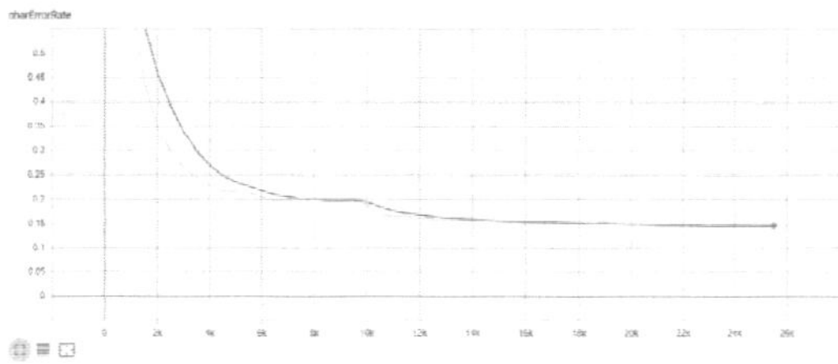

Figure 4. Error Rate - Testing

5. Conclusion and Result Analysis
Handwritten text detection: We implemented a word classifier to recognize words from word images. IAM handwriting database is used as the dataset. We got 65.96% accuracy and 14.73% loss for the word classification. Figure 3 shows the word classification accuracy on testing. X-axis denotes iterations and Y-axis denotes accuracy rate. Figure 4 shows the word classification character error rate on testing. X-axis denotes iterations and Y-axis denotes error rate.
Answer Book Valuation: The answer book valuation have conducted using Universal Sentence Encoder (USE) and Bidirectional Encoder Representation for Transformers (BERT). USE showed an overall accuracy of 73.82% when 15 answer books are evaluated while BERT showed 74.53% accuracy for the same set of answers. Table 1 shows the results obtained.

Table 1. Answer Book Valuation using USE and BERT

Reg.No.	Original Marks	USE		BERT	
		Marks	Accuracy%	Marks	Accuracy%
0001	13	10	78	11	90
0002	10	9	91	11	90
0003	3	7	42	9	31
0004	15	11	75	12	80
0005	11	10	96	12	88
0006	11	13	80	15	68
0007	15	9	63	11	79
0008	19	20	95	20	94
0009	9	11	77	14	62
0010	7	12	56	16	42
0011	17	10	63	15	89
0012	18	12	71	16	91
0013	9	13	67	15	58
0014	6	9	65	10	54
0015	11	8	80	11	93
Overall Accuracy			73.82		74.53

Measuring semantic similarity of text or documents or concepts is a crucial component in many applications. Handwritten text recognition also a complex functionality in NLP. We implemented an answer book valuation system using semantic similarity. It used neural network containing CNN, RNN and CTC for detecting text from word images. We have obtained 65.96% accuracy and 14.73% loss in text detection. More elaborate training and testing on wider database and new technology classifiers may give better results in handwritten text recognition for the future designs. USE and BERT give exceptional performance in text similarity based applications. The automatic valuation methods will replace the conventional valuation by using the advancements in machine learning and natural language processing in near future.

References

[1] Bhakti Chaudhari Rishabh Gada Aakash Mishra Shweta Bhanushali Avani Sakhapara, Dipti Pawade. Subjective answer grader system based on machine learning. *Springer Conf, Feb-2019.*

[2] Chandrabose Aravindan D. Thenmozhi. Paraphrase identification by using clause-based similarity features and machine translation metrics. *The Computer Journal, Sep-2016.*

[3] Henry A. Rowley Li-Lun Wang Victor Carbune Daniel Keysers, Thomas Deselaers. Multi-language online handwriting recognition. *IEEE Trans., May-2016.*

[4] Carlos A Iglesias Ganggao Zhu. Computing semantic similarity of concepts in knowledge graphs. *IEEE Trans, Sep-2016.*

[5] Bhabatosh Chanda Jija Das Gupta, Soumitra Samanta. Ensemble classifier-based off-line handwritten word recognition system in holistic approach, iet image process. *IEEE Trans., Aug-2018.*

[6] Zepeng Gu Ahmed Zeeshan Ming Liu, Bo Lang. Measuring similarity of academic articles with semantic profile and jointword embedding. *Tsinghua Science and Technology, Dec-2017.*

[7] Mourad Oussalah Muhidin Mohamed. A hybrid approach for paraphrase identification based on knowledge-enriched semantic heuristics. *Springer, Apr-2019.*

[8] Avik Bisai Alok Ranjan Pal Riya Goswami, Somik Karmakar. A knowledge based approach for long answer evaluation. *IEEE Conf, Jan-2018.*

[9] Jinghan You Yulun Xiao Xiangping Wu, Qingcai Chen. Unconstrained offline handwritten word recognition by position embedding integrated resnets model. *IEEE Signal Processing Letters, Jan-2019.*

D-App for providing geriatric aid and to collect and handle periodic health survey data

Chinnu Joseph[1,2], Chinchu Joseph[1,3], Atul Krishna P[1,4], Amal George Mathew[1,5] and Fabeela Ali Rawther[1,6]

[1] Department of Computer Science and Engineering, Amal Jyothi College of Engineering, Kerala, India
[2] chinnujoseph29@gmail.com
[3] chinchujoseph2020@cs.ajce.in
[4] atulkrishnap2020@cs.ajce.in
[5] amalgeorgemathew2020@cs.ajce.in
[6] fabeelaalirawther@amaljyothi.ac.in

Abstract. The system aims to build a health survey platform designed to collect basic health information about residents of particular area. Block chain is used to store the collected information making sure only authenticated persons can have access to crucial information. This system also provides information about government health policies or schemes that the members of a family are eligible for and can have easy access to these facilities. This system can also be used as a platform to host health survey in case of an unexpected outbreak of a disease .Machine learning algorithms may be used predict health issues like the possible occurrence of epidemics etc. The system efficiently reduces the paperwork and inefficiency of traditional door to door health surveys by enabling each family to directly feed in their information.

1. Introduction

Machine learning is one of the most thriving technologies that the world has come across. This technology is still in its developing stage. It will undoubtedly continue to attract more people in the coming years as well as make headlines. Machine learning is a tool for turning information into intelligence. In recent decades, there has been an acceleration of data. Such data are worthless until we study them and extract the pattern within them. Machine learning is used to identify useful fundamental correlations efficiently within complicated data that we might otherwise fail to discover. Secret trends and question information can be used for potential activities and for sensitive decisions of any kind [1]. The use of a shared cloud consequently leads to security problems in data storage and processing. So, we need some sort of security algorithm that grants a scalable, key-based encryption system that can resolve delays, security forces, and storage threats.

The history of any digital ledger is irreversible and transparent, through the use of decentralization as well as cryptographic hashing. While generating the first block of a chain, nonce produces the cryptographic hash. The data in the block is characterized as signed and linked forever to the nonce and hash unless it is extracted. Each block comes with its distinct nonce and hash but also relates to the hash of the previous block in the chain which makes it difficult to mine a block particularly on big chains. Decentralization being the most significant principle of block chain, no one machine or organization can own the chain [1]. Every node has its block chain replica, and any freshly mined block requires the network to update, authorize, and test the chain algorithmically. Since these distributed ledgers are transparent, any change in the ledger can be quickly checked and viewed. A specific ID number representing the transaction is given to each user which is alphanumeric. The fusion of digital records with a network of checks and balances aims to preserve transparency and create consumer trust [3].

As machine learning and block chain converge, people will benefit from ML's ability to automate the handling of a massive amount of data. In fact, it would be possible to create a completely new paradigm by putting the two together. By using ML to manage the chain, there is also an opportunity to drastically improve security. In addition, since ML enjoys working with a lot of data, this offers the opportunity to develop better models by utilizing the decentralized nature of block chain which also promotes data sharing. Sometimes you might end up with a qualitatively new data set that is also a better data set when all the data from the silos converge [3]. As a result, it will lead to the creation of a qualitatively new model where new insights can be derived which, in turn, can provide new opportunities to build cutting-edge business applications for next-gen. This can be a game-changer for the majority of industries as it can be used as a tool for fraud detection [1]. Thanks to a common ledger framework.

2. Proposed System

This research was initiated as a way to increase public awareness of the numerous government services and initiatives for the welfare of the elderly [4]. We wondered why not develop apps to bring the specifics of the organizations and student bodies operating health camps and inspections to the geriatric crowd. Further brainstorming led us to a study network idea that allows everybody to be involved in their health care [5]. This platform can provide the public with strong support for the government in relation to health issues and in turn, helps the ministry study people's general health. The second reason for this idea is that the government is not prepared to have an electronic survey forum during an infectious disease epidemic, and it would be of great benefit in our country to create a website that could implement all the above criteria in a limited period [6, 7].

The government struggles to gather its people's data in an incredibly populated country like India [4]. The gathering of data proceeds in conventional formats, e.g. door-to-door surveys. Only with regular and accurate surveys can the government analyze the health, needs, and statistics of its people [8]. The major issues that should be taken into consideration for improved surveys for keeping track of trends in India's disease burden are the inability to collect data on non-communicable disease, injuries and the interval between the survey rounds, another area that needs attention is the delay in making the data available in the public domain [9]. Currently, there is no notification system for the public. We often do not know many of our Government's incentives, which lead to a lack of efficiency in many of our plans. We do not have a platform for an online survey to gain data to study it for medical conclusions even following an outbreak of a disease or a pandemic. These issues should be solved by innovations in data communication and processing technology. The latest computational methods and tools for machine learning are used to effectively enhance our country's health quality by providing an online health database.

The whole network is running on a block chain network, which is an advanced data protection system commonly used. The data storage is unlike the relationship database but as a chain, i.e. one connects to the other; all new data or any data updates made are added as a next block in the chain. Every activity in the network is logged in and every attempt at unauthorized access is strictly controlled by block chain technology [10].

Our system uses an open-source block chain. It provides smart contract functionality, decentralized block chain. Now, the smart contract is a set of rules to be followed by all participants in the block chain network [11]. It is this contract that enables us to specify what actions need to be carried out and who all are authorized to do so. The contract is deployed as the first transaction in the block chain and therefore each node has the same version of the code. We used the language of solidity programming [12] to write our smart contracts, the language has features from C++, python, and JavaScript and is specifically designed for the Ethereum virtual machine.

Truffle development [13] and testing framework have been used for our prototype. This framework allows you to compile and deploy your smart contract on a block chain network. It also provides testing for the behavior of your contracts and thus saves you from unexpected errors after network setup. Since the aim was to build a working prototype of the proposed system, we have used Ganache block chain, which is a personal block chain for the rapid development of Ethereum applications. Ten block chain accounts are provided so that we can develop, deploy, and test our Dapp in a safe environment. You can easily connect the truffle network to the Ganache. Therefore, we are freed from having to set up nodes in our network.

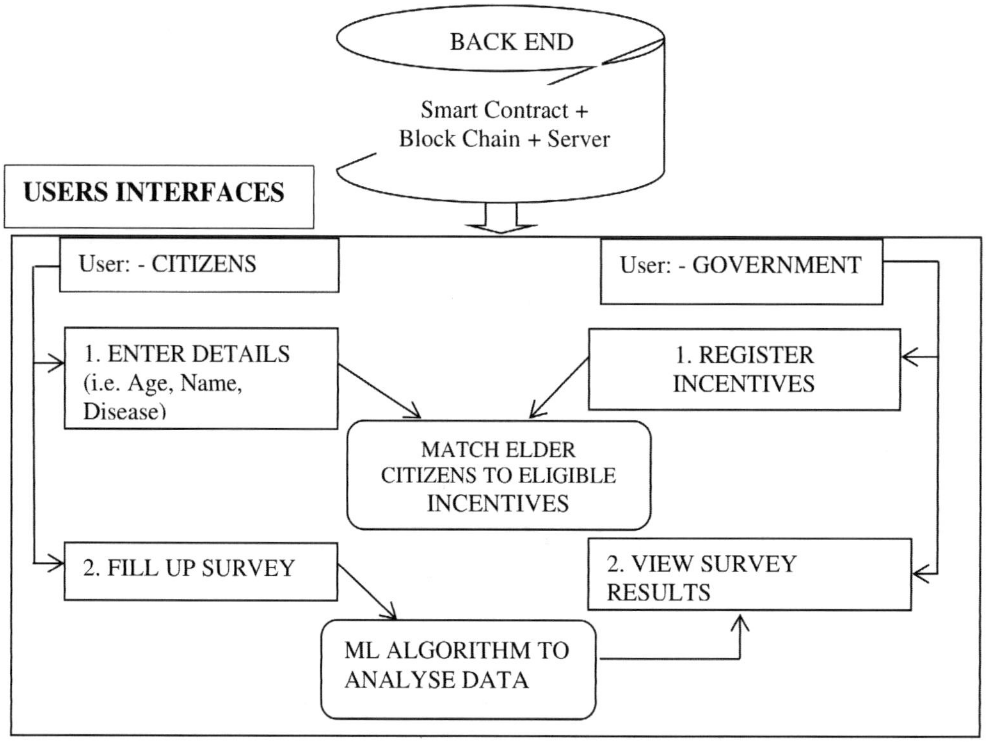

Figure 1 The figure demonstrates the logical architecture of the proposed system.

Web3.js is a library set that will communicate with the block chain using the HTTP-connection (this server is operated by a Ganache). It will access user accounts, send transactions; interact with smart contracts, and more. Extension Meta mask in our browser permits Dapp block chain access, propose transacting, and reads user account address. Every transaction must be checked by the account owner before being executed.

The user can access the website through their block chain account via the Meta mask. If the account is already registered the account holder profile is displayed, otherwise registration. They are provided with a registration page where they register themselves under their block chain account. They have to specify their age, which lets the system separate its users into geriatric and non-geriatric. Once the users are registered, they can view the incentives or schemes they are eligible for and approach the ministry to avail it. If the government sets up a survey, they can give their responses by logging into the system. In case you develop a new disease, you can update the same on your profile. The user can also view a general analysis of the situation in case of a pandemic or epidemic thus staying updated with the correct information.

On the other hand, the government account holder can view all the details of the system users whenever required. Another main responsibility of the government account is to add incentives and schemes and their related information. These are then matched to all the eligible users of the system. The government also gets to see the survey results and thus get an overall analysis of the major diseases among its citizens.

This system can also be used as a host health survey in the event of an unexpected disease outbreak. Machine learning is used to analyse and link symptoms to illnesses. N number of symptoms and n number of illnesses are fed to the machine to train the model. The model is tested to reach the required accuracy. When the user completes the survey, the symptom survey response is registered and mapped to the disease using the trained and tested machine learning algorithm. The output of this study is the likelihood of diagnosing a particular disease to a given person. All the data and results from this step are stored in the block chain. Machine learning algorithms can be used to predict health issues such as future outbreaks (fig 1).

3. Experimental Results

The predominant objective of this system is to perform a survey among a large number of people and to derive an analysis of the data that may be useful to the government [2]. With the help of a data set of 132 symptoms and 41 diseases commonly occurring around a year, we have mapped out a predicted disease after the completion of the survey. The algorithms used for the designed system are decision tree and KNN [2]. In our system, symptoms that give the better classification are assigned higher importance and are thus concentrated near the root node, and other towards the leaf nodes of the tree, providing a more accurate prediction based on the more prominent symptom related to that disease. As already known, the output of a decision tree is a single class, thus our system accurately predicts one of the prominent diseases from those symptoms provided by the user, but chances of presence of multiple diseases to be detected from a survey is also high, thus we need another algorithm which can detect a less evident disease also. K Nearest Neighbour (KNN) is used complementary to the decision tree, in our system. KNN is the most effective pattern detection lazy classification algorithm [2]. The closer neighbours contribute more to the resulting class than the more far off ones. The number of neighbours which are assigned is 10 for the developed system. With even the lower number of neighbours, we can detect a less evident disease from the survey. Since the platform surveys are taken at regular periods the presence of more than two undetected diseases in a person is very less likely to occur. Given four-five prominent symptoms of a disease, we can achieve a prediction with high accuracy from our system. The combination of both the algorithms effectively provides disease that is more obvious along with comparatively less obvious. Hence combining the predictions from the decision tree and KNN we get an effective result of predicted diseases. The table snippet (fig. 2) shows the prediction results from a sample of 34 survey responses. This predicted disease of each individual for a survey is stored under a common serial number in the block chain. The government can view a graphical representation of the count of diseases detected in a region for that survey, thus aiding them to take timely action.

The time series prediction of valuable data using Long Short-Term Memory (LSTM) networks is another component of this system where power machine learning is deployed. During times of a pandemic, like COVID-19, it is the key duty of the government to provide accurate data of the number of confirmed cases on a daily basis to its citizens. Along with this, the government should be able to get an approximation of number cases that might develop in the coming days. Our platform has developed a model that predicts the total number of confirmed cases from all the countries with the presence of COVID-19, up to 10 days with approximate range accuracy (fig. 3).

IOP Conf. Series: Materials Science and Engineering 1085 (2021) 012019 doi:10.1088/1757-899X/1085/1/012019

Survey Number 1 Get Analysis

No. of people predicted to be having the disease from the overall survey responses

DISEASES	COUNT
Fungal infection	1
Allergy	1
GERD	0
Chronic cholestasis	1
Drug Reaction	1
Peptic ulcer diseae	4
AIDS	0
Diabetes	0
Gastroenteritis	1
Bronchial Asthma	0
Hypertension	3
Migraine	0

Figure 2 The table shows the list of diseases and the corresponding no. of people predicted to be having those diseases. Every time the particular survey result, predicts a disease, the count is incremented; thus, giving the government quick analysis of the trending disease.

The data set used for the prediction was compiled by the Johns Hopkins University Centre for Systems Science and Engineering (JHU CSSE),supported by ESRI Living Atlas Team and the Johns Hopkins University Applied Physics Lab (JHU APL) [14].The model was trained with 15 hidden layers and 1500 epochs to achieve the required accuracy over a test data of 60 days. The available and predicted data are graphically displayed to provide the user with improved interpretation or analysis. The LSTM can learn long-term dependencies without the problem of the vanishing gradient, which affects their performances, and can

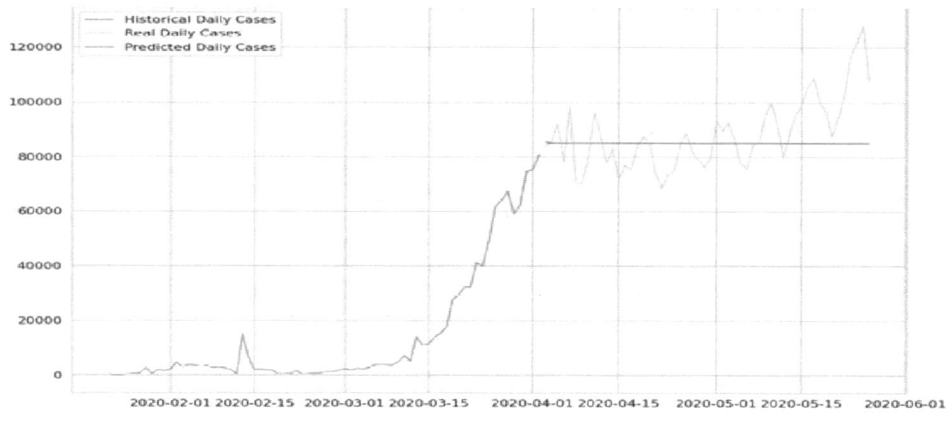

Figure 3 The figure shows the plot of historical daily COVID-19 cases used for the training of LSTM model along with comparative plot of predicted number of cases and the test data of real number of cases.

tackle sequential and temporal problems. LSTM has a chain-like structure, an RNN equivalent, except there's four, composed of a cell, input gate and output gate and a forgot gate consisting of Sigmoid and Tanh functions, rather than providing a single layer of the neural network.

4. Results

The system prototype shows a decentralized web application that will allow every citizen, distinctively the elder citizens, to be aware of the government-funded special monetary and nonmonetary medical schemes and check-up activities [4]. We successfully mapped the right health incentives (which are made available by the govt and NGO's) to the elderly that requires it using solidity datatypes. The survey page allows easier access to mass health information in short time notice. As described in the experimental section above, ML algorithms were used to process the collected survey data to give us insight into the predominant and trending disease among the citizens.

5. Conclusion

The website will make the annual health survey digital and provide a much easier way to analyse the survey responses. Health surveys can be conducted in shorter intervals with minimum human resources and government authorities responsible for conducting them will become more answerable about the task assigned to them. It will greatly aid timely action on healthcare. Another aspect of this system involves mapping the diseases of an individual to available incentives from the government or other NGOs especially for the elder members of the society enabling them to receive their rightful aid from these agencies, which in turn, we strongly believe, would improve their life at adulthood [4]. In case of a sudden outbreak staging an instantaneous survey to an outsized group of citizens now become realizable. Collection of data can now be done remotely, thus reducing the likelihood of disease spread among the concerning data collection authorities. The system briefly brings down a busy task of data collection to the citizen's fingertips.

The future scope includes the security aspects of the web site and improvement of the machine learning algorithm for disease prediction. Also, a cloud service might be connected to the system, to collect together all the incentives and schemes found out by the government or NGOs for the people.

6. References

[1] S. Vyas, M. Gupta and R. Yadav 2019 *Amity International Conference on Artificial Intelligence (AICAI)* pp. 709-711

[2] D. Dahiwade, G. Patle and E. Meshram, 2019 *3rd International Conference on Computing Methodologies and Communication (ICCMC)* pp. 1211-1215

[3] B. Narendra Kumar Rao, B. Bhaskar Kumar Rao,Vellingiri J 2019 *International Journal of Innovative Technology and Exploring Engineering (IJITEE)* **Volume-8 Issue-8** June ISSN 2278-3075.

[4] Abhishek Verma, Sunita Aggarwal, Sandeep Garg and Pooja Anand 2019 *Journal, Indian Academy of Clinical Medicine* **Vol. 20**, No. 3-4

[5] Angell BJ, Prinja S, Gupt A, Jha V, Jan S 2019 The Ayushman Bharat Pradhan Mantri Jan Arogya Yojana and the path to universal health coverage in India: Overcoming the challenges of stewardship and governance (PLoSMed16(3): e1002759).

[6] J. Li, P. Ray, H. Seale And R. Macintyre, 2012 *45th Hawaii International Conference on System Sciences* (MAUI, HI) PP. 2800-2809

[7] J. Li and P. Ray, 2010, *The 12th IEEE International Conference on e-Health Networking, Applications and Services* pp. 391-398.

[8] Dandona R, Pandey A, Dandona L. 2016 *Bull World Health Organ.* **94(4):** 286-96A.

[9] Devadasan N, Ghosh S, Nandraj S, Sundararaman T 2014 “Monitoring and Evaluating Progress towards Universal Health Coverage in India.” PLoS Med 11(9): e1001697.

[10] Jones M, Johnson M, Shervey M, Dudley JT, Zimmerman N 2019 *J Med Internet Res*;**21(8)**: e13600

[11] https://www.kaleido.io/blockchain-blog/what-are-smart-contractsand-how-do-theywork/ “What are Smart Contracts and How Do They Work?”

[12] https://solidity.readthedocs.io/en/v0.6.8/, “Solidity Documentation”

[13] https://www.trufflesuite.com/docs, “Truffle Suite Documentation”

[14] Dong E, Du H, Gardner L. An interactive web-based dashboard to track COVID-19 in real time. *Lancet Inf Dis*. 20(5):533-534.

Automatic Detection of Anomalies in Video Surveillance using Artificial Intelligence

Sreedevi R Krishnan[1*], P Amudha[1], and S Sivakumari[1]

[1]Department of Computer Science & Engineering, School of Engineering, Avinashilingam Institute for Home Science and Higher Education for Women, Coimbatore, Tamil Nadu, India.

[*]Email id sreedevirkrishnan@gmail.com

Abstract The significance of security in day to day life is increasing, and hence the use of a video surveillance system is becoming commonly accepted in almost every public places. Even though by placing surveillance system in public places will help to trace out the culprits of the anomalous situation, it is hard to trace it out. The knowledge of the incidence time is needed to identify the event, and also, it requires an enormous data handling task. To identify the anomalous situation in real-time will help the authorities to decrease the consequences and loss during the anomalous event. The paper proposes an in-depth study of various automatic anomaly detection techniques which helps to reduce the loss occurred of the anomalous situation. The advancements in Artificial Intelligence help in the quick and automatic identification of nominal and anomalous events. The sequential and incremental learning approach in feature extraction will help to generate a model that will provide much accurate classifications and predictions of anomalies.

1. Introduction

The need for security to the life and property of ordinary people is increasing. As the violence during theft increases, for guaranteeing the security of citizens in public places need an authorised, and the behaviour of the people needs to be analysed. Nowadays, most public areas are under Closed Circuit Television (CCTV) surveillance for providing security to people and property. Using CCTV surveillance, a large amount of video data needs to be analysed and to handle this Big Data is tedious.

A smart CCTV is needed for the identification and behaviour analysis of the people in public places. Various Neural Networks algorithms are useful for the automatic identification of abnormal situations in a security system. Using Artificial Intelligence, we can identify the features and frequency of occurrence of abnormal events. Besides, the study provides techniques to reduce the amount of data storage by removing the videos, which is under normal conditions, and this will solve the main problem related to Big Data management. The motivation to select this topic is to ensure the safety and security of people in the public area in a fast and effective way. The immediate identification of abnormal events will help to reduce the casualty of the situation or even avoiding the situation, thereby enforcing the security of people.

2. Automatic Anomaly Detection in Video Surveillance

Through Artificial intelligence, the machine can mimic a human being and also may have human cognisance without being manually coded. Artificial Intelligence performs a systematic study of algorithms and statistical analysis to build a model using artificial neural network. It relies on data patterns and deductions for making predictions using computers. AI algorithms build a model based on the information known as *training data*; this model can predict the future without explicitly coded to execute out the task. During the Learning procedure, certain information which is known from previous experience is given to the machine called training. During this training, the machine will consequently distinguish the knowledge. Using the model built for classifying nominal and anomalous events using Artificial Intelligence, the various anomalies occurring in the real-life scenario can be detected. By providing proper alerts, the after-effects of abnormal events can be reduced to an extent. Some of the anomaly detection methods are discussed in the following subsections.

2.1. Smart Processing and Storage Utilisation method in Surveillance System

Z. Shao, J. Cai, and Z. Wang [2] present an innovative smart processing and storage utilisation method, in a video surveillance system for event identification and alert message. The strategy incorporates three sections: smart pre-alert for rare events, sufficient storage for surveillance video, and quick access to suspected videos, which ultimately discovers the temporal-spatial relation analysis of rare events in various checkpoints. Initially, an irregular behavioural database is established for a smart surveillance system that stores and manages alerts in case of crisis. Then, selected videos of the surveillance system are stored according to the alert information and avoid other videos; this significantly decreases the storage space consumption. Finally, when video evidence is required, the videos related to irregular behaviours are outlined and retrieved accordingly.

The characteristics of criminal behaviour are studied from previous experiences and recorded, to distinguish the anomalous practices from ordinary conduct. Anomalous behaviours can be decided by the frequency of occurrence, the identity of the target, and the surrounding area of an event. Anomalous activity database is initially created by storing the vivid data of anomalous activities captured from various video cameras. This metadata comprises of several types of anomalous conduct, name of checking premises, the details of video camera including device ID and location, starting and finishing time of the event, Etc. The anomalous behaviour database gives the metadata for further examination.

The Event- Abnormal Behaviour Correlation Model [2] is used to identify and co-relate the events. The model can identify abnormal activities with temporal auto-correlation greater than the previously set observational limit as an unsafe one. When an anomalous behaviour alert is activated from a specific video surveillance device, the smart surveillance system will analyse the temporal correlation regarding previous records in the database. A multi-site relation is examined regarding the activity of the criminal suspect.

The criminals usually follow a practice of studying the premises by roaming before doing a criminal activity; this spatial association can be identified from the various surveillance cameras placed in the smart surveillance system. The surveillance system is shown in figure 1, in which people who are commonly roaming during an event can be recognised and re-identified [4]. If the person is identified using the face recognition system [1] and found as walking suspiciously, it will be marked as a notified event. If the roaming occurs in various destinations at the same time: the smart observing camera tries to identify the persons by checking the multi-site checkpoints using spatial relevance and then raise the alert. The reoccurrence at a specific time interval and temporal auto-correlation is taken for the significance of the conduct. If two or three roaming occasions are recognised at a time by the surveillance system in various areas, the roaming conduct will be fixed as a high-risk one. If the same person visits multi-site and reports events of wandering: then identify that he is part of an activity that is closely related to time and space risk. In the event, if the roaming reoccurrence in the examination time frame is deeply found a lot higher than the ordinarily reported information at the hour, then this event will be a suspicious one.

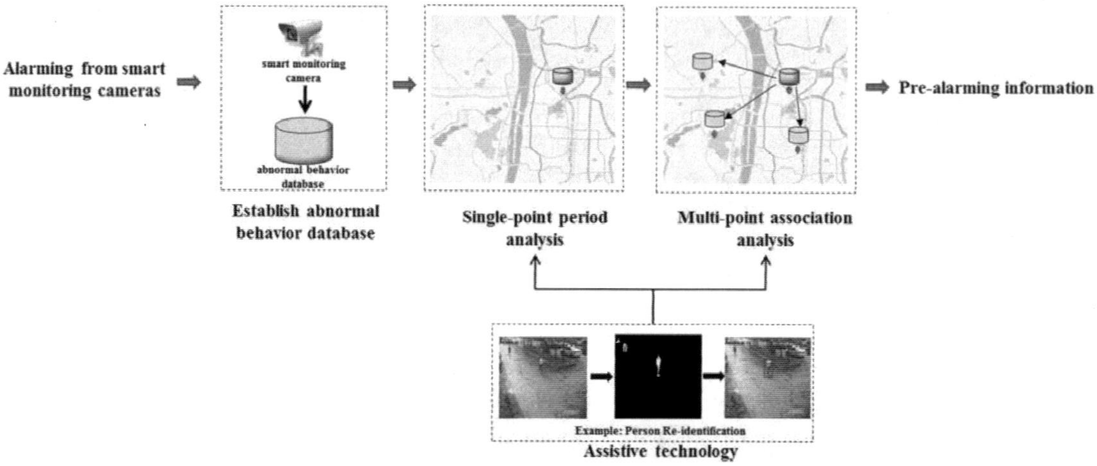

Figure 1. Pre-alarming based on abnormal behaviour association analysis

When an accident or abnormal condition occurs, the model will be refreshed and updated. From the abnormal characteristics database, recurrence, and various abnormal behaviour are considered for the last seven days, and the security system can be implemented more or less accordingly. Sometimes the same reoccurrence may happen for abnormal behaviours, and then the model features out the more significant weight. The study of the above paper gives knowledge about the smart monitoring of Big Data in Surveillance Video and memory utilisation. Even though the system can quickly identify and update the anomalous situation, this model will be more useful for identification of culprits of an event occurred than avoiding the abnormal condition.

2.2. Unsupervised learning and Knowledge Attaining in Surveillance System

Nawaratne, Bandaragoda, Adikari, Alahakoon, De Silva, and Yu [3] present a methodology that consolidates significant level video investigation with self-learning, incremental knowledge attaining, and self-governing smart surveillance system. This methodology is fit for identifying variations that happen after some time and isolating anomalies from re-events, without the essential of a labelled data set. The analysis of the behaviour is a beneficial methodology for finding out the primary usage of irregular and regular behaviour patterns. A constant irregular movement detection method is introduced in [3] for the analysis of a group of people, which helps to analyse regular, irregular, or emergency in the group stream. In this methodology, the behaviour of movement is taken from movement vectors in the video clips. Depending on the world-wide group movement behaviour, and movement vectors, the unexpected conduct in social groups can be identified based on the approach given in [6]. The blend of an optical stream and background elimination procedures are utilised to extract movement at each picture outline and discard small movement vectors having noise, issues of compression, and non-stationary foundation pixels.

Using Microsoft Cognitive Service [7] the characterisation and automatic labelling of videos based on semantic matters can be performed. For example, the activities of a person, objects, or complex events in the video frame can be characterised and labelled as such. The framework is given by Google Cloud Vision [8], and the Tensorflow open-source provides a very significant level of video analysis like emotion, face, optical character recognition, logo identification, object detection, and unsuitable content identification [10].

The learning and thinking in human cognisance is a progressive method. A human builds up fundamental knowledge and information through real-time surroundings and gradually refines the understanding as new data shows up. Learning, revising, and unlearning is a usual and constant procedure in human cognisance [11]. The unsupervised-learning algorithm, Incremental Knowledge Acquisition and Self-Learning (IKASL) tend to update the labelled dataset by continuous learning, thereby keeps up reliability and versatility across the iterations of learning [11]. The IKASL algorithm has procedures for generalisation and learning which help in knowledge acquisition and self-study. By the unsupervised learning method, the data set and features learned from the IKASL algorithm will be exploiting the video exploration capacity of Microsoft Cognitive Services API. The algorithms working process is explained in the following subsections.

The recorded labels created by Microsoft Cognitive Services API for frames of videos are used to build the feature space. From the video frames, the details of feature space comprise of keywords for identifying the activities, behaviours, and various objects are present. Figure 2 shows two video frames taken at time intervals from a typical checkpoint. According to [3], the keywords are identified in figure 2(a) using Activity Recognition [12]. Figure 2(b) is having the video frame taken from the same video checkpoint after 8 seconds, and the keywords are identified. Different methods are used for identifying this feature space like object recognition method [13] to infer the idea of the video frame, computer visual service provider [8] that study the video frames to generate a description or its combinations.

(a) (b)

Figure 2. Two video frames are taken at time intervals from a typical checkpoint

IKASL algorithm uses a layered structure of n layers, and each layer consists of learning and generalisation sub-layers. The clustering of the input feature vector is included as a Growing Self-Organising Map (GSOM) [13] component in the sub-layer called the learning layer. The encoded summarised representation of the next learning layer is done in the generalisation sub-layer, and this layer will become the initial layer of the following learning layer, as shown in figure 3.

In the basic structure and model of the IKASL algorithm [3], the initial layer consists of 4 nodes with the random initialisation of weights. Initially, the sequences of data sets are given as tuples, in which the selection of a successful node is decided based on distance measurements like Euclidean or Cosine distance. The weight adjustment of a successful node is made according to the input tuple, which will be further adjusted by the nearby nodes. The learning in every layer is done by the GSOM algorithm [13]. After an iteration of learning, every successful node in a layer is identified and generalised up to an extent, and the outcome will return two significant results in the learning procedure. The learning knowledge is passed to neighbouring nodes as successful nodes of weight vectors. In the generalisation sub-layer, the fuzzy integral will consolidate the weight vectors into a single one. The knowledge obtained from the learning and self-organisation process is provided in the generalisation sub-layer. Several iterations of learning are performed in the data set using learning and generalisation sub-layers.

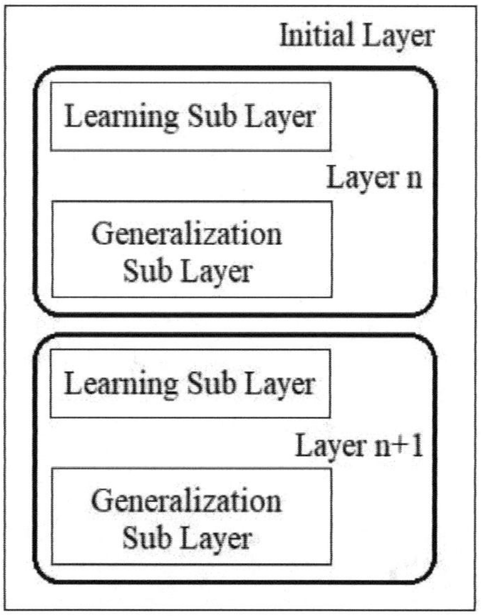

Figure 3. IKASL Model

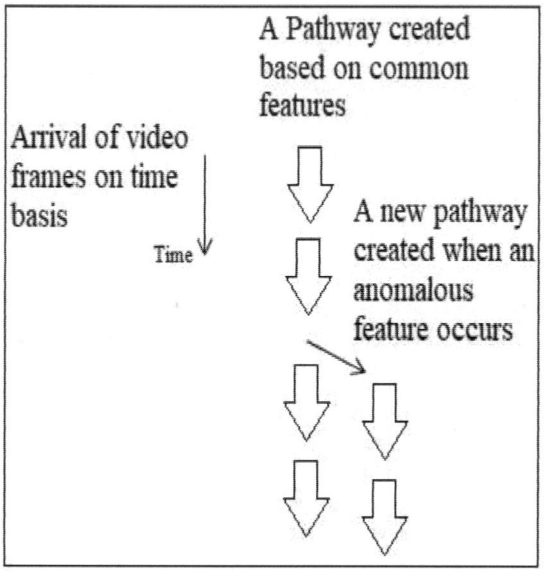

Figure 4. IKASL Pathway Generation

Using the IKASL algorithm, incremental learning of a new pathway is dependent on clusters recognised in learning sub-layers as in Figure 4. New pathways might be produced if there are new features in the input given to the algorithm and to classify the cluster according to the input tuples or feature vectors. For instance, consider figure 2, in which video frames are given to the IKASL algorithm in the order of occurrence, and each pathway represents the feature vector based on the keywords which form a cluster. If the feature vector has common features, then all the video frames will be in the same cluster, and if there is a substantial change in a feature vector, the new pathway will be created. For instance, if a video frame consists of danger features like fire, the IKASL algorithm will create a new pathway with new feature vectors and keyword sets. If the danger feature is cleared in the subsequent frame, the feature vector will move back to the original pathway using the IKASL algorithm.

The new pathway creation using IKASL algorithm can be used for various video surveillance applications like anomaly detection, identification of unknown objects, intrusion detection, observation of road accidents, and safety alerts generation when needed. The model can be extended by adding more dataset to extract more feature vectors.

2.3 Continual Learning for Anomaly Detection

Keval Doshi and Yasin Yilmaz in [5] propose an Anomaly detection method using transfer learning [14] and continual learning [15] which helps to reduce the training complexity. The model uses feature extraction based on neural networks for transfer learning, and for decision making continual learning statistical k-nearest neighbour (kNN). In transfer learning, a model which is developed for one task may be reused for another related task. Whereas Continuous learning with automatic incremental development of models is termed as Continuous Learning, and this helps to solve complex problems.

This Machine Learning model will updates the Prediction model smoothly using task and data thereby enables reuse and retrain useful knowledge and skills. The use of transfer learning will considerably reduce the training complexity. The statistical framework for sequential anomaly detection can perform continual and few-shot learning of video data, which help to evaluate video anomaly detection datasets and real surveillance video inputs.

The anomalous events are usually circumstantial, for example, a problem in the practical situation a riding a bike through the road is a normal situation whereas riding through pedestrian way will be an anomalous situation. So in practical, the nominal event needs to be updated continually, and the decisions are updated using Deep Neural Networks (DNNs) [16], which uses the sequential nature of video anomaly detection.

The features from the video are evaluated, and feature vectors of each objects having attributes like motion, location and appearance are created. The weights are applied to features which adjust according to the relative importance of each feature category during feature selection. To obtain the features form a real-time video stream, a pre-trained real-time object detection system like You Only Look Once (YOLO) [17] is used. The proposed model uses YOLOv3, which gives a bounding box (location) and class probabilities (appearance) for each video frame. The centre of the box is monitored for obtaining the area and location features. From the test videos, the objects deviating from nominal paths or new classes will help to identify anomalies related to spatial information. For temporal information, the hypothesis used is, for any motion anomaly the probability distribution of the optical flow of the frame would alter. For this, the mean, variance and high order statistics skewness, and kurtosis is extracted, which is used to represent irregularity and sharpness of the probability distribution. A feature vector is constructed for each time by combining the motion, location and appearance features [5]. For Anomaly Detection with minimum delays, a nonparametric sequential anomaly detection algorithm [5] is used. The Euclidean k Nearest Neighbor (kNN) distance is used to identify the local interactions between nominal data points. The hypothesis of the statistical distance between the nominal features and anomalous instances will be further away from the k nearest neighbour is used here.

During the testing phase, the feature vectors are extracted for each object, and kNN distance is computed. The instantaneous frame-level anomaly is evaluated, and running decision statistics for each time is calculated. For nominal data, the value of decision statistics will always be a negative value, and anomalous data will get a positive value. If the anomalous data persists to exist for n frames for a successive time periods the value of the decision statistics will grow. Furthermore, if it exceeds the threshold, the frames will be labelled as anomalous. The test will continue with another set of frames and label the frames.

The continual learning is used to test statistical results and labelling the frames. The model proposes a human-in-the-loop approach to classify the label as true or false alarms from time to time. This sequential updates will continually learn new data and prevents from wrong classification of abnormal alarms. This model avoids retraining of deep learning methods from scratch and disastrous forgetting. The model can be extended to work effectively on challenging situations like various weather conditions occurring dynamically.

2.4 Ensemble Learning using Bagging and Inception-v3 for Anomaly Detection

Yumna Zahid, Muhammad Atif Tahir and Muhammad Nouman Durrani in [9] propose an Anomaly Detection method by extracting video features using Inception-v3 deep learning network which avoids segmentation. The model is implemented by transfer learning the Convolution Network Neural (CNN) [19]. The Homogeneous bagging ensemble of 3-Layer Fully Connected (FC) Network is used for classification of anomalous events.

For unsupervised feature extraction of video, a widely used pre-trained Inception-v3 image recognition model is used. The Inception-v3 network provides video level and frame-level feature representations. Bagging Ensemble is used for randomly selecting data points from the training data, and to build the training model. A Fully Connected 3-Layer Neural Network is used as the base model for training. The Fully connected base classifier consists of ReLu function and Sigmoid function in the first and last layer respectively for the prediction score calculation. The confidence score from various models is integrated using an average operation called Bagging Ensemble algorithm [9].

The Fully connected 3- Layer Neural Network [9] is used for the calculation of the prediction score with an Adagrad optimizer [9]. The model will binary classify the video input into nominal or anomalous according to the probability score. A high probability score indicates the anomalous events and a low score indicates nominal events. The model also eliminates the necessity of segment videos for feature extraction. The videos are randomly selected, and each iteration is having a batch size of 60. The learning rate is kept as 0.001, and the hinge loss function is used in binary Support Vector Machine (SVM) classification. The dataset used for training consists of real world anomalous and nominal data. This model provides a good accuracy level for real time video anomaly detection and can be extended for fine-grained classification.

3. Conclusion

This paper studies some of the novel techniques used for automatic Anomaly Detection in a video surveillance system. In this paper, a comprehensive description of various Anomaly Detection Methods that exists in real video streaming is discussed. Anomaly detection in real video streaming needs a fast and novel model, which uses the previously learned knowledge and continual learned knowledge. For developing a fast and automatic model, novel techniques in Neural Networks is an elegant solution. Convolution Neural Networks [19] and Convolutional Long Short Term Memory (LSTM) [18] will provide motion and appearance feature identification effectively. The Convolutional LSTM can be used for sequence prediction problems with spatial inputs like images or video. Generative Adversarial Networks can generate internal scene representation from a given frame. Convolutional Autoencoder (CAE) [21] is another widely used neural network used in real-time video anomaly detection. The CAE apply the reconstruction error of each frame with an anomaly score.

By implementing a reliable model for automatic Anomaly Detection, the smart security system can work more efficiently and ensure the safety of people effectively. The applications of Neural Networks are developing rapidly and are yet to explore more in the area of computer vision and automatic anomaly detection in real-time videos.

4. References

[1] W Heng, T Jiang, and W Gao 2019 How to Assess the Quality of Compressed Surveillance Videos Using Face Recognition *IEEE Transactions on Circuits and Systems for Video Technology* **29** 2229-43

[2] Z Shao, J Cai, and Z Wang 2018 Smart Monitoring Cameras Driven Intelligent Processing to Big Surveillance Video Data *IEEE Transactions on Big Data* **4** 105-16

[3] R Nawaratne, T Bandaragoda, A Adikari, D Alahakoon, D De Silva, and X Yu 2017 Incremental knowledge acquisition and self-learning for autonomous video surveillance *IECON 2017 - 43rd Annual Conference of the IEEE Industrial Electronics Society* 4790-95

[4] L Bazzani, M Cristani, A Perina, and V Murino 2012 Multiple-shot person re-identification by chromatic and epitomic analyses *Pattern Recognit. Lett.,* **33** 898–903

[5] K Doshi and Y Yilmaz 2020 Continual Learning for Anomaly Detection in Surveillance Videos *IEEE/CVF Conference on Computer Vision and Pattern Recognition Workshops (CVPRW)* **10** 1025-34

[6] I R de Almeida and C R Jung 2013 Change Detection in Human Crowds Proceedings of the 2013 XXVI Conference on Graphics, Patterns and Images 63–69
[7] Cognitive Services—Intelligence Applications —2020 *Microsoft Azure [Online]* Available link https://azure.microsoft.com/enin/services/cognitive-services [last Accessed 14-Feb-2020]
[8] Vision API - Image Content Analysis 2020 *Google Cloud Platform [Online]* Available link https://cloud.google.com/vision/ [last Accessed 14- Feb-2020]
[9] Y Zahid, M A Tahir and M N Durrani 2020 Ensemble Learning Using Bagging And Inception-V3 For Anomaly Detection In Surveillance Videos *IEEE International Conference on Image Processing* 588-92
[10] Comparing image tagging services 2020 *Google Vision, Microsoft Cognitive Services, Amazon Rekognition, and Clarifai, Filestack Blog,* [Online] Available link https://blog.filestack.com/thoughts-andknowledge/comparing-google-vision-microsoft-cognitive-amazonrekognition-clarifai/ [Last Accessed 15-Feb-2020]
[11] D D Silva and D Alahakoon 2010 Incremental knowledge acquisition and self-learning from text *The 2010 International Joint Conference on Neural Networks* (IJCNN) 1–8
[12] Collective Activity Dataset 2020 [Online] Available link http://vhosts.eecs.umich.edu/vision/activity-dataset.html [Last Accessed: 15-Feb-2020]
[13] D Alahakoon, S K Halgamuge, and B Srinivasan 2000 Dynamic self organising maps with controlled growth for knowledge discovery *IEEE Trans. Neural Networks* **11** 601–14
[14] Jason Brownlee 2017 [Online] A Gentle Introduction to Transfer Learning for Deep Learning *Deep Learning for Computer Vision* Available link https://machinelearningmastery.com/transfer-learning-for-deep-learning/ [Last Accessed 29-Nov-2020]
[15] Vincenzo Lomonaco 2017 [Online] Why Continual Learning is the key towards Machine Intelligence *Continual AI* Available link https://medium.com/continual-ai/why-continuous-learning-is-the-key-towards-machine-intelligence-1851cb57c308 [Last Accessed 29-Nov-2020]
[16] Michael Nielsen 2015 Neural Networks and Deep Learning (Determination Press) Available link http://neuralnetworksanddeeplearning.com/faq.html [Last Accessed 30-Nov-2020]
[17] Redmon, Joseph and Farhadi, Ali 2018 YOLOv3: An Incremental Improvement *arXiv* Available link https://pjreddie.com/darknet/yolo/ [Last Accessed 30-Nov-2020]
[18] Jason Brownlee 2017 CNN Long Short-Term Memory Networks *Machine Learning Mastery - Long Short-Term Memory Networks* Available link https://machinelearningmastery.com/cnn-long-short-term-memory-networks/ [Last Accessed 29- Nov-2020]
[19] Brownlee 2017 How Do Convolutional Layers Work in Deep Learning Neural Jason Networks? *Machine Learning Mastery- Deep Learning for Computer Vision* Available link https://machinelearningmastery.com/convolutional-layers-for-deep-learning-neural-networks/ [Last Accessed 29- Nov-2020]
[20] Jason Brownlee 2019 A Gentle Introduction to Generative Adversarial Networks (GANs) *Machine Learning Mastery-Generative Adversarial Networks* Available link https://machinelearningmastery.com/what-are-generative-adversarial-networks-gans/ [Last Accessed 29- Nov-2020]
[21] Manasses Ribeiro, Andre Eugenio Lazzaretti and Heitor Silverio Lopes 2018 A study of deep convolutional auto-encoders for anomaly detection in videos *Pattern Recognition Letters* **105** 13-22

Analysis of distance measures in spatial trajectory data clustering

Sharmila S [1,2] and Sabarish B A [1,2]

[1]Department of Computer Science and Engineering, Amrita School of Engineering, Coimbatore
[2]Amrita Vishwa Vidyapeetham, India

E-mail: cb.en.p2cse19022@cb.students.amrita.edu (Sharmila S)

E-mail: ba_sabarish@cb.amrita.edu (Sabarish B A)

Abstract. Tremendous growth of Location-based technologies resulted in the generation of a huge volume of spatial data, which needs to be analyzed to get potentially important patterns. The spatial patterns extracted can be used to design a better infrastructure ensuring reliable service coverage. Trajectory data is one variant of spatial data that are generated by moving objects travelling across. It is represented as a sequence of spatial coordinates (latitude, longitude) of a location. Trajectory clustering tries to group similar spatial data points to extract the most common movement behaviors. Trajectory data poses major challenges including uncertainty, sampling rate, representation, relationships, spatial autocorrelation, serialization, redundancy, and triviality, which makes it hard to apply traditional clustering algorithms over trajectory data. In this paper, K-Means and DBSCAN (Density-based spatial clustering of applications with noise) clustering algorithms are analyzed using different similarity measures like Euclidean, Hausdorff and Haversine distances with the help of index measures say Adjusted Rand Index (ARI) and Fowlkes-Mallows scores (FMS). Experiment is carried out over two different trajectory datasets and it is proved that usage of Haversine distance for clustering is efficient than Euclidean and Hausdorff distances in terms of spatial trajectory data.

1. Introduction

Due to the rapid technological advancement in various fields using location-based information, data accumulated from multiple resources and services like military, banking, bio-informatics, transportation (e.g., taxi GPS data, logistics tracking), retail, health care units etc. The data collected in huge amount has to be processed to extract useful knowledge or behaviour patterns, which is used to mend the productivity with regard to cost and efficiency. Clustering groups objects together based on similarity to gain knowledge. [1-4]. The process of data mining and clustering of trajectory data is difficult when compared to the clustering of numerical or categorical data by traditional algorithms. Due to the factors like, relationships, spatial autocorrelation, serialization, redundancy and triviality, traditional algorithms fail to make optimum trajectory clusters [5]. In data collection process, data collected are usually of high-dimensions. Due to high dimensional feature space, data becomes sparse and clustering of data into different groups becomes harder to separate one cluster data from others. Because of high-dimensional data, search space and complexity of model increases, dimensional reduction techniques becomes as integral part in trajectory clustering [6-9].The following paper is organised with related works in Section 2, datasets description in Section 3, Analysis of different distance measures on two different trajectory datasets is mentioned with its results in Section 4, the paper is concluded and future scope is mentioned in Section 5 and References.

2. Related Work

As there exists a huge variety of data, the different types of clustering algorithms should be analysed properly before usage. Major groups into which multiple clustering algorithms are aggregated based on: Partition, Density, Hierarchical, Model and Grid [10]. Partition-based or hard clustering algorithm is K-Means, used in automatic text data categorization and external document plagiarism detection [11, 12]. Based on the type of data like Text, Multimedia, Stream, Uncertain, Time Series, different types of clustering algorithms should be analysed before building a model, otherwise, it will result in loss of efficiency [13]. Near Point with Index Ratio (NPIR) algorithm [14] deals with the searching of a nearest neighbour data point by not considering all the data points at once like K-Nearest neighbour (KNN) algorithm, instead by considering only few neighbour points based on election process using distance vector. DBSCAN clustering is dependent on the type of data. Multiple steps of pre-processing should be made before using DBSCAN on spatial-temporal data. To overcome these multiple pre-processing steps, ST-DBSCAN was proposed in [15]. Due to the reduced number of steps of processing, the efficiency of ST-DBSCAN is high when compared to DBSCAN. Data point may tend to be a member of different clusters based on the type of data. So, a comparative study between soft and hard clustering algorithms are done [16] and concluded that for a generic dataset like iris data [17], it is proved that K-NN outperforms Fuzzy C-Means (FCM) in-terms of efficiency as it incorporated huge fuzzy steps. This study proves that the selection of a clustering algorithm based on data type is trivial.

Choosing perfect Intra-cluster and Inter-Cluster similarity measure and Rand Index for analysing the quality of clusters is considerable. The analysis of multiple such similarity measures and the Rand Index are made with varying dimensions of data as inputs [18]. This study comes out with major findings like the Pearson correlation coefficient is not suitable for datasets which are low dimensional and centroid dependent, instead, it is suggested to be used on high-dimensional data which uses hierarchical approaches. Finally, the same study reveals an important finding of doing ANOVA (Analysis of Variance) test for proving the correctness of the similarity measure used in a model. The selection of similarity measure should also be compatible with data types. [19,20] used sematic similarity measures for documents to detect plagiarism. The common distance measures like Euclidean, Manhattan, Cosine distances demands the same length of trajectories. Since trajectory length varies, Haversian and Hausdorff distance measures are studied against trajectory data [21]. Hierarchical trajectory clustering algorithm is proposed in [22] uses Dynamic Time Warping (DTW) to identify the resemblance between trajectories. Multiple trajectory mining algorithms for future location prediction like HMM (Hidden Markov Model), String matching algorithms, Semantic-based prediction methods, Pattern matching algorithm-based prediction methods are analysed in [23]. The Index measures are used for cluster validation as it measures the agreement between two partitions. Adjusted rand index and Fowlkes-Mallows score are verified in [24,25] and concluded that the measures are suitable for measuring the quality of trajectory clustering.

The trajectory data is collected every single second and the amount of data and its dimensionality is huge. The general dimensionality reduction techniques like PCA, FA, LDA are not relevant in this case. Sampling of data should be done to reduce its size by retaining its feature and underlying information. Vertex sub-sampling algorithms which are used to select certain vertices in a line or trajectory are Douglas-Peucker, Visvalingam–Whyatt. Reumann–Witkam, Opheim simplification algorithms [26-29]. These algorithms aim to reduce a curve or line with huge data points to fewer data points without affecting its shape. These algorithms are used for data reduction in many data-type specific applications. In particular, the Douglas-Peucker algorithm is applied on ship trajectory data in [30] to reduce trajectory data and proposed a modified DBSCAN algorithm for marine traffic pattern recognition.

3. Datasets

TN291 and Microsoft T-Drive datasets are used in this analysis

3.1 TN291

It contains the Latitude and Longitude of a locality as data points. It is collected from 9 different districts of Tamil Nadu, India and contains about 2910 data points of 291 trajectories. It is already pre-processed using the Douglas-Peucker algorithm.

IOP Conf. Series: Materials Science and Engineering 1085 (2021) 012021 doi:10.1088/1757-899X/1085/1/012021

3.2 Microsoft T-Drive [31]

It is collected by taxis in Beijing, China for Research purposes. It contains the Latitude and Longitude of a locality as data points along with taxi id and time stamp. One-week trajectories of 10,357 taxis are collected. The data set contains about 15 million data points and it covers the distance of about 9 million kilometres.

4. Analysis of Distance Measures on Trajectory clustering

The popular clustering algorithms K-Means and DBSCAN will be applied on two trajectory datasets with Euclidean, Hausdorff and Haversine distance measures. The clustering validity is calculated with the Adjusted Rand Index (ARI) and Fowlkes-Mallows scores (FMS).

4.1 Data Reduction using Ramer-Douglas-Peucker (RDP) algorithm

TN291 is already pre-processed to have only 2910 data points. First 500 taxi data from Microsoft T-Drive trajectory dataset is chosen and RDP is applied with $\varepsilon = 0.0005$, such that 7,89,543 data points are reduced to 5,86,512 latitude longitude pairs. Figure 1 visualizes the spread of datapoints of both TN291 and T-Drive datasets.

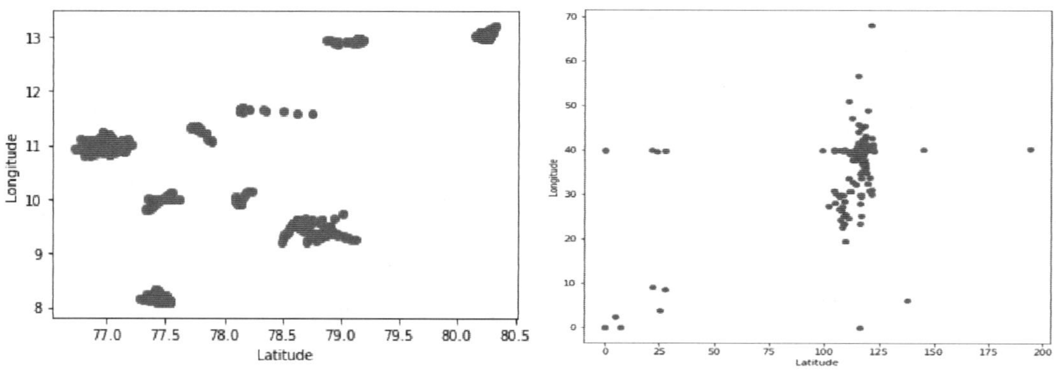

(a) (b)

Figure 1. Latitude and longitude pair of a) TN291 dataset and b) T-Drive dataset.

4.2 Analysis of hyper-parameter for clustering

K-Means clustering algorithm faces a problem of identifying number of clusters initially. Elbow method is used to define optimum K by calculating Within Cluster Sum of Squared Error (WSS) for different values of K and by choosing the K for which WSS value becomes the first start to diminish as in Figure 2.

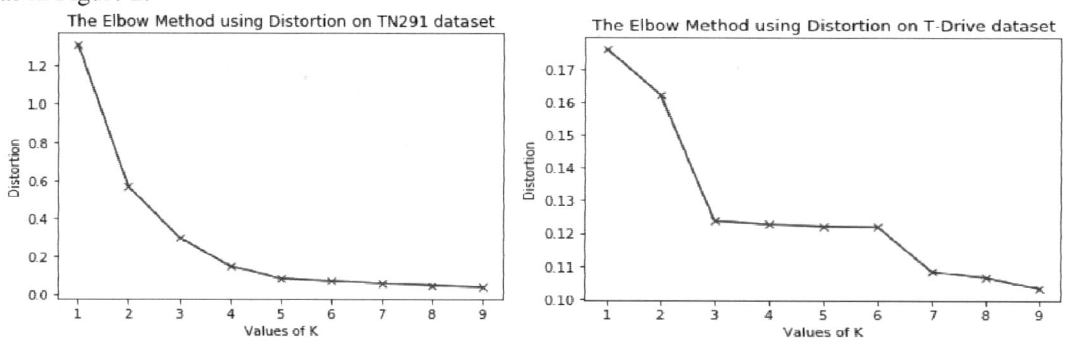

(a) (b)

Figure 2. Distortion is plotted against different K values for a) TN291 dataset and b) T-Drive dataset

It is clear from the above Fig 2(a)that, optimum number of centres (i.e) K value for TN291 dataset is either 3 or 4 and from Fig 2(b), optimum K value is either 2 or 3. In this case, further analysis is done with K=3 for both TN291 and T-Drive dataset.

K-means clustering algorithm

STEP 1: Initialization – K initial centroids are generated in random.
STEP 2: Assignment – Distance between data points and centroid will be calculated and K clusters are formed by associating each data points with the nearest centroid.
STEP 3: Update – The cluster centroid are updated with the mean of each cluster data points.
STEP 4: Steps 2 to 3 will be repeated until the new assignment is equal to the previous assignment.

DBSCAN clustering algorithm

STEP 1: Initially the algorithm starts with selection of data point (not visited)
STEP 2: Distance between points is calculated and the neighbourhood of a data point is extracted using epsilon (ε).
STEP 3: If there exist minPoints of neighbour points around a point, then clustering starts by marking that point as visited, else it will be marked as noise (temporarily, until it gets allotted to a cluster).
STEP 4: If a point is noted to be a part of a cluster, then its ε neighbourhood points is also a part of the cluster. Steps 2 to 3 will be repeated until all the points in a cluster is determined.
STEP 5: New unvisited data point will get retrieved and processed, resulting in the discovery of cluster point or noise.
STEP 6: Steps 2 to 5 will get repeated until all the data points are marked as visited.

K-Means and DBSCAN with different distance measures like Euclidean, Hausdorff and Haversine are implemented and analysed as follows.

4.3 Euclidean distance

Let D_i represents the spatial co-ordinates, where D_1 is Latitude and D_2 is the Longitude of a locality. Consider the randomly generated centroid as C_i where C_1 is centroid latitude and C_2 is centroid longitude. The general formula for finding Euclidean distance d(D,C) between latitude longitude pair and centroid is represented in Equation 1

$$d(D,C) = \sqrt{\sum_{i=1}^{2}(C_i - D_i)^2} \quad (1)$$

4.3.1 K- means clustering using Euclidean distance

STEP 1: Generate random centroids
STEP 2: Assignment and update stages are repeated until the cluster reassignment is not possible using the Euclidean distance measure. Figure 4 represents final clusters visually differentiated with different colours.

IOP Conf. Series: Materials Science and Engineering 1085 (2021) 012021 doi:10.1088/1757-899X/1085/1/012021

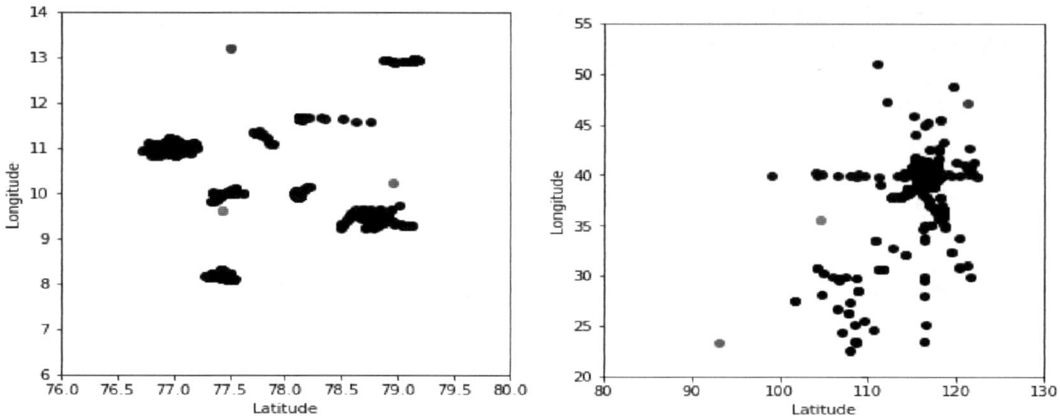

(a) (b)

Figure 3. represents the initial indication of randomly generated K Centroids in a) TN291, b) T-Drive

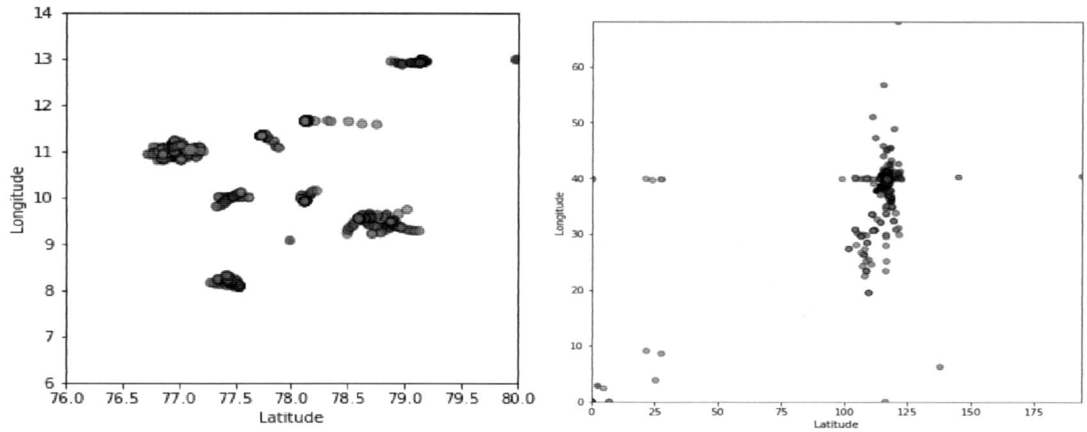

(a) (b)

Figure 4. Final clusters assignment a) TN291 dataset with 3 clusters and b) T-Drive dataset with 3 clusters

4.3.2 DBSCAN clustering using Euclidean distance

STEP 1: Optimal epsilon (ε) value can be found using K-Distant Graph. Figure 5 represents the K-Distant graph on TN291 dataset, the optimum Epsilon value where the maximum curvature resides is at 0.05. So, from this graph ε = 0.05 for TN291 dataset

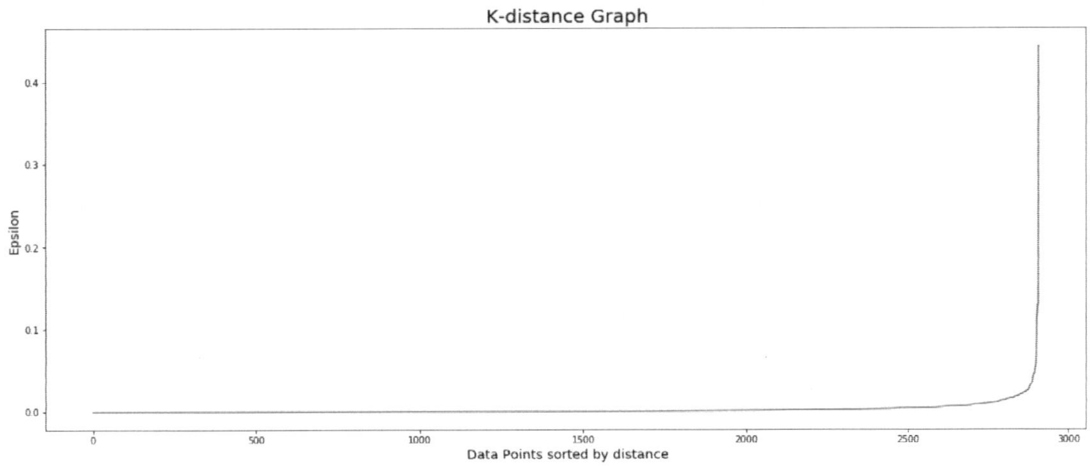

Figure 5. K-distant graph for finding optimum epsilon value

STEP 2: There exists no standard method to choose optimum minPoints. So, minPoints=40 is chosen using the trial and error method. Figure 6 represents final clustered data points using the DBSCAN algorithm with ε =0.05 and minPoints = 40.

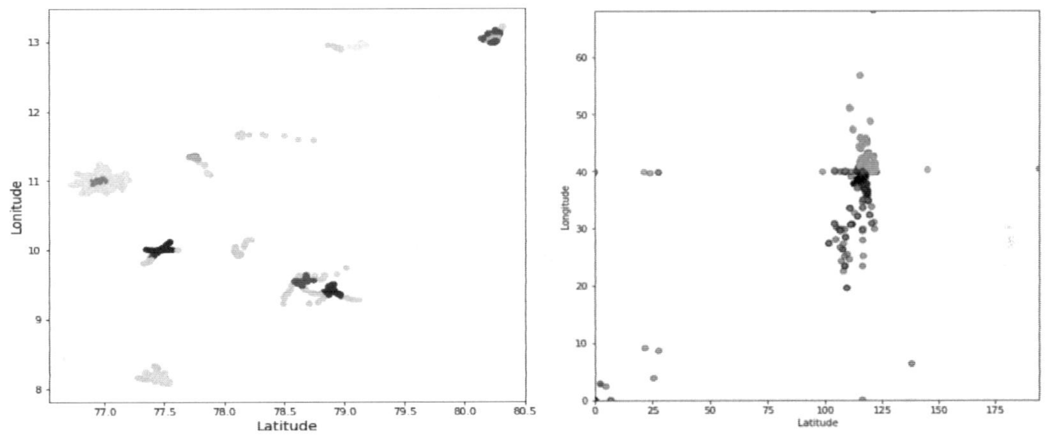

(a) (b)

Figure 6. Clustered using DBSCAN on a) TN291 dataset with 13 clusters and b) T-Drive dataset with 4 clusters

4.4 Hausdorff distance

The Hausdorff distance calculates dissimilarity amongst two different groups of data points. In other words, it returns the longest distance from one group of points closest to the next group. Let D_i represents the latitude, longitude pair of a locality. Consider centroid as C_i. The general formula to find Hausdorff distance H(D,C) between spatial co-ordinates and centroid is represented in Equation 2

$$H(D,C) = H(C,D) = \max\{\, h(D,C), h(C,D)\,\} \quad (2)$$

where $h(C,D) = \max c \in C\,\{\min d \in D\,\{\,d(c,d)\}\}$,
$h(D,C) = \max d \in D\,\{\min c \in C\,\{\,d(d,c)\}\}$ and
d(d,c) and d(c,d) is distance between d,c and c,d using any distance measure.

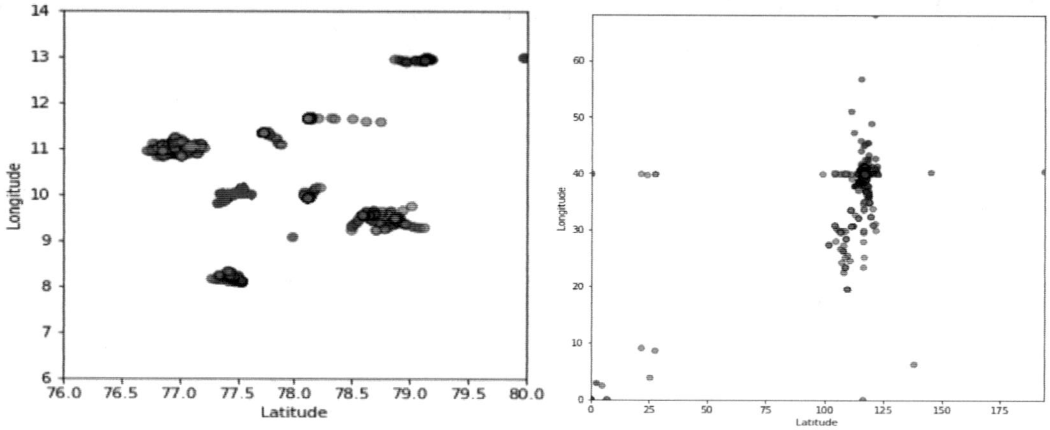

(a) (b)

Figure 7. represents Clustering using K-Means and Hausdorff distance measure on a) TN291 dataset with 3 clusters and b) T-Drive dataset with 3 clusters.

4.4.1 DBSCAN clustering using Hausdorff distance

The final clustered partition with DBSCAN and Hausdorff distance measure on both TN291 and T-Drive datasets is mentioned in Figure 8.

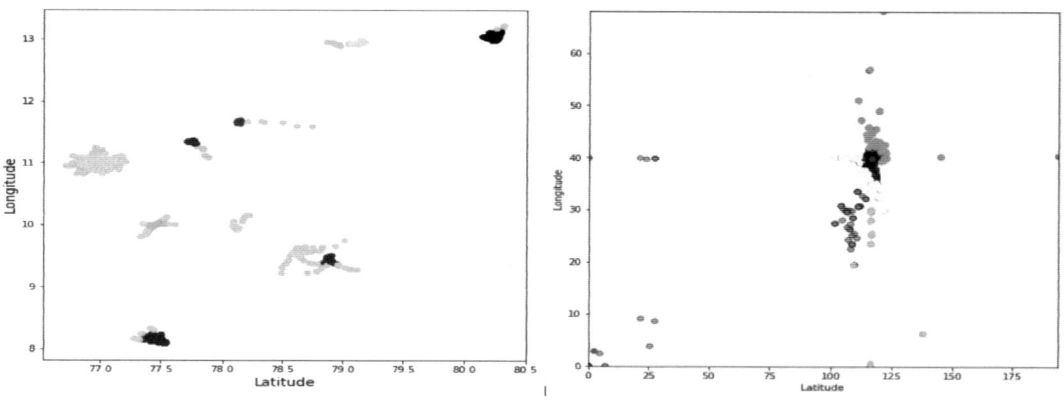

(a) (b)

Figure 8. represents Clustering using DBSCAN and Hausdorff distance measure on a) TN291 dataset with 11 clusters and b) T-Drive dataset with 6 clusters.

4.5 Haversine distance

The haversine distance is accurate in terms of measuring distance in spherical object say the Earth. The equation to calculate distance d(D,C) between the data point and a centroid is mentioned in Equation 3

$$d(D,C) = 2\,\gamma\ \sin^{-1}\sqrt{sin^2\left(\frac{D_1-C_1}{2}\right) + \cos(D_1)\cos(C_1)sin^2\left(\frac{D_2-C_2}{2}\right)} \quad (3)$$

where D_1 and D_2 are data point, C_1 and C_2 are centroid. γ is the approximate radius of Earth, $\gamma = 6371$ km.

4.5.1 K-MEANS clustering using Haversine distance

The final clustered partition on both TN291 and T-Drive datasets with K-Means and Haversine distance measure is mentioned in Figure 9

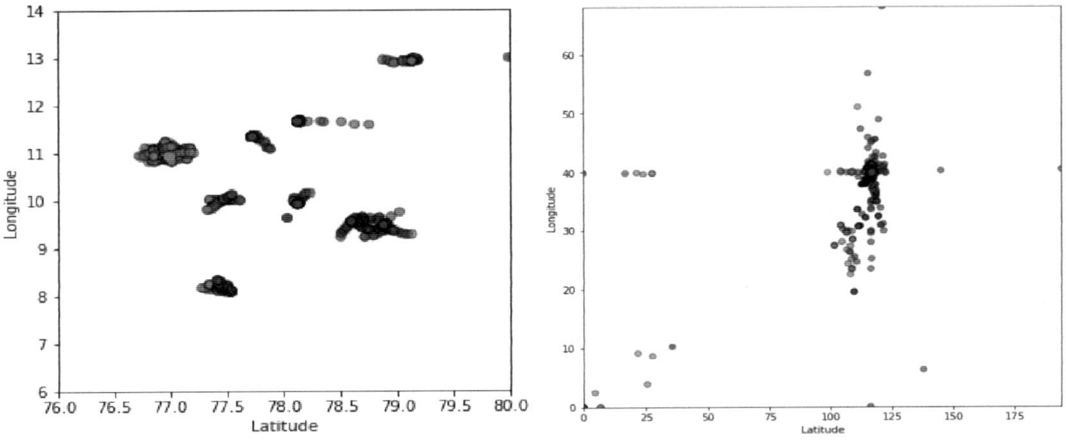

(a) (b)

Figure 9. represents Clustering using K-Means and Haversine distance measure on a) TN291 dataset with 3 clusters and b) T-Drive dataset with 3 clusters.

4.5.2 DBSCAN clustering using Haversine distance

The final clustered partition with DBSCAN and Haversine distance measure on both TN291 and T-Drive datasets with is mentioned in Figure 10

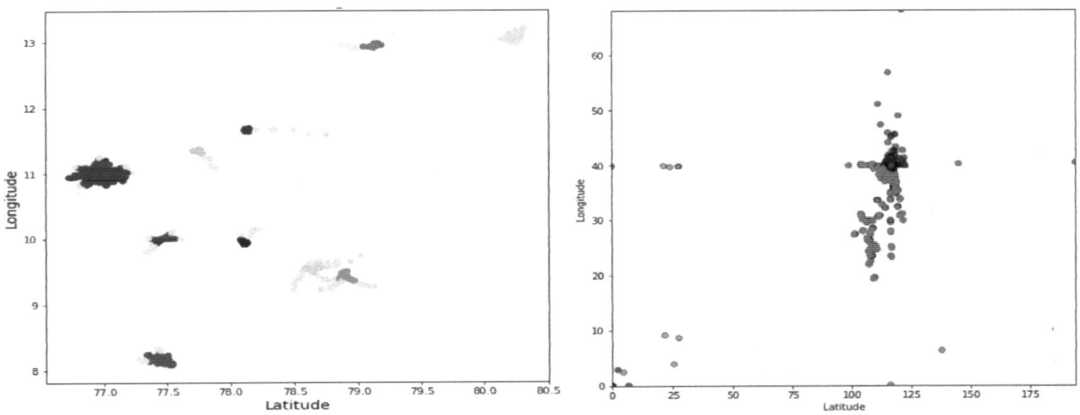

(a) (b)

Figure 10. represents Clustering using DBSCAN and Haversine distance measure on a) TN291 dataset with 11 clusters and b) T-Drive dataset with 3 clusters.

4.6 Performance Analysis and Findings

Adjusted Rand Index (ARI) and Fowlkes-Mallows scores (FMS) are the popular cluster validity measurement standard as both measures agreement and similarity between two clusters. These measures are used for validation with reference to [18,19], as these two index measures are suitable for trajectory data. Table 1 statistically represents both ARI and FMS values for the applied K-Means and DBSCAN algorithms with different distance measures on two different datasets.

Table 1. ARI and FMS values for K-Means and DBSCAN algorithms with Euclidean, Hausdorff and Haversine distances on TN291 and T-Drive datasets

Algorithms	Dataset	**Euclidean**		**Hausdorff**		**Haversine**	
		ARI	FMS	ARI	FMS	ARI	FMS
K-Means	**TN291**	0.55	0.610	0.50227	0.61042	0.67782	0.68313
	T-Drive	0.545112	0.57443	0.45019	0.42143	0.72389	0.72626
DBSCAN	**TN291**	0.66553	0.27095	0.62294	0.32520	0.74662	0.78872
	T-Drive	0.52419	0.43951	0.67752	0.69067	0.82511	0.79801

From Table 1, the distance measure which gives high values for both ARI and FMS can be concluded as the optimal distance measure for Trajectory data. It can be seen that the Haversine distance reported having comparatively high ARI and FMS values that other distance measures. That is, ARI = 0.67782 and FMS 0.68313 for K-Means algorithm on TN291 dataset, ARI = 0.72389 and FMS = 0.72626 for K-Means algorithm on T-Drive dataset, ARI = 0.74662 and FMS = 0.78872 for DBSCAN on TN291 and ARI =0.82511 and FMS = 0.79801 for DBSCAN on T-Drive dataset.

5. Conclusion and Future Work

The different distance measures like Euclidean, Hausdorff and Haversine distances are applied and validated with K-Means and DBSCAN algorithms on both TN291 and Microsoft T-Drive datasets. Thus, the Haversine distance measure outperforms both Euclidean and Hausdorff distances and it ensures to have better validity and accuracy results with respect to moving objects Trajectory data. The Haversine distance resulted to be optimal because it is specifically for measuring distance on spherical objects say Earth.

Further, it is noted that when K-Means and DBSCAN algorithms are used for clustering Trajectory data, both ARI and FMS values are at most 82.0 which is lesser than even 90. This can be improved with the concept of fuzzy clustering as there exist chances for one data point to belong to multiple clusters in trajectory data and with a better way to predict the optimum number of clusters before building the model. Because, it is obviously seen from TN291 dataset that it contains 9 clusters visibly, but the popular elbow method reports a lesser number of clusters than it actually contains, which may have resulted in lesser ARI and FMS scores. The future work is to propose a fuzzy concept for clustering trajectory data with an improved way to foresee the optimum number of clusters beforehand.

References

[1] Gupta M K and Chandra P *International journal of Information Technology* **12** 1243-1257.
[2] Benabdellah A C, Benghabrit A, Bouhaddou I 2019 *Procedia Computer Science* **148** 291-302.
[3] Rokach L 2009 *A survey of Clustering Algorithms, Data Mining and Knowledge Discovery Handbook* (Boston, MA:Springer)
[4] Rui Xu and Wunsch D 2005 *IEEE Transactions on Neural Networks* **16** 645-678
[5] Yuan G, Sun P and Zhao J 2017 *Artificial Intelligence Review* **47** 123–144
[6] Dou W, Dai G, Xu C and Zhang Z 2010 *Machine Learning and Knowledge Discovery in Databases, ECML PKDD 2010* **6321**
[7] Sivasankar E, Selvi C and Mala C 2017 *Advances in Intelligent Systems and Computing* **556**
[8] Kavilkrue M and Boonma P 2015 *Advances in Intelligent Systems and Computing* **361**
[9] Assent I 2012 Clustering high dimensional data URL 10.1002/widm.1062; https://doi.org/10.1002/widm.1062
[10] Mehta V, Bawa S and Singh J 2020 *Artificial Intelligence Review* **53** 5995-6023
[11] Vani K and Gupta D 2014 Using K-means cluster based techniques in external plagiarism detection *International Conference on Contemporary Computing and Informatics (IC3I)* pp 1268-1273.
[12] Unnikrishnan S, Sreelakshmi S and Deepa G 2016 *International Journal of Control Theory and Applications* **9** 7619-7626

[13] Oyelade J 2019 *International Conference on Computational Science and Its Applications (ICCSA)* pp 71-81
[14] Qaddoura R, Faris H and Aljarah I 2020 *International Journal of Machine Learning & Cybernetics* **11** 675–714 URL 10.1007/s13042-019-01027-z;https://dx.doi.org/10.1007/s13042-019-01027-z
[15] Birant D and Kut A 2007 *Data & Knowledge Engineering* **60** 208-221 URL 10.1016/j.datak.2006.01.013;https://dx.doi.org/10.1016/j.datak.2006.01.013
[16] Bora D J, Gupta D and Kumar A 2014 *International Journal of Computer Trends and Technology* **10** 108-113
[17] Fisher R URL https://archive.ics.uci.edu/ml/datasets/iris
[18] Shirkhorshidi A S, Aghabozorgi S and Wah T Y 2015 *PLOS ONE* **10** e0144059 URL 10.1371/journal.pone.0144059; https://dx.doi.org/10.1371/journal.pone.0144059
[19] Nedungadi P, Raj H 2014 *Smart Innovation, Systems and Technologies* **1**
[20] Gupta D, Vani K and Singh C K 2014 Using Natural Language Processing techniques and fuzzy-semantic similarity for automatic external plagiarism detection *International Conference on Advances in Computing, Communications and Informatics (ICACCI)* pp 2694-2699
[21] Wai K P and New N *2017* Measuring the distance of moving objects from big trajectory data *IEEE/ACIS 16th International Conference on Computer and Information Science (ICIS)* pp 137-142.
[22] Sabarish B A, Karthi R and Gireesh K T 2018 *Lecture Notes in Computational Vision and Biomechanics* **28** 215-226
[23] Sabarish B A, Karthi R and Gireesh K T 2014 A Survey of Location Prediction Using Trajectory Mining, Artificial Intelligence and Evolutionary Algorithms in Engineering *Systems Proceedings of ICAEES 2014* vol 1 (Springer) pp 119–127
[24] Lin Y, Huang H, Simon G E and Liu S 2016 *Mathematical Biosciences* **282** 191-203 URL 10.1016/j.mbs.2016.10.008;https://dx.doi.org/10.1016/j.mbs.2016.10.008
[25] Yu Q, Luo Y, Chen C and Chen S 2019 *Applied Intelligence* **49** 2315–2338 URL 10.1007/s10489-018-1385-x;https://dx.doi.org/10.1007/s10489-018-1385-x
[26] Saalfeld A 1999 Topologically Consistent Line Simplification with the Douglas-Peucker Algorithm *Cartography and Geographic Information Science* **26** 7-18 URL 10.1559/152304099782424901;https://dx.doi.org/10.1559/152304099782424901
[27] Visvalingam M and Whyatt J D 1993 Line generalisation by repeated elimination of points *The Cartographic Journal* **30** 46-51 URL 10.1179/caj.1993.30.1.46:https://dx.doi.org/10.1179/caj.1993.30.1.46
[28] Reumann K and Witkam A P Optimizing curve segmentation in computer graphics
[29] Opheim H Smoothing a digitized curve by data reduction methods
[30] Zhao L and Shi G 2019 A trajectory clustering method based on Douglas-Peucker compression and density for marine traffic pattern recognition URL 10.1016/j.oceaneng.2018.12.019:https://dx.doi.org/10.1016/j.oceaneng.2018.12.019
[31] Zheng Y 2011 URL https://www.microsoft.com/en-us/research/publication/t-drive-trajectory-data-sample/

Intelligent Handwritten Character Recognition For Malayalam Scripts Using Deep Learning Approach

Bineesh Jose[1], **K. P. Pushpalatha**[2]
[1] Research Scholar, School of Computer Sciences, M. G. University, Kottayam, Kerala, India bineeshjose@gmail.com
[2] Associate Professor & HoD, School of Computer Sciences, M. G. University, Kottayam, Kerala, India pushpalathakp@mgu.ac.in

Abstract. Machine Learning, especially Deep Learning has been incorporated into Pattern Recognition and Image Processing for the Handwritten Character Recognition (HCR) research which is always a hot area of research. Foreign scripts are enriched with Handwritten Character Recognition (HCR) studies. But very less research can be seen for HCR in Indian scenario, especially in Malayalam script. This paper provides an overview of different feature extraction and classification techniques used for Handwritten Character Recognition of Malayalam Scripts using Deep Learning Approaches. The accuracy measures of classification of the different algorithms have been given so that the researchers can select a particular algorithm with high performance.

1. General Character Recognition System:- An Architecture View

The recognition of characters contains preprocessing, segmentation, feature extraction, and classification as shown in Fig.1.

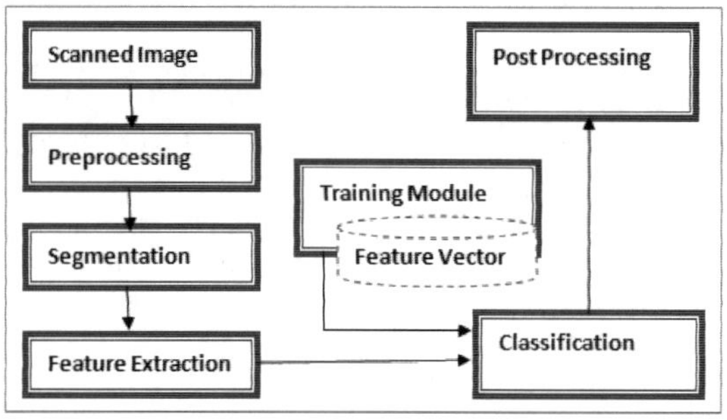

Figure 1. Architecture of a character recognition system

1.1. Preprocessing

Preprocessing contains the following sequence of operations[11]

- Noise Removal

- Thresholding
- Skeletonization

1.2. Segmentation
It is having a major role in handwriting character recognition problem which requires line, character and word segmentation. White space and pitch, projection analysis and connected component labeling are the certain techiniques used for character segmentaion[11].

1.3. Feature Extraction
Feature Extraction is one of the important step towards handwriting character recognition. Salient Features are extracted from segmented images and it is captured in feature vectors. Lot of feature extration techiniques are available for character recognition system[11].

1.4. Classification
This is the final step towards the handwriting character recognition. Wide varity of classifiers are available in pattern recognition and image processing area. Any of the classification model can be used to map features to corresponding character. Supervised and Unsupervised methods are used here. Slicing of single data set into training set and test set is an important concept in Classification [6].

2. Malayalam Handwritten Character Recognition:- A Literature Review
Malayalam is one of the popular languages of South India which is mainly used by Kerala and Union territories of Lakshadweep and Mahe as official language. But very less research can be seen for HCR in Indian scenario, especially in Malayalam script. It is very difficult to have a complete handwritten text recognition system in Malayalam because of its huge character set.

2.1. HCR using Stacked Long Short Term Memory (LSTM)
2.1.1. Proposed work Jino, et al. in [1] introduced a stacked LSTM model for isolated Malayalam handwritten character recognition. 18000 samples of ninety Malayalam characters are included in the dataset. LSTM model consists of two layers and prediction is done by a final output layer. More than 90% of accuracy was achieved. Improvement can be done on dataset by adding more samples to the dataset. In this model a version of RNN which is capable of handling machine learning application called Long Short Term Memory (LSTM) was used.

LSTM Cell Architecture LSTM Cell contains three gates namely input gate, forget gate and output gate [1] as in Fig.2 which is able to solve the vanishing gradient problem of RNN.

Stacked LSTM Architecture The architecture Contains 3 layers, namely LSTM Layer1, Layer2 and a third Output Layer for prediction. Softmax function is used as activation function at Output prediction Layer.

2.1.2. Result Experiment results obtained by using Stacked LSTM architecture was really good. Better results can be achieved by using more samples per class. By fine tuning the architecture parameters, this approach can lead to state of the art result. Word recognition also can be implemented by using this model.

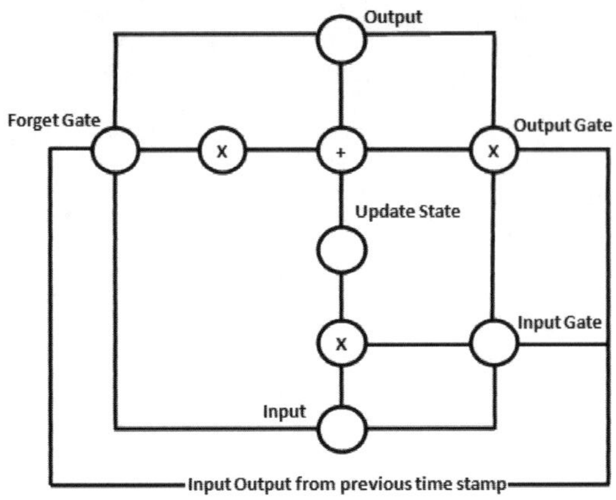

Figure 2. Block Diagram of LSTM Cell

2.2. Online Handwritten Bangla Character Recognition using CNN

2.2.1. Proposed Work Shibaprasad in [2], proposed a online handwritten isolated Bangla characters recognition system using Convolutional Neural Network (CNN) architecture. The author was tried to analyse CNN architecture with various parameters like kernel variations, pooling strategies and activation functions. He has used a dataset with 10000-character samples. The total dataset was sliced into test set and train set with a percentage of 30 and 70 respectively. An accuracy of 99.4% has been achieved on test dataset using this technique. As shown in Fig.3, the proposed system consists of 2 convolutional layers, 2 pooling layers and one fully connected classification layer.

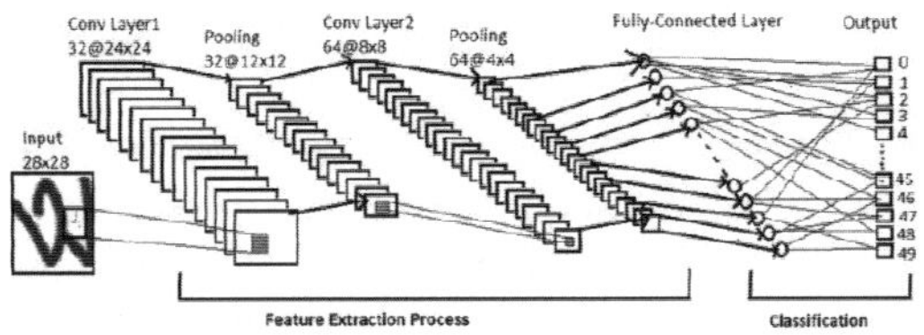

Figure 3. Architecture of CNN used in the present work

2.2.2. Result In this paper, they have analyzed different parameters like kernel size, pooling scheme, and activation function for online handwritten Bangla character recognition

2.3. HCR using wavelet energy and extreme learning machine

Binu, et.al. in [3] employed wavelet energy feature (WEF) for feature extraction and extreme learning machine (ELM) for classification of handwritten Malayalam characters. The proposed model makes use of a new parameter called wavelet energy (WE). Characters can be represented in terms of WE distribution which returns WEF. Character images can be recognised by using WE of different decomposition levels as shown in Fig.4. These features will act as input to classifiers of handwritten character recognizer.

Feature extraction using wavelets Different parameters such as pixels, shape data, or mathematical properties act as features to the recognizer. In this paper WT act as a feature extractor. Pattern recognition and image processing uses wavelet transformations for the past two decades [3].

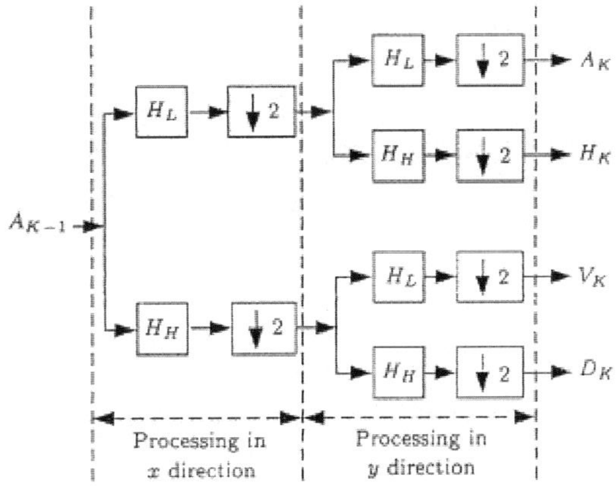

Figure 4. 2D discrete wavelet transform

Extreme learning machine Feed-forward neural networks is consuming long time for learning which is the major disadvantage of such methods in their applications during the last 20 years. Iterative changing of parameters will lead to this issue while training such models [3].

2.3.1. Result High learning speed and low training error are the major features of ELM with different activation functions. Wavelet energy based feature vectors are used for pattern classifications. Multiresolution analysis can be done by using WEF which is defined in terms of wavelet. More experiments can be conducted based on different features with good classifiers for better result on Malayalam character recognition.

2.4. HCR using Gradient Based Features and Run Length Count(GBF-RLC)

2.4.1. Proposed Work Raju, et.al. in [4] proposed a model with gradient-based features and run length count (GBF-RLC) as feature extractor for handwritten character recognition system. Simplified Quadratic Classifier (SQDF) and Multi-Layer Perceptron (MLP) were used as classifiers here.

Feature extraction The feature extraction methods are applied to each character image in the database. Character codes can be obtained from the binarized and thinned images. Binarized, resized (72x72) and thinned images are used for centriod calculation. Different steps in the feature extraction is given in Fig. 5.

Feature extraction Algorithm It is used for extracting suitable features by applying each image in the database.

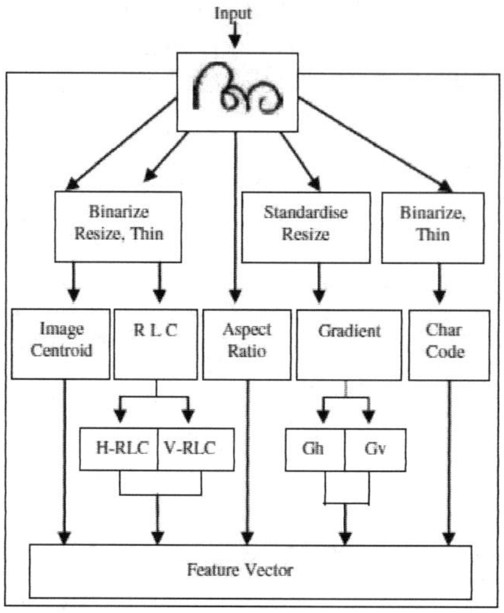

Figure 5. Feature extraction in GBF-RLC

2.4.2. Result The proposed model use a combination of gradient based features as well as run length count (GBF-RLC) for hand written character recognition system. For classification, SQDF and MLP are used. In order to achieve higher recognition rate, we can use combination of fearutes as input. The experimental result showed by SQDF and MLP classifiers are 99.66% and 99.78% respectively. Performance of the different methods are shown in Fig. 6.

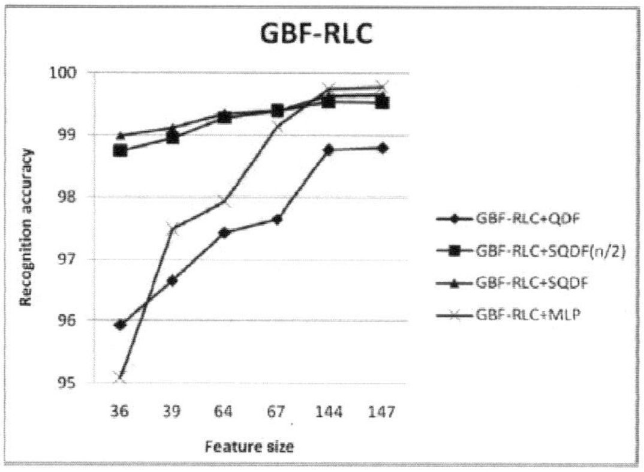

Figure 6. Performance of Different Methods

2.5. HCR Based on Chain Code Histogram
2.5.1. Proposed Work Jomy, et.al. in [5] proposed a model for handwritten character recognition with CCH as feature extractor and Neural Network as classifier. The biggest challenge in the Malayalam Character recognition system is the largest number of characters.

The problem is more complex when we are considering different writing patterns of different people.

Chain code calculation of Handwritten characters Significant features are extracted from boundary points of binary segmented character image. Fig. 7 Shows the four directional and eight directional coding of segmented image. Chain code representation of the contour can be used for CCH calculation. Reconstruction of Character image using boundary points is shown in Fig. 8.

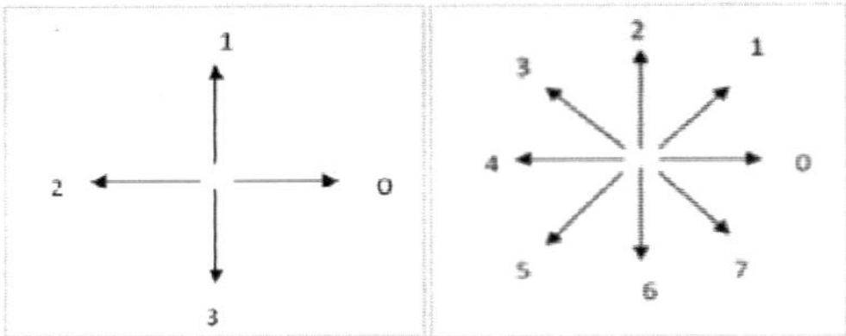

Figure 7. Directions of four connected and eight connected chain code

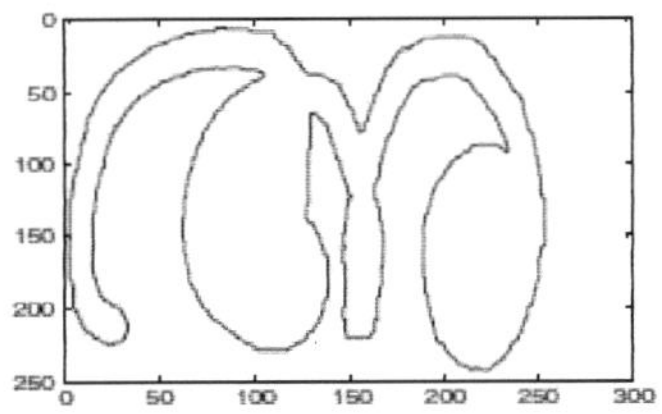

Figure 8. Reconstruction of Character image using CCH

Classification For classification they used a Neural Network Model which is having one input layer, one output layer and many hidden layers.

2.5.2. Result The dataset is sliced into three sets as train, test and validation with percentage 70, 15 and 15 respectively. Feature set IV with 8dir CCH, 8dir NCCH and Centroid got highest average accuracy.

2.6. HWR Using a Deep Architecture

2.6.1. Proposed Approach Jino, et.al. in[7] came up with a offline model for Malayalam handwritten word recognition. He has developed a bench mark database for offline handwritten word samples. A deep convolutional neural network (CNN) architecture as in Fig. 9 was introduced for feature extraction. As usual Support vector machine (SVM) was used for classification in the above work.

The following sequence of operations are essential for the proposed word recognition model [7].

- Preprocessing

- Convolutional Neural Network
- Training of the CNN
- Support Vector Machine

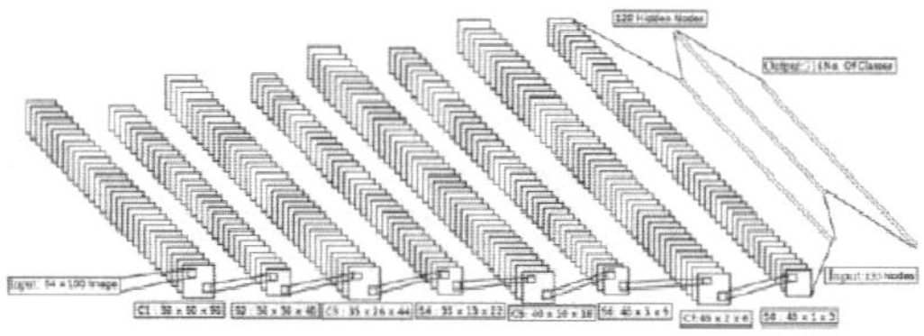

Figure 9. Proposed Architecture of Convolution Neural Network(CNN)

2.6.2. Result In this work the author is mainly concentrating on offline handwritten Malayalam word recognition. He is also trying to extent the same on some existing databases and got good result on this databases.

2.7. CNN based common approach to HCR

2.7.1. Proposed Work Durjoy, et.al. in [8] proposed a 5-layer model which is similar to LeNet-5 with two convolution layer, two subsampling layer and one fully connected layer as feature extractor. Support Vector Machine (SVM) was used as the corresponding classifier in this paper.

Accuracy can be incresed by applying some preprocessing methods on the input images for handwriting character recognition. The model do not require any preprocessing to remove noises present in the input samples. The CNN architecture itself will take care of several sources of variation among the data set.

5 Layer Convolutional Neural Network The architecture proposed here consists of two convolutional layer, two subsampling layer and one fully connected layer [8] as shown in Fig.10.

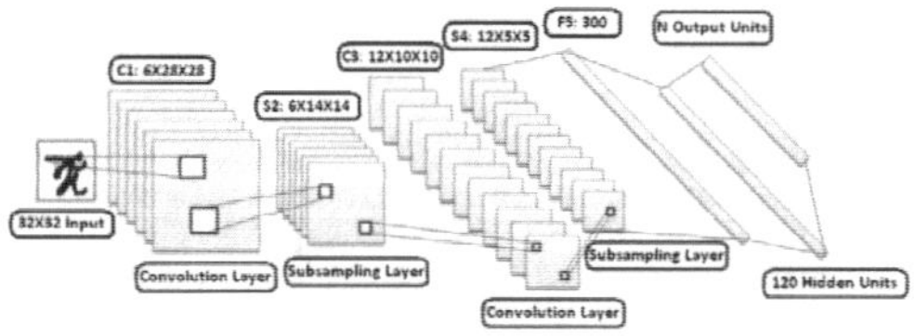

Figure 10. 5 Layer Convolutional Neural Network Architecture

2.7.2. Result Six different character databases are used to test handwritten character recognition system in this proposal. The Proposal has shown a good recognition accuracy on train and test dataset. Any character recognition problem can reuse the feature extraction strategy that they used here. In the given proposal, they used the popular SVM classifier. By the aid of this multilayer perceptron, a moderate recognition rate is attained.

2.8. HDR Using an Improved Deep CNN Architecture

2.8.1. Proposed Work Chandrika, et.al. in [9] proposed a model for Bangla handwritten digits recognition by the aid of Deep Convolutional Neural Network (D-CNN). The D-CNN model consists of total seven layeres in which three are convolution layers, another three are average pool layers and one fully connected layer for classification of Bangla handwritten digits.

2.8.2. Result The model was given 96.80% accuracy on a LeNet-5 like five-layered architecture. The total dataset for taining was divided into 80 batches of 50 images each.

2.9. HCR using Neural Networks: Feature based Classification

2.9.1. Proposed Work Shreya, et.al. in [10] has done a detailed review of Handwritten Character Recognition using Neural Network by back propagation algorithm. Character recognition techniques recognize the characters written on a paper documents and convert it in digital form.

Neural Network Classification and Recognition tasks in this model are carred out by an Artificial Neural Network. In a feed forward neural network, nodes are organized in layers; each over other in a stack structure. The neural network consists of one input layer, one or more hidden layers and an output layer. Here Back propagation algorithm is used for training of multi-layer feed-forward neural network. Improved Deep CNN Architecture is shown in Fig. 11.

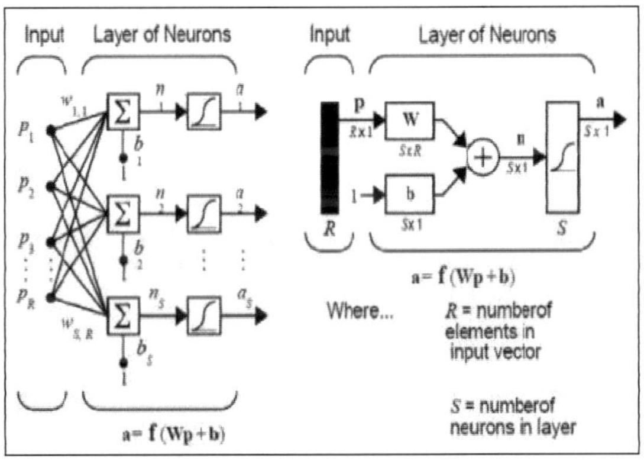

Figure 11. Improved Deep CNN Architecture

2.9.2. Result In this paper, the author was trying to analyze various feature extraction and classification techniques used for offline handwritten character recognition. Due to the high noise tolerance capability of ANN(Artificial Neural Network), it is used here to perform handwritten character recognition. They got excellent recognition results by using the above-proposed model.

3. Technical Comparison

Here we have given a technical comparison of different methodologies used for feature extraction and classification in TABLE I. The accuracy measures of classification of the above algorithms have been also given so that the researchers can select a particular algorithm with high performance. Among the different proposals that we have discussed in this paper, GBF-RLC with SQDF or MLP has given the highest performance.

Table 1. Technical Comparison

Author	Script	Feature Extractor	Classifier	Accuracy
Raju et al.[2014]	Malayalam	GBF&RLC	SQDF Or MLP	99.66
Shibaprasad et al.[2018]	Bangla	CNN	CNN	99.40
Jino et al. [2017]	Malayalam	Stacked LSTM	LSTM	97.00
Jino et al.[2019]	Malayalam	10 layer CNN	SVM	96.90
Chandrika et al.[2019]	Bangla	7 layer CNN	CNN	96.80
Durjoy et al.[2015]	Bangla	5 layer CNN	SVM	95.60
Binu et al.[2012]	Malayalam	WE	ELM	95.59
Jomy et al.[2011]	Malayalam	CCH	NN	72.10

4. Conclusion

In this paper, handwritten recognition systems for characters and words from different languages are discussed. Various feature extraction and classification techniques are discussed here with deep learning platform. Big character set, similarity among characters and complex structers of compound characters are the major challenges that the researchers are facing in this area. We believe that our survey will be helpful for researchers in this field.

References

[1] Jino P J,Jomy John,Kannan Balakrishnan 2017 Offline Handwritten Malayalam character Recognition using stacked LSTM *ICICICT-IEEE* **17** 1587

[2] Shibaprasad Sen 2018 Online Handwritten Bangla Character Recognition Using CNN: A Deep Learning Approach *Springer Nature* **695** 413

[3] Binu P Chacko, VR Vimal Krishnan, G Raju, and P Babu Anto 2012 Handwritten character recognition using wavelet energy and extreme learning machine *Springer* **3(2)** 149

[4] G Raju, Bindu S Moni, and Madhu S Nair 2014 A novel handwritten character recognition system using gradient based features and run length count *Sadhana-Springer* **39(6)** 1333

[5] Jomy John, Pramod K. V, Kannan Balakrishnan 2011 Offline Handwritten Malayalam Character Recognition Based on Chain Code Histogram *ICETECT-IEEE* **11** 736

[6] Abhishek Hazra, Prakash Choudhary, Sanasam Inunganbi and Mainak Adhikari 2020 Bangla-Meitei Mayek scripts handwritten character recognition using Convolutional Neural Network *Applied Intelligence-Springer Nature* **20** 1901

[7] PJ Jino, K Balakrishnan, U Bhattacharya 2019 Offline Handwritten Malayalam Word Recognition Using a Deep Architecture *Soft Computing for Problem Solving-Springer Nature* **816** 913

[8] Durjoy Sen Maitra, Ujjwal Bhattacharya and Swapan K. Parui 2015 CNN Based Common Approach to Handwritten Character Recognition of Multiple Scripts *ICDAR-IEEE* **15** 1021

[9] Chandrika Saha, Rahat Hossain Faisal and Md. Mostafijur Rahman 2019 Bangla Handwritten Digit Recognition Using an Improved Deep Convolutional Neural Network Architecture *ECCE-IEEE* **19**

[10] Shreya Girdhar, Sachin Gupta, Bhaskar Kapoor 2018 Handwritten Character Recognition using Neural Networks: A Study of Various Feature based Classification Techniques *IJSRD* **6** 2163

[11] Jomy John, K.V. Pramod, and B. Kannan 2011 Handwritten Character Recognition of South Indian Scripts: A Review *National Conference on Indian Language Computing* **1** 11

An Early Disease Prediction and Risk Analysis of Diabetic Mellitus using Electronic Medical Records

Rutuja A Gulhane[1] **and Sunil R Gupta**[2]

[1]Research Scholar, PRMIT&R, Badnera, Maharashtra, India
[2]Assistant Professor, PRMIT&R, Badnera, Maharashtra India

E-mail: gulhanerutuja@gmail.com, sunilguptacse@gmail.com

Abstract. In the world today, the fourth leading disease is Diabetes that could lead to other serious complicating diseases. Diabetes is one of the most common chronic disease which can also be the cause of death in many cases. An efficient system for early disease prediction and risk analysis of diabetic mellitus is very much needed as it has the major adverse effects. The large amount of medical data is collected by healthcare industry in the form of Electronic Medical Records. The Electronic Medical Records is communal database for clinical disease and risk prediction that are useful in accurately predicting multiple medical events using machine learning approach. Therefore, this research presents an efficient technique for early prediction and risk analysis of diabetic mellitus disease to improve accuracy and precision using Electronic Medical Records.

1. Introduction

The changing habits of individuals has various adverse effects on human health. Diabetic Mellitus (DM) has become an essential global health concern that has no models for its appearance and at any age of life it can impact humans [1]. According to WHO Diabetes was seventh leading cause of death in 2016, and was estimated to have caused four million deaths globally in 2017 [2]. According to International Diabetes Federation Diabetes Atlas, by 2019, an estimated 463 million people were living with diabetes; and by 2030, this figure is estimated to rise to 578 million and by 2045, to around 700 million [3]. DM is a metabolic disease characterized by high blood glucose (BG) and causes abnormal BG regulation that may lead to death if not properly managed. Diabetes is a kind of chronic disease and often accompanies with other disease that might result in short and long-term health complications [4, 5].

DM is generally divided into two types of diabetes: type-1 (T1DM) & type-2 (T2DM) diabetes. In type 1, DM occurs when insulin can no longer be created in the pancreas. Kids, teens and adults will be diagnosed with type-1 diabetes at a younger age. Type 2 diabetes happens when body does not efficiently use pancreas to prevent production of glucose and promote the use of glucose. Type-2 commonly happens in older adults & affects more people who are overweight [6, 7]. Type 1 diabetes does not have complications at the start, as the pancreas remains moderately useful. The disease is perceived only when 80-90% of pancreatic insulin-producing cells are dead. The uncertainty involved in type 2 diabetes brings lifestyle interruption, psychosocial transition, cost for medical treatment. This causes Hypoglycemia & many health problems [8, 9]. About 10% of people are having type-1 & about 90% of people are having type-2 diabetes [10]. DM can affect blood vessels, which increases the risk that the heart, eyes, kidneys, and nerves will be harmed by serious health issues. Diabetic retinopathy, diabetic neuropathy, diabetic nephropathy, and strokes, etc. are the most prevalent diabetes complications [11]. The effects of diabetes can be suppressed when it is identified in the earlier stages because it cannot be cured, but it can be prevented, detected, and managed [12, 13]. Therefore, early detection of diabetes is the only solution for prevention of the diabetes mellitus disease [14].

Many types of medical information sources are present, such as, hospital information systems (HIS), electronic health records (EHR), or electronic medical records (EMR). EMR is a collection of medical records of individual patients or many people. EMR contains many kinds of healthcare data such as structured, unstructured and semi-structured data. Prediction of disease becomes quick and easier if data is precise, consistent and free from noise. Chronic diseases can be detected and prevented early with current medical techniques. For early diagnosis, patient's data consisting of various features and diagnostics related to disease should be entered with utmost care to provide quality services [15, 16]. The data mining methods, algorithms and its application are more significant for healthcare services [17]. Data stored in medical databases may contain missing values and redundant data, so it is difficult to extract medical data. As it can impact mining outcomes, before implementing mining algorithms, it is important to have good data planning and data reduction. For disease prediction and health risk analysis, effective machine learning algorithms are therefore used to provide accurate results [18].

To perform DM prediction, it needs history of various diabetic patients. Such records are collected from different medical organizations and named as diabetic data set. For detection of diabetes mellitus fuzzy logic, Artificial Neural Network (ANN), Support Vector Machine (SVM), Random Forest (RF), Convolutional Neural Network (CNN), CBR classifiers are used [19]. It is important to select the correct extraction technique as this helps to extract vital information from the signals, thereby enhancing the efficacy of the classifier. [20]. Many techniques are introduced to improve performance of diabetic detection system. But the improvement is still needed for diabetic detection based on accuracy metrics, so this paper presents an efficient early detection and risk level analysis technique.

The research paper of an early disease prediction and risk analysis of diabetic mellitus using electronic medical records is organized as follows: Section I introduces the need of predicting of diabetes mellitus. Section II provides detailed literature review on machine learning based risk prediction model. Section III presents the methodology of early prediction and risk analysis of diabetes mellitus disease through electronic medical records using machine learning technique. Section IV presents summary and discussion. Finally, Section V concludes the paper.

2. Literature Review

Machine learning has significant contribution in building predictive models in the diagnosis and forecasting of the diseases using electronic medical records (EMR) is being made by recent advances [21]. The emergence of machine learning and electronic medical records for prediction of disease viewed as analogous to the development of imaging, genetic, or some other new cause of highly informative clinical data [22]. Machine learning can be used in numerous ways to achieve rising health data challenges in predicting diabetic mellitus disease and make the system more efficient in practice. The analytic processes such as data mining can solve different problems through data analysis which are based on the category of data that needs to be handled [23]. Recent progress of machine learning methods in designing and applying it on EMR has shown promising outcomes as discussed below:

Cui S *et al.* [24] presented improved support vector machine (SVM) based method to predict diabetic readmission. There were 3 stages: preprocessing, feature selection and classification. In order to fix imbalanced data and to improve prediction, hybrid feature selection mechanisms were developed to select important features. Synthetic Minority Over-Sampling Technique (SMOTE) data preprocessing were implemented. Finally, using improved SVM-based learning method, classification was carried out. Experimental result has shown that SVM method achieves 81.02% accuracy. Authors claimed that other common methods for readmission predictions were outperformed by decision tree and SVM process.

Zhu C *et al.* [25] designed data mining model for early prediction of diabetes using principal component analysis, k-means and logistic regression methods. PCA used for dimensionality reduction, k-means technique used for clustering the data and logistic regression used for classification technique. Logistic regression model has been improved by integrating k-means and PCA for prediction of diabetes at an early stage. The authors claimed that the approach can also model a new dataset effectively. The results showed that PCA improved the logistic regression classifier accuracy of 1.98% higher and the k-means clustering output of 25 more correctly classified data as compared to results of other studies. Also, the approach achieved accuracy of 97.4% for diabetes prediction.

Devi R D H *et al.* [26] presented hybrid method of farthest first clustering algorithm and support vector machine (SVM) classification algorithm for diagnosing Diabetes Mellitus. Clustering algorithm farthest first was used for gathering data into number of clusters and the time require for computation was reduced significantly as original data diminished. The output of clustering algorithm was given as input to classification algorithm for classification of patients into positive & negative. Results showed that classification accuracy of 99.4% has achieved for diagnosing patient with diabetic and non-diabetic.

Beloufa F and Chikh M A [27] proposed reliable and improved Artificial Bee Colony (ABC) algorithm for diabetes diagnosis by designing optimal fuzzy classifier. Decision-making process of doctor supported by the fuzzy classifier made with ABC approach and provided more information by using optimal fuzzy rules with membership function. The improved ABC algorithm used Pima Indian diabetes dataset for demonstration. Results showed that classification method achieves highest accuracy of 84.21% by using fuzzy rules with improved ABC algorithm in diagnosis of diabetes disease.

Singh N and Singh P [28] have developed ''NSGA-II-Stacking'' a stacking-based multi-objective evolutionary ensemble system for predicting of Type-2 diabetes mellitus. Initially, data pre-processing step is carried out where the missing values and outliers were identified and the values assigned with the median. The multi-objective optimization algorithm was used to increase classification accuracy and decrease ensemble difficulty for base learner selection. Also, for combine predictions of base learner selection, k-nearest neighbor (KNN) was used as meta-classifier. Experiment results showed that the system achieved 83.8% highest accuracy, 96.1% sensitivity, 79.9% specificity, 88.5% f-score & 85.9% area under ROC curve. Authors claimed that datasets having different features and risk factors can be investigated further for early prediction of DM by modifying NSGA-II stacking method.

El-sappagh S et al. [29] proposed framework for diabetes prediction based on a heterogeneous ensemble classifier. The approach used an ensemble framework by considering classifiers based on random subspace and bagging techniques. The framework uses different data mining techniques, such as naive Bayes, k-nearest neighbors, fuzzy decision tree, decision tree, support vector machine, logistic regression, and artificial neural network, each with separate set of suitable features. The framework accurately selects for each sub-dataset, accurate classifier and appropriate feature set for diabetes prediction. The result showed that framework achieved 90% of accuracy, 94.9% of precision, and 90.2% of recall, which shows that ensemble framework significantly outperforms all other classifiers.

Alama T M et al. [30] introduced early prediction model for diabetes disease. The approach used machine learning & data mining techniques. Important attributes were selected & association of attributes were categorized to predict diabetes. This was achieved by principal component analysis (PCA) technique. It is observed that glucose level & BMI are main features for diabetes prediction which was excavated by apriori method in association rule mining. Random forest (RF), artificial neural network (ANN), and K-means clustering techniques were applied to predict diabetes at an early stage. Among all, ANN technique has valuable treatment results by achieving the best accuracy of 75.7%.

3. Methodology

Early prediction of preventable diseases and classification of risk level is very significant for more efficient health-care resource allocation and better disease management. The aim is to make use of significant features, and design an efficient system to predict DM at early stage and analyze risk level using electronic medical records. The methodology is split into three phases, such as,

PHASE I: Diabetes Prediction
PHASE II: Risk Analysis
PHASE III: Early Prediction.

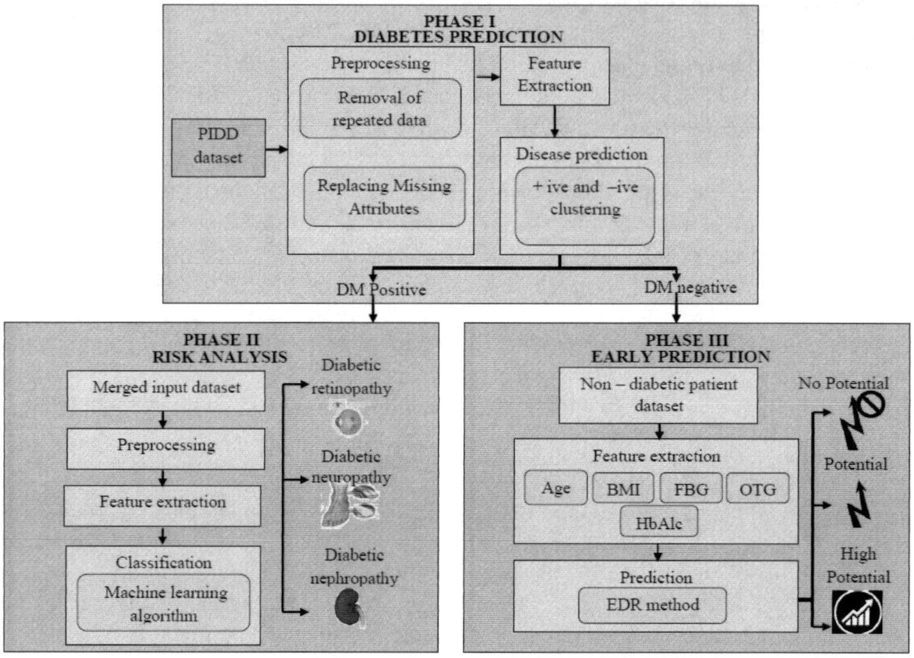

Figure 1: Work flow of Methodology

PHASE I: Diabetes Prediction

Data preprocessing is an essential step in disease prediction, because of the quality of data, to a large extent, affects the result of prediction. In this step, the preprocessing includes two steps for simplifying the dataset. They are the removal of repeated data and replaced the missing attributes.

i. Removal of repeated data

In this stage, repeated data are removed from dataset using Hadoop Distributed File System. It provides software framework for dispersed storage and processing of big data by using the Map Reduce programming model that consists of map() & reduce() functions to remove duplicate data.

- *Map () stage:* The input file is given to the mapper function line by line. The mapper processes the data and makes several minor chunks of data.

$$m_i = map(D_i) \tag{1}$$

Where, $map(\bullet)$denotes the map () function and $D_i$denotes the input dataset.

- *Reduce () stage:* The Reducer's job is to process the data that comes from the mapper. It removes the repeated data and produces a new set of output, which will be stored in the HDFS.

$$R_i = red(m_i) \tag{2}$$

Where, $red(\bullet)$denotes the reduce () function, $R_i$is the reduced collection of data.

ii. Replace the missing attributes

If any records encompass unrecorded values, then these values will be filled by replacing the missing value for a particular attribute by the average value for that attribute.

$$M_i^a = \frac{(\sum D_i)}{n} \tag{3}$$

Where, M_i^adenotes the missing attribute, n represents the number of data.

iii. Feature Extraction

In this phase, the important features are extracted from preprocessed data. In DM prediction system, the important features, like Plasma glucose concentration, Diastolic blood pressure (mm Hg), Two-hour serum insulin, Body mass index, Diabetes pedigree function, etc. Extracted features are represented as,

$$\vec{F}(e)_i = \{\vec{F}(e)_1, \vec{F}(e)_2, \ldots\ldots\ldots\ldots, \vec{F}(e)_n\} \tag{4}$$

Here, $\vec{F}(e)_i$ signifies extracted feature set and $F(e)_n$ denotes n number of extracted features.

iv. Diabetes Disease Prediction utilizing Clustering Algorithm

The clustering algorithm is utilized for categorization of patient results as positive and negative classes. Positive class means, the output of the patient data is a diabetic patient. Here, the value is signified as, 1 and 0, 1 means positive result and 0 implies a negative result. The negative result means the output of the particular data is a non-diabetic patient. The less variation within clusters, the more homogeneous (similar) the data points are within the same cluster. The algorithmic steps of clustering are given below:

Step 1: Specify number of clusters $K_i = 2(K_1, K_2)$ then initialize the K number of class values as $\vec{c}_i = \{\vec{c}_1, \vec{c}_2, \ldots\ldots, \vec{c}_n\}$, here $\vec{c}_i$is the class value dataset D_i and set of cluster centroid as $c_i = \{c_1, c_2, \ldots\ldots, c_n\}$.

Step 2: Select the number of K_i cluster centers randomly to cluster the dataset.

Step 3: Calculate the distance between each class values and cluster centers.

$$E_{d_1} = |\vec{c}_i - c_1| \tag{5}$$

$$E_{d_2} = |\vec{c}_i - c_2| \tag{6}$$

Step 4: Assign datapoint to cluster center whose distance from cluster center is minimum of all centers.

Step 5: Repeat steps 2, 3, and 4 until the same points are assigned to each cluster in consecutive rounds.

The final clustering result contains two classes ('0', '1'). Here, '0' indicates DM negative and '1' indicates DM positive. In the case of DM positive, Phase II is performed to analyze the future risks in diabetes mellitus. Otherwise, Phase III is carried out for the early prediction of DM.

PHASE II: Risk Analysis

Diabetes can lead to problems that affect many parts of the body, including brain, heart, eyes, kidneys, and nerves. Thus, proper assessment of patient's prognosis plays a central role in the management of diabetes. This phase is utilized to ascertain Risk Analysis of diabetic disease after 3 years.

The risk analysis phase consists of four steps, that are, merge input dataset, pre-processing, feature extraction, and classification. First, data from diabetic disease is merged by utilizing common attributes, like sex, age, etc. to form a new dataset. Then, the data is arranged into a standard format for obtaining a better result, which is done in the pre-processing step. After that, the important features, such as BP, Anemia, numbness, Blood urea, and Blood sugar, etc. are extracted from dataset for efficient analysis. The extracted feature set is denoted as, F_i^d. In this step, extracted features F_i^d are given as input to machine learning algorithm to ascertain risk value for diabetic neuropathy, retinopathy, and nephropathy after 2 or 3 years. The machine learning algorithms classify the risk level of diabetes patients effectively.

PHASE III: EARLY PREDICTION OF DM

In this phase, early detection of diabetes mellitus is carried out. Early diagnosis of diabetes and pre-diabetes is important so that patients can start to manage the disease early and possibly prevent or delay the serious disease complications. The steps involved in early prediction of DM is given as follows:

i. Non-Diabetic Patient Dataset

From the hospital laboratory, the method gathers patient's previous health information about blood glucose tests, age, fasting blood glucose (FBG), Oral glucose Tolerance (OTG), HbA1c examination.

ii. Feature extraction

In this phase, important features are extracted from dataset. Technique extracts following features:

Table 1. Extracted Feature set

Features	Extracted Feature Set
Patient's age Body Mass Index (BMI)	Patient's age (F1)
Family History of Diabetes Physical Stress	Body Mass Index (BMI) (F2)
Blood pressure Smoking	Fasting blood glucose test results (mg/dL) (F3)
History of polycystic ovary syndrome (PCOS) Being of African, Latin American, or Asian-Pacific	Oral glucose Tolerance test results (mg/dL) (F4)
Fasting blood glucose test results (mg/dL) Oral glucose Tolerance test (mg/dL) HbA1c examination (Percent)	HbA1c examination (percent) (F5)

Above mentioned features are chosen for DM early prediction. Extracted feature set is expressed as,

$$F_i^{nd} = \{F_1^{nd}, F_2^{nd}, \ldots\ldots\ldots\ldots, F_n^{nd}\} \qquad n = 5 \tag{7}$$

Here, F_i^{nd} signifies the extracted feature set and F_n^{nd} denotes the n number of extracted features.

Table 2. Diagnosis level of Diabetes Ranges

	Age	BMI	HbA1c	FPG (mg/dL)	OGT (mg/dL)
Diabetes	-	-	6.5 or above	126 or above	200 or above
Pre-diabetes	Above 45	25 to 29.9	5.7 to 6.4	100 to 125	140 to 199
Normal	-	18.5 to 24.9	Above 5	99 or below	139 or below
F_i^{dp}	45	20	5	99	139

iii. Euclidean Distance Ranking (EDR) Method for Early Prediction of DM

In this phase, the early prediction of DM is carried out utilizing the Euclidean Distance Ranking method. The EDR method is a basic and simple technique that measures the distinguishing between two feature set with real values. The techniques use two feature set for EDR calculation, one is extracted feature set F_i^{nd} and another one is "diabetic-prediction feature set", which is denoted as F_i^{dp}. The real value of each feature in F_i^{dp} is shown in table 2. The Euclidean Distance between two features is calculated as:

$$Eucl_{dis}(F_i^{dp}, F_i^{nd}) = \sqrt{\sum_{i=1}^{n}\left(F_i^{dp} - F_i^{nd}\right)^2} \tag{8}$$

$Eucl_{dis}(F_i^{dp}, F_i^{nd})$ measures the numerical difference for each corresponding features of feature set F_i^{dp} and F_i^{nd}. The stages are found centered on the $Eucl_{dis}$, which means that the method fixed some threshold if the $Eucl_{dis}$ is below the threshold, then the stage is denoted as the 'risk'. Otherwise, the stage is denoted as 'no risk'. After calculating each feature stage, the final result is a decision on whether a person has potential against DM or not is predicted using following rules.

Table 3. Rules involved in Early Prediction of DM

Rule Details		Early Prediction Output
Features (F3 or F4 or F5)	Features (F1 or F2)	
No risk	No risk	No potential
Risk	No risk	Potential
No Risk	Risk	No potential
Risk	Risk	High potential

Based on the above-mentioned rules, the patient's potential against DM is predicted at an early stage. This early prediction system can help to reduce the risk of serious complications, such as premature heart disease and stroke, blindness, limb amputations, and kidney failure.

4. Summary and Discussion

Availability of huge amount of clinical data in the form of electronic medical records is the origin of the research which has ability to transform healthcare industry. In many of the research, the most important challenge is to build models that accurately diagnose and predict diabetic mellitus disease by developing different algorithms and methods for processing and analysis of biomedical data. On analyzing existing techniques, it is observed that more diabetes mellitus cases are caused due to delayed detection. So, early recognition of diabetes is needed to treat diabetic mellitus and also diabetes patients will have so many risks in the future that may damage brain, heart, eyes, kidneys, and nerves. It is also needed to provide efficient and accurate system to enhance workflow, clinical tasks, streamlining work of early disease prediction and risk level classification. Machine learning models yield better performance in many tasks and require less manual feature engineering. Thus, the system will be useful to improve quality of life and avoid health risks.

5. Conclusion

The research focuses on applying machine learning methods and its applications in prediction of diabetic mellitus disease and health risks for healthcare systems using electronic medical records (EMR). A continuously growing large amount of EMR has to be processed and analyzed using data mining to improve accuracy and precision of disease prediction and future health risks. The objective is to analyze electronic medical records to efficiently classify sets of patient data in different categories (e.g. normal and disease affected) by investigating machine learning based approaches to improve efficiency and accuracy of risk analysis. It helps to improve performance of risk level classification for different future stages of disease affected patients. Further, machine learning algorithms is applied to achieve better results in healthcare systems. The approach is designed in such a way that it will give greater results than previous diabetes prediction model. Various machine learning algorithms are used in state-of-art works but low accuracy problem still exists. So, this paper presents an efficient and robust technique to improve accuracy and precision of early disease prediction and risk level analysis of Diabetic Mellitus. In future work, finding from this paper in changing medical practices, leading to more cost effective by considering unstructured data like family health history, previous health check, etc. The further research can be carried out on how to improve performance of risk prediction for different types of diseases.

References

[1] Baiju B V and Aravindhar D J 2019 *1st international conference on innovations in information and communication technology (ICIICT) IEEE* pp 1-6
[2] Health Topics, "Diabetes", World Health Organization, [Online] Available https://www.who.int/news-room/fact-sheets/detail/diabetes
[3] Saeedi P, Petersohn I, Salpea P, Malanda B, Karuranga S, Unwin N, Colagiuri S, Guariguata L, Motala A A, Ogurtsova K, Shaw J E, Bright D and Williams R 2019 *Diabetes Research and Clinical Practice* vol **157**
[4] Woldaregay A Z, Arsand E, Walderhaug S, Albers D, Mamykina L, Botsis T and Hartvigsen 2019 *Artificial Intelligence in Medicine* pp 109-134
[5] Zheng H, Park H W, Li D, Park K H and Ryu K H 2018 *5th NAFOSTED Conference on Information and Computer Science (NICS), IEEE* pp 110-113
[6] Qalhati N S J M Al, Hussain S A and Singh A V 2018 *7th International Conference on Reliability, Infocom Technologies and Optimization (Trends and Future Directions) (ICRITO), IEEE* pp 840-846
[7] Fabris C, Facchinetti A, Sparacino G and Cobelli C 2014 *36th Annual International Conference of IEEE Engineering in Medicine & Biology Society* pp 6643-46

[8] Yahyaoui A, Jamil A, Rasheed J and Yesiltepe M 2019 *1st International Informatics and Software Engineering Conference (UBMYK) IEEE* pp 1-4
[9] Sumangali K, Geetika B S R and Ambarkar H 2016 *International Conference on Control, Instrumentation, Communication and Computational Technologies (ICCICCT), IEEE, (Kumaracoil, India)* pp 389-392
[10] Undre P, Kaur H and Patil P 2015 *international conference on pervasive computing (ICPC) IEEE: Advance Communication Technology and Application for Society* pp 1-4
[11] Fiarni C, Sipayung E M and Maemunah S 2019 *Procedia Computer Science* vol **161** pp 449-457
[12] Kannadasan K, Edla D R and Kuppili V 2019 *Clinical Epidemiology and Global Health* vol **7** pp 530-535
[13] El-Sappagh S, Ali F, El-Masri S, Kim K, Ali A and Kwak K S 2019 *IEEE* vol **7** pp 21917–47
[14] Rosyadi I, Ramadhona E, Wahono A T U and Hijrati Y N 2018 1st *International Conference on Bioinformatics, Biotechnology and Biomedical Engineering, BioMIC* vol **1** pp 1-4
[15] Lan G, Lee C, Lee Y, Tseng V, Chin C, Day M, Wang S, Chang C and Cheng S 2012 *Conference on Technologies and Applications of Artificial Intelligence* pp 27-32
[16] Jain D and Singh V 2018 *Egyptian Informatics J* vol **19** pp 179-189
[17] Jothi N, Rashid N A and Husain W 2015 *Procedia Computer Science Elsevier in Information Systems International Conference (ISICO)* vol **72** pp 306-313
[18] Shinde S A and Rajeswari P R 2019 *International Journal of Engineering and Technology* vol **7** pp 1019-23
[19] Han L, Luo S, Yu J, Pan L and Chen S 2015 *IEEE Journal of biomedical and health informatics* vol **19** pp 728-734
[20] Lekha S and Suchetha M S 2017 *IEEE Journal of biomedical and health informatics* vol **22** pp 1630-1636
[21] Battineni G, Sagaro G G, Chinatalapudi N and Amenta F 2020 *Journal of Personalized Medicine* vol **10**
[22] Adkins D E 2017 American Journal of Psychiatry vol 174 pp 93-94
[23] Islam M, Hasan M, Wang X, Germack H and Noor-E-Alam M 2018 *Publication of Healthcare (Basel)* vol **6**
[24] Cui S, Wang D, Wang Y, Yu P and Jin Y 2018 *Computer methods and programs in biomedicine* vol **166** pp 123-135
[25] Zhu C, Idemudia C U and Feng W 2019 *Informatics in Medicine* vol 17
[26] Devi R D H, Bai A and Nagarajan N 2020 *Obesity Medicine* vol **17**
[27] Beloufa F and Chikh M A 2013 *Computer Methods and Programs in Biomedicine* vol **112** pp 92-103
[28] Singh N and Singh P 2020 *Biocybernetics and Biomedical Engineering* vol **40** pp 1-22
[29] El-Sappagh S, Elmogy M, Ali F, Abuhmed T, Islam S M R and Kwak K S 2019 *Electronics Article Special Issue Internet of Things (IoT) Based Wireless Health: Enabling Technologies and Applications* vol **8** pp 1-29
[30] Alama T M, Iqbala M A, Alia Y, Ijazb A W S, Baigb T I, Hussainc A, Malikb M A, Razab M M, Ibrarb S and Abbas Z 2019 *Informatics in Medicine, Elsevier* vol **16**

Object Identification Model using Deep Reinforcement Machine Learning Concept for Image

Saurabh Tiwari[1], Dr S Veenadhari[2], Dr Sanjeev K Gupta[3]

[1]Phd Research Scholar, Department of Computer Science and Engineering RNTU, Bhopal 462003, Madhya Pradesh India
[2]Associate Professor, RNTU Department of Computer Science and Engineering RNTU, Bhopal 462003, Madhya Pradesh India
[3]Professor, RNTU Department of Electronics and Communication Engineering RNTU, Bhopal 462003, Madhya Pradesh India

Email: saurabh_tietech@rediff.com

Abstract. This paper presents a model which gives the detailed process of object identification. We need to identify class and location of object in image for completing process of objet identification .Proposed model works on the principal of reinforcement learning which takes action on the basis of rewards and experiences. Normally methods in literature uses sliding window which moves in same direction but proposed algorithm provides a variable mask which moves 360 degree for identifying object using action history vector proposed with RL also not only this work focuses on localization like other work but also used class information with Softmax classification able to classify multiple object in single image with efficient time which is novel. Proposed mask acts as agent and focuses on proposed candidate reason this saves time and works in efficient manner for identification. Agent depends on transformation action and by applying top down reasoning it gives location of object. Classification is done using Softmax classification as we are having features of image by CNN. Reinforcement learning concept used for training of agent and Pascal voc dataset used for testing. Analysis of only 10 to 25 regions is sufficient with proposed work to identify first instance of object. Experiment and performance evaluation shows the efficiency of proposed work.

1. Introduction

Now days deep machine learning is widely used for object identification it become more crucial if there is more than one object in single image we need to repeat the counter and process for identification. Definition of object says Objects [1] are standalone things which having well define boundary and centre and separates itself from background. Most common approach for object detection is to define problem as classification and with convolution or mask used on feature map for region classification. Two commonly used training algorithm known as supervised and unsupervised trainings are used for training model but proposed work used reinforcement training. Many techniques [3]-[4] try to improve computational time and space complexity with different algorithm which work well. Faster R CNN uses variable mask over feature map for region proposal with trained neural network.

Proposed approach follows strategy that different path for different object. Also we not proposed candidate region with low level cues but high level concept of logical reasoning are used for candidate region selection. We proposed action strategy with Reinforcement learning which works by paying attention on current region and transform bounding box in a way that it can be more focused to target object. We define reward function which is equivalent to how much current bounding box covers the ground box. Deep-Q-network algorithm is used for learning policy for localization. We calculate precision and recall for Pascal Voc Dataset and obtain competitive results. Factor to compute processing time for an image we define IAET image average evaluation which gives time required for complete process of identification.

2. Related Work

Several work proposed which define different algorithm and methodology for object detection. The research method given in [5] depends on neural network for identifying market products. For high accuracy and fast training inception v3 proposed in tensor flow library. Alex Krizhevsky proposed model named Alex net [6] in 2012 which uses CNN for image classification it consist five convolution and three fully connected layer. Within image RCNN [7] was first successful attempt to detect object efficiently but due to its slow speed and high computational cost different improvement proposed for detection process. FAST RCNN was next version of RCNN which uses ROI pooling for reducing amount of forward passes by passing it to sub regions. FASTER-RCNN [8] replaced selective search used in fast RCNN with region proposal network (RPN) and reduced time taken up to half a second. independent region proposal for detection of objects B. Schiele et al[9] given a detailed and structured analysis of eight region proposal on the basis of recall, ground truth annotation, precision keeping objective in mind that they want to give method among eight which is feasible to use for other projects. Every researcher in region proposal is working to reduce the number of evaluated region so detection process can be made fast and efficient. M. B. Blaschko et al proposed a theory using concept of branch and bound [11] which use few locations to predict region with high score. Aleksis [10] et al proposed deep reinforcement learning based object detection which replaces greedy ROI selection process with sequential attention mechanism using object detector with RPN.

Reinforcement learning [12] work on the experiences objective is to define a policy that decide sequential action by enhancing future rewards. Recent trends [13]-[15] in reinforcement learning is to combine deep neural network with RL and obtaining policy from RL to solve the problem normally RL model defines policy or value function by evaluating deep features RL can successfully produced algorithm to play many Games like Atari[16] and others. Figure 1 shows the view of simple Reinforcement Learning concept to understand things clearly. Also by employing Deep RL many problems related to computer vision like object identification, localization, object tracking has solved efficiently.

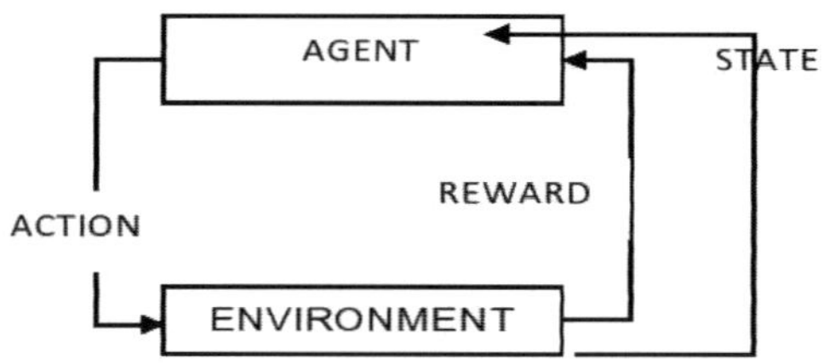

Figure 1 Reinforcement learning concepts

Object identification problem solved by reinforcement leaning will follow normally two algorithm named policy gradient and Q-learning. Proposed work follows the concept of Q learning for policy framing David Silver [14] et al proposed policy gradient algorithm for RL with continuous action, comparative to stochastic policy deterministic gradient descents evaluated more efficiently and produce better computation results. We used Softmax classifier for class label because it gives probabilities for each class we compare Softmax with SVM classifier and we found that interpretation of probabilities is easier than marginal scores and we are calculating probabilities thus we elected Softmax classifier for proposed work.

3. Object Identification Model and scheme

This section will give detail about model and scheme used for region proposal, features extraction, action and reward of RL and proposed model with algorithm.

3.1. Object identification model

We proposed an improved and boosting model for identifying object figure 2 will give theoretical view of proposed model .we named Model ODDRL-NET that is object detection deep reinforcement learning model. It having three convolution layer based on CNN responsible for features generation also we are using kernel Relu and strides as shown in figure to generate features map. Input of model is a pre-processed image taken from Pascal Voc dataset. Model has two fully connected layers named FC4 and FC5 .FC4 is responsible to processed input from features map and provides output with 512 nodes after dropout it concatenated with action history vector which finally provides the state representation. After this we use Q-learning for correct localization on the basis of probabilities calculated Softmax classifier will predict class information of object.

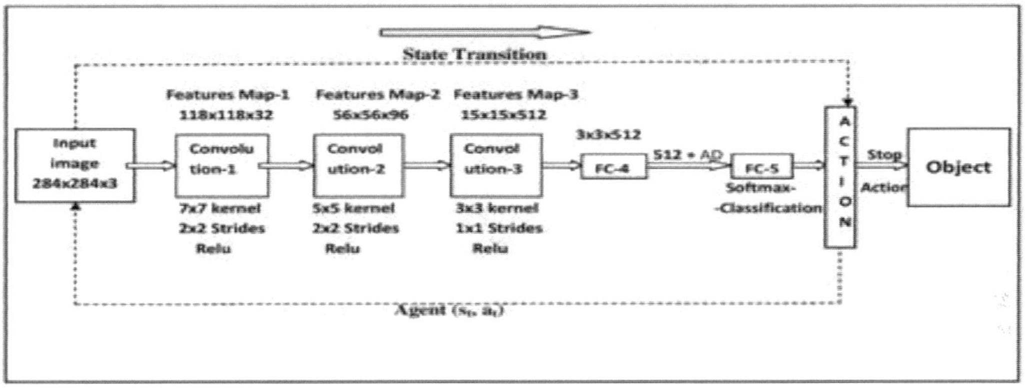

Figure 2 Proposed Object Identification Model

3.2. Problem statement with action dynamics

We defined problem statement as Markov Decision Process (MDP) which helps work with decision making. Reinforcement learning concept works on environments, state, action, agent and reward we have image as environment and on different state with action history agent will try to transform a tight mask called bounding box around object with the help of different action history. The agent makes transition from one state to another it marks current region which is visible and past region. Agent mark rewards as positive and negative as well for getting visible it mark 1 and other for 0 which helps the algorithm to calculate score function for further accuracy of algorithm.

Figure 3 Eleven Action including translation, scale, stop

Proposed work uses eleven action for movement which includes scaling , translation and special action stop which finishes the first pass of algorithm if image having single object. Translation consist of four moves (left right up down) with fast moves with double capacity for each four, scale changes includes (up Scale, down Scale) which maintain and set aspect ratio. Bounding box in implementation simply represented by $B = (x_1, y_1, x_2, y_2)$ it changes shapes with a factor β given by following equation

$$B_w = \beta^* (x_2 - x_1) \quad \beta h = \beta^* (y_2 - y_1)$$

Where the value of β lies in range of 0 and 1 given by β € (0, 1) new position of bounding box is defined by adding or eliminating β_w or β_h from x or y co-ordinates. Finally stop action is the only action which stops the process and indicates successful identification of single object and terminates process and place the mask on initial position for multiple objects present in the environment it will restart process . With the stop action mask will overlap ground truth and intersection over union [17] (IoU) computed, which is used in many visual attention model to calculate the effective identification.

3.3. State transition and Reward calculation

State is factor which move process further we define some parameter like observed region vector (Or) and action history vector (Ha).transition of two states depends on (Or, Ha) we used state generalization for effective transition. Feature vector or is the vector extracted with current region using first three convolution layer of ODDRL-Net model acting as pretrained model for vector extraction as proposed by Girshick et al. Any region attended by agent deformed to match the required input size (284x284) by using technique provided by [18] J. Donahue et al.

Function which lead to search towards objective refer as reward its count one gain when agent identifies an instance of object after having one appropriate action. Measuring gain or improvement depends on intersection over union (IoU) calculated on predicted and target bounding box. we get reward when state changes mathematical computation of reward will done with moving one state to other we can define it say current state s is changes to s' while applying action and every state have current box b than reward will be defined as…

$$R_a(s, s') = \text{sign}(I_oU(b', g) - I_oU(b, g))$$

If reward is improving with current to next state than we get reward positive and otherwise negative. The reward scheme is binary because without quantization difference in IoU will be very small and hard to predict which action is bad or good. When we found target object in visible region it means time to have stop action and identification phase. The stop action is a special action because it lead the box in terminal state which doesn't change in this case we get IoU zero so at termination we can define reward function as

$$R(s, s') = \begin{cases} 1 & \text{if IoU}(b, g) \geq 0.6 \\ -1 & \text{otherwise} \end{cases}$$

Where IoU (b, g) is the overlap ratio between terminal box b and ground truth g. Proposed agent follows Greedy strategies which provide minimum number of steps which helpful to make proposed scheme efficient and systematic.

4. Training and Testing

Training and testing is done using Pascal Voc Dataset for the very first step we pre-processed images using a python script to get suitable dimension and space. We already know that an image is collection of neuron and if we reduce size it means less neuron to handle. We created XML file with label image software than a CSV file which contains map with that we can classify images easily. We implemented our model in tensor flow [19] using anaconda python version 3 this make easy this model to use with different architectures devices.

The objective of the agent is to find a policy to transform bounding box in a manner that it find appropriate action which enhance the reward when it interact with the environments. We Framed policy for training with reinforcement learning using Q-learning algorithm and solved problem of Q-learning with algorithm given by D silver et al [20]. We used shallower faster RCNN concept in our features and proposals which is trained in a way that is acts as pretrained network for Q network it having advantages that learning of Q function is relatively faster and RCNN will acts as feed forward for Q network. Also we are having full features learning as we using customize Pascal-Voc Dataset learned features improved performance of Q network. We trained agent by initializing Q-network parameter randomly and agent interacted with the environment with Greedy strategy. Environment having information about ground truth box and for calculating Reward we use IoU with the current bounding box so here we get action with positive and negative record. We used stochastic gradient descent and back propagation algorithm also for handling dropout we use [21] to regularized dropout.

Now when we completed phase of training it's time to test model for identification .behaviour of agent is more important whether it's learned to find region out of proposed region by algorithm. Initially we don't know how much object are present in single image so we allow the agent to run with 200 steps it means we are going to evaluate 200 region per image . The search will be restart in two conditions if agent has used stop action or 30 steps passed without stop it means agent is trapped in obscure region we decided 30 steps because we getting results in fewer than thirty steps also it make the search time efficient.

5. Performance Evaluation and Discussion

We setup performance evaluation based on detection quality, precision and recall. Every object identification algorithm works to improve these parameters; evaluations shows that proposed work have improved technique compared to other methods in literature. We evaluated proposed scheme with customize Pascal VOC dataset and also added many other images which used for training and testing purposes for proposed model .we implemented this work with Anaconda python 3.5 combined with tensor flow. All operation executed in 8th generation Intel core I7 8557U processor, 8 GB DDR-4 RAM and hard drive with 500 GB capacity with windows10. Implementation details of this work shown in figure 4 and figure 5 which gives identification of two sample test image for giving the quality of project. As shown in figure 4 and 5 the value of detection is more than 80 percent which clears the proposed box and ground truth box having more than 70 percent intersection over union which shows mask is properly placed by agent and based on score class calculated by Softmax algorithm is correct.

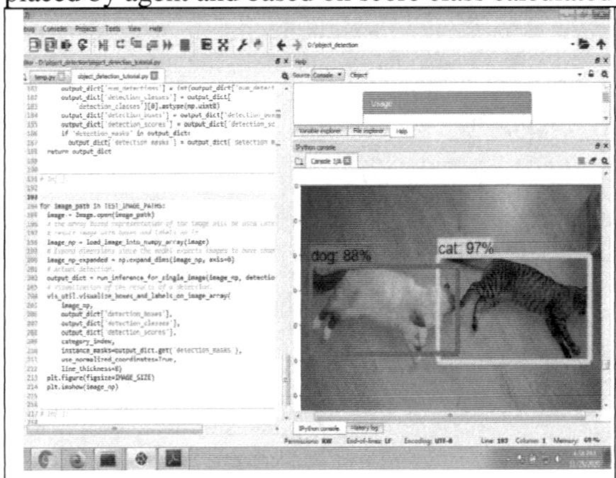

Figure 4 Test Image Dog and Cat

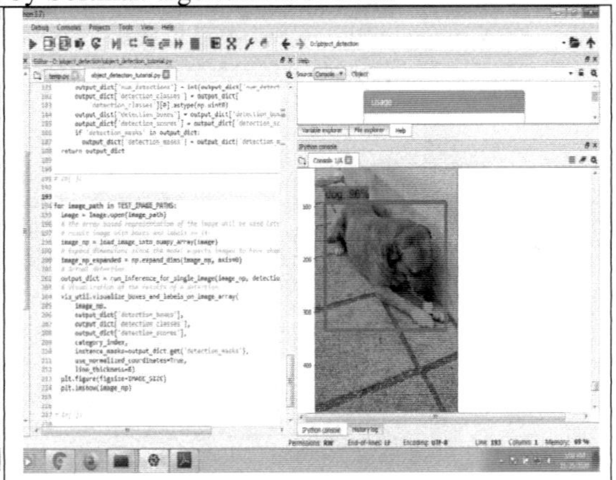

Figure 5 Test Image Dog

5.1. Estimation of precision

Precision tells how accurate is our prediction proposed approach will considers all attended region which predict object we have a detector which scores all region on single search step. Normally in object detection precision depends on intersection over union (IoU) given by ratio between area of intersection and area of union of predicted bounding box and ground truth box. Precision means positive predicted value which given for class in classification process. Precision is calculated by

$$Precision = \frac{\text{True positive}}{\text{True Positive} + \text{False Positive}}$$

Table 1 Calculation of AP (Performances with bold font present best results)

Methods	Aero	Bike	Bird	Boat	Bottle	Bus	Car	Cat	Chair	Cow	Table	Dog	Horse	Bike	Person	Plant	Sheep	Sofa	Train	Tv	MAP
R-CNN[10]	64.2	69.7	50.0	41.9	32.0	62.6	71.0	60.7	32.7	58.5	46.5	56.1	60.6	66.8	54.2	31.5	52.8	48.9	57.9	64.7	54.2
Region lets[38]	44.6	55.6	24.7	23.5	6.3	49.4	51.0	57.5	14.3	35.9	45.9	41.3	61.9	54.7	44.1	16.0	28.6	41.7	63.2	44.2	40.2
RPN-SS[39]	77.0	78.1	69.3	59.4	38.3	81.6	78.6	86.7	42.8	78.8	**68.9**	84.7	82.0	76.6	69.9	31.8	70.1	**74.8**	80.4	70.4	70.4
RPN-300[39]	87.4	83.6	76.8	62.9	59.6	81.9	82.0	91.3	54.9	82.6	59.0	89.0	85.5	84.7	84.1	52.2	78.9	65.5	85.4	70.2	75.9
Proposed method	**87.7**	**84**	**78**	**64**	**59.9**	**83**	**82.6**	**92.1**	**55.2**	**83.1**	60.2	**89.4**	**86.0**	**85.1**	**84.5**	**55.0**	**79.3**	66.2	**86.2**	**71.0**	76.6

Precision tells how many selected items are relevant and for many class we take MAP that is mean average precision. Above table 1 will give comparison of precision and average of **20 class** images taken from Pascal Voc Dataset. Also proposed method compared with different methods and bold font showing best results.

5.2. Recall

In simple words recall defined as how many relevant items are selected Recall formula for a given class in classification is given by

$$Recall = \frac{True\ positive}{True\ positive + False\ negative}$$

Region attended by agent can be consider as object proposal candidate for evaluating these we follow approach given by B. Schiele et al. [22] .we process 20 category and running 200 steps per category it means 4000 region will be evaluated and get 78 percent recall. Evaluation of recall is done by Q values predicted by agent to score attended region, when agent initiate stop we add constant to score those region so they can clearly get position in top for clearly identify region. Figure 6 will show the plot between recall vs. number of proposals also we compared proposed method with other discussed earlier. With ten proposals we having more than 40 percent recall while other methods reached only 20-30 percentage.

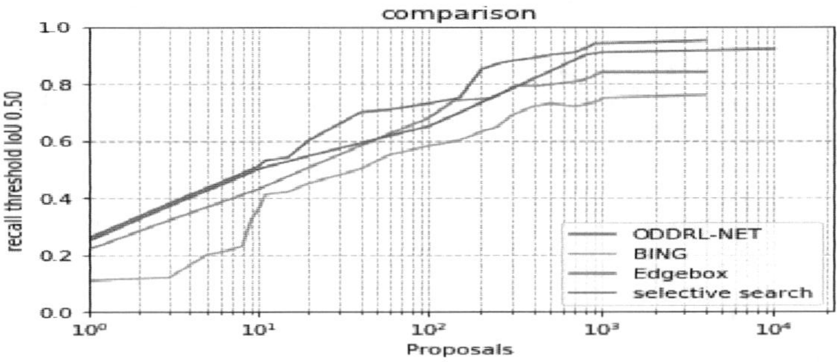

Figure 6 Plot Recall @threshold IoU 0.50

For better illustration we plot distribution among correctly identified object and number of steps required to identify object correctly. Proposed method identify more than 85 percent object in less than 40 steps and average of 18.

5.3. Image Evaluation time

Image evaluation time refers to complete time required to process one image for identifying object. Single Image Average Evaluation Time (SIAET) [23] and storage used by Carlos Augusto et al. to show the time taken to execute an image. Following table will compare the different method with proposed method for image evaluation time and storage requirement.

Table 2 Calculation of image evaluation time

Method	Storage Required	IET
R-CNN	Data Dependent	47302ms
FAST R-CNN	477 MB	2185ms
FASTER R-CNN	523 MB	385ms
FLODNET + SLIDING WINDOW	105 MB	6304ms
FLODNET+CONVOLUTION FEATURE MAP	95 MB	113ms
ODDRL-NET	96 MB	110ms

6. Conclusion and Future work

Thus this work present a novel object identification model based on faster RCNN concept and Reinforcement Learning. We used the action history vector to calculate reward and used Q learning policy for placing the box at correct location. Reinforcement learning is used in literature for object localization but proposed work not only localize object but defines class information as well. Precision and recall gives the quality of identification very high with this scheme. All results implemented in python which is the language of machine learning now days. Overall method gives better results and time efficiency for identification.

References

[1] Deselaers T et al what is an object *Proceedings IEEE Conference Computer Vision and Pattern Recognition* (CVPR), Jun 2010, pp 73-80.
[2] Lapte I Improvements of object detection using boosted histograms in *Proceedings BMVC*, vol. **3**. 2006.
[3] Dollár P et al Edge boxes, Locating object proposals from edges in *Proceedings European Conference Computer Visi* 391-405.
[4] Hoffman S C and Thiagarajan D Continuity report Revisiting grocery recognition using tensor flow publication *vision.cornell.edu.*
[5] Krizhevsky A and Hinton G E Image Net classification with deep convolutional neural networks in *Proceedings Adv. Neural Infor. Proce. System*, 2012, 1097-1105.
[6] Darrell T and Malik J et al Rich feature hierarchies for accurate object detection and semantic segmentation in *Proced. IEEE Confer. Comp Vi Patt Reco* Jun. 2014, 580-587.
[7] Girshick R et al Faster R-CNN Towards real-time object detection with region proposal networks in *Proced. Advance Neural Info. Proce. System* 2015 pp. 91-99.
[8] Schiele B and Hosang J et al How good are detection proposals really *arxiv* 1406-6962, 2014.
[9] Aleksis P and Christian Deep Reinforcement Learning of Region Proposal Networks for Object Detection *IEEE conferen. CVF* 2018.
[10] Hofmann T et al Beyond sliding windows objects localization by efficient sub window search *CVPR IEEE*, 2008.
[11] Sutton R S Introduction to reinforcement learning volume – 135 books on *Reinforcement Learning.*
[12] Silver D and Wierstra D et al Playing Atari with deep reinforcement learning *arxiv pre print* 1312-5602.
[13] Lever G et al Deterministic policy gradient algorithms *Intern. Confere. On Machine Learning* ICML, 2014.
[14] Guez A and VanHasselt H Deep reinforcement learning with double q-learning *CORR*, 1509.06461,2015.
[15] Silver D et al Playing Atari with deep reinforcement learning *arxiv preprint arxiv*-1312-5602, 2013.
[16] Itti L et al. Computational modelling of visual attention *Nature reviews Neu. Scie.* 194-203, April 2001.
[17] Donahue J, Girshick R et al Rich feature hierarchies for accurate object detection and semantic segmentation *CVPR* 2014.
[18] Abadi M, Paul B et al. Tensor Flow Large-scale machine learning on heterogeneous distributed system *arxiv.org/abs*/1603.04467.
[19] Silver D, Veness J et al. Human level control through deep reinforcement learning *Nature* 518-529-533, Feb 2015.
[20] Hinton G, Salakhutdinov R et al Dropout A simple way to prevent neural networks from over fitting *The Journal of Machine Learning (ML) Research*, 15-1929-1958, Jun 2014.
[21] Hosang J et al How good are detection proposals really *arxiv* 1406-6962, 2014.
[22] Augusto Carlos et al. Fast and Lightweight Object Detection Network Detection and Recognition on Resource Constrained Devices *IEEE* 2169-3536 2018.

IOP Conf. Series: Materials Science and Engineering 1085 (2021) 012025 doi:10.1088/1757-899X/1085/1/012025

Enhancing the efficiency of continuous integration environment in DevOps

Jetty Benjamin[1] and Juby Mathew[2]

[1]Research Scholar, APJ Abdul Kalam Technological University, Kerala, India

[2]Associate Professor, Amal Jyothi College of Engineering, Kanjirappally, Kerala, India

E-mail:jettybenjamin@amaljyothi.ac.in

Abstract. Faster release of features to the production environment is the need of the hour of every IT organizations. This faster and continuous software delivery is made possible through DevOps. DevOps is a set of practices that bridges the gap between development team and operations team. The different aspects of DevOps came from agile methodology. Continuous integration is the paramount of DevOps life cycle. In a continuous integration development practice, the development code and the applications code are continuously integrated several times a day. Developers follow the practice of committing changes frequently into the version control system such as git. A change in the version control system triggers the continuous integration system such as Jenkins. This exploratory study presents the ways that support the developers during the build break in continuous integration environment and thereby improve the efficiency of continuous integration process. Our findings contribute that the selected list of plugins in the Jenkins support the developers during the build break, thus developers can come up with faster recovery of build failures which in turn increase the efficiency of continuous integration environment.

1. Introduction

DevOps aims to bring together development and operations. The primary focus of the development team is to bring maximum number of changes that are pushed to production environment whereas the IT operations team aims to make the system stable and available. Adoption of DevOps practice in software companies has made a positive and significant impact. The usage of wide variant of tools accelerated the deployment pipeline activities.[1]. Frequent changes to the production environment take place through the two core processes continuous integration and continuous delivery.

Continuous integration is the process in which the developers continuously integrate or merge the code changes into a shared repository. Continuous delivery is the process in which new changes are released to the customers frequently. Every change that passes all stages of continuous integration/continuous deployment pipeline is released to production. Although the continuous delivery approach increases the delivery of features, there are numerous challenges that needs to be considered during the adoption of DevOps. Categories of problems including build design, integration, testing, release when adopting continuous delivery is addressed and the number of cases per testing problem is found to be greater [2].

When a complete transformation occurs from the traditional software development to continuous development a change will be reflected in every environment such as development, testing, deployment, operation and maintenance. A set of strategies to overcome the challenges while adopting the continuous delivery is described in [3]. The strategies include convincing the stakeholders about the advantage and benefits of implementing continuous delivery system, establishing a team with experts in different areas and taking a gradual shift to continuous delivery by adding small features, parallelly obtaining the feedback from the team.

Integrating the features continuously will shorten the cycle of release and improves the quality of the software. However, the entire process will come to a halt when there is break in the automated build in continuous integration. Analyzing the build failures, finding the reason for the build failures and quickly fixing the build break is a prior need of continuous integration system.

2. Background and related work

The IT organizations are trying to release small sets of features through a series of release trains instead of big features. The smaller change sets results in lesser complexity. Frequent updation without compromising the quality, results in high customer satisfaction. The collaboration between the development team and the operation team obtained through the DevOps practice helps to create a positive environment.

Table 1. Lists the continuous delivery features and the stakeholders' gain

Continuous Delivery Features	Stakeholders gain
Frequent release of features.	Observe continuous improvement
Continuous integration of small fragment of code	Make the changes smaller, thereby less likely to break build.
Continuous testing through automated testing tools.	Lesser bugs at the production environment
Automated build and deployment.	Ensures repeatable, reliable process and saves a lot of time.
Visibility of build status to the entire team	Increased visibility and transparency.
Faster feedback loop	Improvement in the quality of application

The usage of toolchain supporting the continuous integration and continuous deployment practice obtained from 2020 DevOps trends survey.[13]

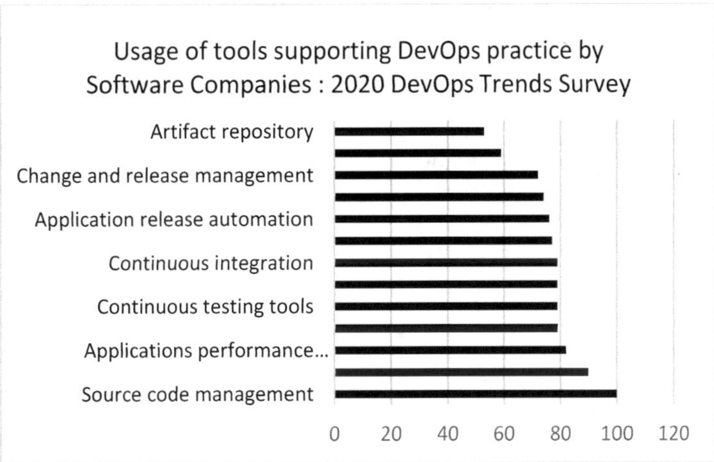

Figure 1. Usage of tools supporting DevOps practice

Continuous integration involves the process of build automation and code testing every time when a team member commits the changes to the version control. A build breakage can occur due to many factors. To focus our study, we have constructed the following concepts.

- Build failures have a significant impact in the continuous integration environment.
- Build failures increase the cycle time of DevOps.
- Quick responds to the build failures increase the efficiency of continuous integration environment.

2.1 Previous related work

The adoption of DevOps practice has a significant impact on the software quality. There are set of features that contributes to the quality attributes. Automation, collaboration and fast feedback are analysed as the key features to improve software quality. [4]

The anti-patterns or poor practices that affects the performance of continuous integration system is analysed. The implemented method automatically detects anti-patterns. A metric which assess the performance of the continuous integration is established. The practices that improve the performance of continuous integration and the practices that degrades the performance of continuous integration system is identified through the metric.[5]

Build Abstraction and Recovery Tool [BART] was created to analyse the build log, extract the keywords and group the keywords into categories as testing, compilation, code quality etc. BART generates the information that is relevant or that cause the build failure from the build log. It identifies the reason for failure and searches for similar errors on developers' forums. It provides developers with proposed solutions. Summaries generated with BART increases the understandability of build log.[6]

A method to reduce the cycle time of DevOps by optimizing the test, finding out the test that covers maximum fault detection and minimum run time execution was proposed. Although there are many tests in consideration in a continuous integration, the work is done with integration testing.[7]

Reinforcement learning method to prioritize the test cases in a continuous integration environment was proposed. Test execution results are also used for learning test case prioritization. The method analyses the important indicators for failing test cases. [8]

Test automation plays an integral part in the continuous development and continuous delivery of features. There are several factors that needs to be considered in test automation in a DevOps environment. In addition to the factors such as selection and integration of test tools, the factors including team effort and agile-oriented process helps in the improvement of test automation process.[9]

Continuous testing is the process of running any type of automated test case as quickly as possible in order to provide rapid feedback to the developers and detecting crucial issues before going to production.[10]

3. Methodology

3.1 Research goal and questions

Continuous integration tool Jenkins is selected for the experimental study. The goal of the study was to analyse the developer supporting plugins in Jenkins. These developer assistance tools help to analyse the build failures thereby facilitates the faster recovery of build failure. This process in turn improves the efficiency of continuous integration. Previous related works [5] deals with the working of BART to summarise the build log for easier understandability. This paper analyses the working of test result analyser plugin in Jenkins. Thus, we propose the following research questions:

- What are the categories of build failures?
- Does the build break that results from test failures have a significant impact on continuous integration?
- Does the automated test result analyser improve the understandability of test failures?
- Does the report obtained from the automated system helps to overcome the future issues related to continuous testing?

3.2 Proposed Model

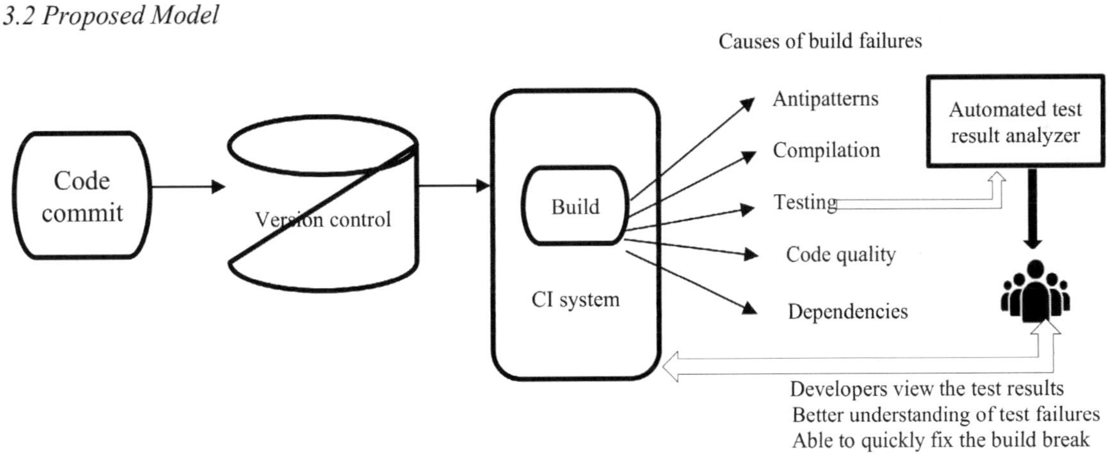

Figure 2. Build history of the build failure due to testing shown by the automated test result analyzer

3.3 Implementation

To perform the experimental study, we use the Jenkins 2.249.2 version [11]. Jenkins is a continuous integration tool that facilitates continuous development, continuous testing and continuous deployment of newly created codes. The entire software development process is accelerated trough Jenkins. Any code changes in the places like GitHub, automatically triggers the continuous integration process. The code is build, quality of the code is analyzed and the code is tested and it is deployed to production environment. Jenkins offers more than 1700 plugins. The software pre-requisites for Jenkins are Java 7 or above. We used Java 8.A maven project [12] with automated test cases is selected for the study. Maven is an automation tool for build that is used primarily for Java projects.

Table 2. Maven build lifecycle

Maven build lifecycle phases	Functions
validate	The project is validated to check whether it is correct and all information's are available
compile	Source code of the project is compiled.
test	Compiled source code is tested using unit testing framework.
package	Package the compiled code in a distributable format.
verify	Check whether the package is valid and met the quality criteria
install	Package is installed into the local repository
deploy	Final package is copied to the remote repository.

IOP Conf. Series: Materials Science and Engineering 1085 (2021) 012025 doi:10.1088/1757-899X/1085/1/012025

Table 3. Software configurations for the study

Plugins installed in Jenkins for performing the study	Version
Git	4.3.0
Maven Integration Plugin	3.8
Test Result Analyzer Plugin	0.3.5

Software's installed in the local system	Version
Java	1.8.0_271
Apache Maven	3.6.3

System Configuration	Windows 10, 64-bit, 4 GB RAM

We performed the required goal and options "**compile test**" in the continuous integration system for the maven project. The required post build action "Aggregate downstream test results" is also executed to obtain the test report.

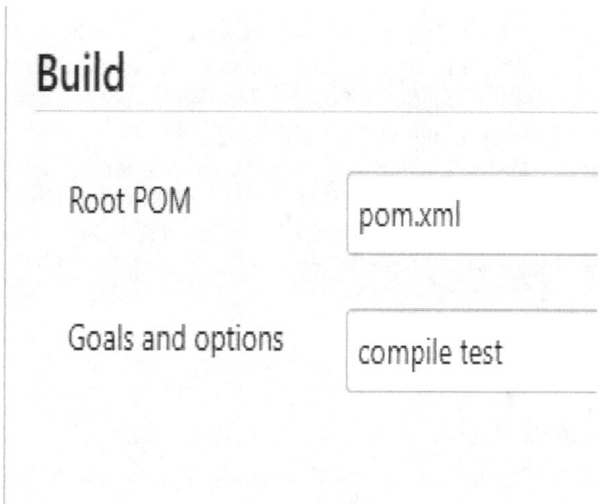

Figure 3. Goals and options

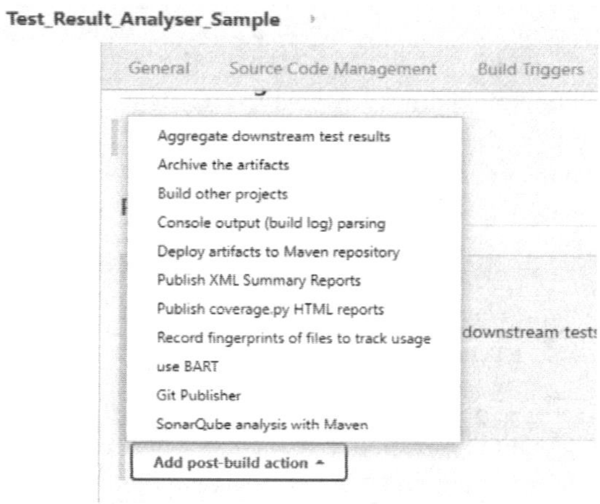

Figure 4. Post build action

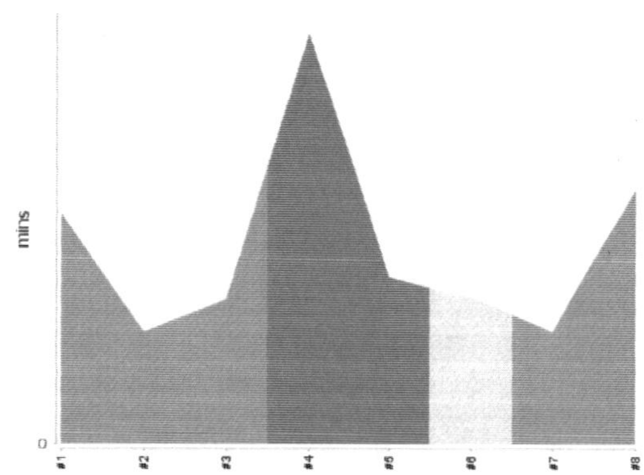

Figure 5. Build Trend

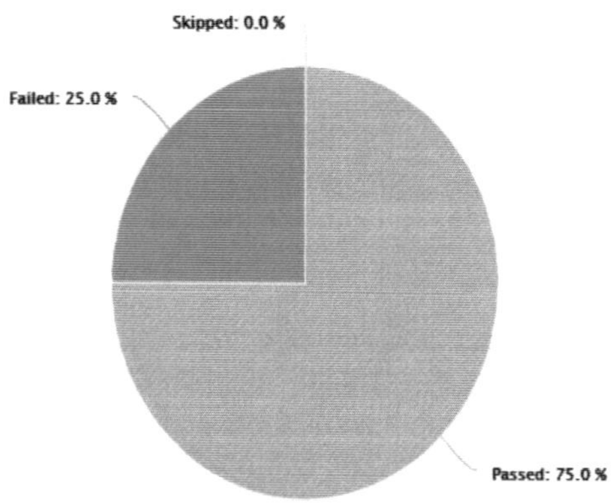

Figure 6. Build status

3.4 Results and Discussions

The obtained test results which are in a tabular tree format helps and assist the users to categorize the results as passed, failed and skipped. We obtain a graphical report of our test execution. The version of the Test Result Analyzer Plugin which we have used to perform the study (version 0.3.5) allow the users to specify the number of builds that are needed to be fetched by the plugin for the reporting purpose. This test results gives a better understanding about the test failures which in turn helps the stakeholders to quickly fix the failure. Continuous testing is an important phase in a continuous delivery process and test failure is one among the major cause of build failure. The cycle time of the continuous integration and delivery process is significantly increased if there is delay in fixing the build. If the developer's understandability in test results can be improved, then we can significantly reduce the cycle time of continuous delivery process.

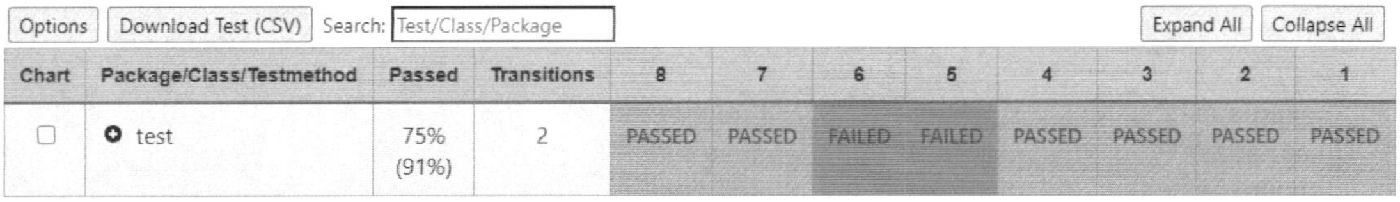

Figure 7. Test Passed and Failed

4. Conclusion

In the continuous integration, the process of continuous testing is followed and when we run a series of test executions, it is a difficult task for the users to move to the each and every build report and analyze the result. The automated test analyzer tool solves the issue by presenting a tabular tree format of the test results, which helps the developers and testers to identify which all tests have passed, which all tests have failed and which all tests have been skipped.

5. Future work

We have discussed the causes of build failures. From the different causes of build failures, we selected and studied the test failures which is one among the cause of build breakage. We worked with a Maven project and obtained the above results. As a future work the test results obtained from the test result analyzer can be further processed to identify the anti-patterns that affects the efficiency of continuous integration environment.

6. References

[1] Lwakatare L E, Kilamo T, Karvonen T, Sauvola T, Heikkilä V, Itkonen J, Kuvaja P, Mikkonen T, Oivo M and Lassenius C 2019 DevOps in practice: a multiple case study of five companies *Elsevier* **114** 217-230

[2] Laukkanen E, Itkonen J and Lasssenius C 2017 Problems causes and solutions when adopting continuous delivery-a systematic literature review *Elsevier* **82** 55-79

[3] Chen L 2017 Continuous delivery: overcoming adoption challenges *Elsevier* 128 72-86

[4] Mishra A and Otaiwi Z 2020 DevOps and software quality: a systematic mapping *Elsevier* **38**

[5] Vassallo C, Proksch S, Gall H C and Penta M Di 2019 Automated reporting of anti-patterns and decay in continuous integration *Int conf on automated software engineering* (San Diego, CA, USA) IEEE

[6] Vassallo C, Proksch S, Zemp T and Gall H C 2019 Every build you break: developer- oriented assistance for build failure resolution *Springer* **25** 2218-2257

[7] Marijan D, Liaaen M and Sen S 2018 DevOps improvements for reduced cycle times with integrated test optimizations for continuous integration *Annual computer software and applications conf* (Tokyo, Japan) IEEE **01**

[8] Spieker H, Gotlieb A, Marijan D and Mossige M 2017 Reinforcement learning for automatic test case prioritization and selection incontinuous integration *Proc of 26th Int Symp on*

software testing and analysis (Santa Barbara, CA, USA) ACM 12-22
[9] Wang Y, Pyh¨aj¨arvi M and M¨antyl¨a M V 2020 Test automation process improvement in a DevOps team: experience report *Int conf on software testing, verification and validation workshops* (Porto, Portugal) IEEE
[10] Mascheroni M A and Irrazábal E 2018 Continuous testing and solutions for testing problems in continuous delivery: a systematic literature review *ISSN* **22** 1009-1038
[11] https://www.jenkins.io/doc/book/installing/
[12] https://maven.apache.org/
[13] https://www.atlassian.com/whitepapers/devops-survey-2020

Energy consumption monitoring in smart home system

R S Hariharan[1], Reema Agarwal[2], Madhurya Kandamuru[3] and H Abdul Gaffar[4]

[1,2,3,4]School of Computer Science and Engineering, VIT University, Vellore, Tamil Nadu, India

Email: haran8861@gmail.com, reemasmu1398@gmail.com

Abstract: Energy Consumption, especially electricity consumption, is one of the serious problems that we are facing in today's world. There is a need for an efficient system to monitor this energy consumption. Internet of Things opens a way to solve these problems by interconnecting hardware, software and cloud. Therefore, we propose an energy consumption monitoring system for home appliances which can be used to calculate the energy consumption of the household and to keep the user informed about the electricity consumption through an android app where he can view the units of electricity used and a prediction of the bill at the end of the month. The system can also be incorporated with the features of controlling the energy consumed by the appliances as desired by the user and the app will notify if the electricity consumption exceeds a threshold value set by the user.

1. Introduction

Electric meters installed in houses take a reading at the end of every month to determine the monthly bill based on the units of electricity consumed by each household. The energy consumption meter calculates the amount of energy consumed by various electrical devices and appliances. The meter counts the number of kilowatt-hours delivered and the annual electricity consumption is measured directly on the meter. The billing unit is typically kWh- kilowatt hour, which is used to take readings at the end of each billing period from the electric meters installed in a household or workplace premises.

But what if we can monitor electricity usage and be notified by a mobile app when the electricity bill for a given month is above a certain threshold? The electricity bill for a household is measured by calculating the electricity consumed in a month. Households belonging to the middle or low-class face difficulties to minimize and keep track of the electricity consumption and end up paying a higher bill amounts due to very few extra units consumed by that household that could have been conserved easily. As a result, an effort was made to propose a solution by taking readings of all devices in the house and predicting the bill of that month by calculating the amount of the bill of consumption of electricity each day until the current date to help people realize how much electricity they consume each day and how to reduce it to keep consumption below a certain threshold. Implementation of IoT in an energy monitoring system can transform the way industries, companies and individuals consume and control electric devices and household appliances and IoT can offer a solution to optimize power consumption.

Once a system such as a house or an office building or a facility is equipped with this IoT technology, it incorporates the particular space with the features of energy monitoring and control overconsumption. Among them are Periodical statistics on power consumption in each day, month and year with the additional functionality of monitoring each appliance in the house/workspace through a mobile app. Prediction of electricity bill by the end of the month on an analysis of data i.e., units consumed in that particular month till the current date and giving an estimate of the bill for the amount of consumption.

2. Literature review

Chupong et al. proposed a home energy monitoring system that uses an application which can forecast the user's electricity bill. The proposed system uses an Application Programming Interface (API) which allows the application to fetch stored data from the hardware unit, predict the user's electricity bill and send a report to the user via email every day. This system uses only the last 7 days energy consumption to predict the electricity consumption and the effect of energy saving with larger sampling size is unknown [1]. Maheepala D C et al. addressed the issues of energy auditing and proposed a system which monitors energy consumption in real time, analyses the measured data, finds out the problems and provides technical solutions to it. The system also generates reports and warns its users in case of any emergency [2].

Another method proposed by Mohammed Abo Zahhad et al. consists of a sensor nodes network in which the nodes are deployed in different locations which will measure the power consumption and send the readings through a multi-hop network to the EMU that has a graphical user interface to send control signals to nodes during peak loads and show runtime data. It is integrated with WSN and the energy management unit records the energy consumption, manages it and predicts ensuing energy consumption [3]. Chooruang and Meekul proposed a system which comprises energy monitoring nodes that make use of Peacefair PZEM-004T, a low-cost energy meter, SD3004 chip and microcontroller for measuring voltage, current, power and transferring that data as JSON via MQTT protocol. This data is interpreted and the user gets an energy consumption pattern [4].

Energy Monitoring System design based on IoT was designed by Luan and Leng for energy conservation by improving energy management levels. The EMS along with IoT can collect, transmit and save a huge amount of data in energy running processes by making use of a database, communication networks, etc. The execution of the system has 3 levels namely perception, transport and application layer. This proposed IoT system has some advantages over traditional methods like strong reliability and practicability [5]. Jinsoo Han et al. designed a smart HEMS architecture using ZigBee and PLC-based renewable energy gateway which considers both energy consumption and generation. The home server collects data on energy consumption and performs analysis for energy estimation and controls the home energy usage schedule [6]. Al-Ali A R et al. proposed a system in which all of the devices in a house are interfaced with an IoT object with a unique IP address in a data acquisition module which results in a large mesh wireless network of devices. A prototype with a smart approach to monitor, control and cut the energy cost has been developed and tested [7].

Woong Hee Kim et al. proposed an energy management system based on a wireless sensor network i.e., ZigBee and an intelligent home gateway where it senses, records and updates electricity data to provide real-time electricity consumption monitoring. The method makes use of three technical components such as information acquisition, information processing and information presenting. The system allows users to detect and control electrical appliances using web and mobile interfaces [8]. Smart energy monitoring using IoT and GSM-based Cloud computing was proposed by Govindarajan

R et al. which makes use of a two-way communication protocol between metering infrastructure and the consumer end. To alert about power utilities, the multiple energy conservation live report is sent to the user [9]. Devadhanishini A Y et al. proposed the idea of a Smart Power monitoring system by integrating Arduino, WIFI and GSM Short Message Service (SMS). The system includes the design of a smart power meter that reduces power consumption and ensures energy saving with a motion sensor which automatically sends signals to cut the power supply when there is no sign of humans in the place [10].

Rishabh Jain et al. proposed an IoT-based energy meter reading system that consists of fault detection and SMS alert to the user. The method makes use of EEPROM as a flash storage memory and consists of Arduino board, buzzer, 16*2 LCD, ESP 8266 Wi-Fi module and power supply. The user can deal with their vitality utilization by knowing the vitality from time to time using this system [11]. Ganesh Shirsat K et al. designed a cost-effective IoT-based energy monitoring system with a low-fee PZEM-004T, the use of non-invasive CT sensors, SD3004 electric powered electricity dimension chip and ESP8266. This system is used as a power billing device and strength monitoring appliance [12]. Collins K et al. addresses the challenges faced by industries in maintaining energy cost and comes up with an energy monitoring and management system which identifies the individual machine's energy consumption, processes the collected data to a fuzzy inference system that helps to reduce energy costs by providing feedback to a graphical user interface [13]. Sarmad Sohaib et al. proposed a low-cost design using ADE7753 energy meter IC, ATmega8 microcontroller, current sensor, etc. for real-time energy-consumption monitoring and control to reduce energy wastage. Compared to the commercially deployed energy meters, their system has high accuracy with a low average error rate of 0.66% in power measurement and led to reduction in average power utilization of about 12% [14].

3. Proposed Methodology

Recently in many places, the government has come up with strict electricity bill calculation procedures to avoid people from overusing and wasting electricity. According to these procedures, the cost per unit of electricity increases with increasing total energy consumption in a month. Due to this, the public needs a product that could help them to monitor and keep the number of kilowatts consumed in a month under control. This product must be able to predict and tell them the details of the expected bill based on their usage from the beginning of the month so that they can control their energy usage and get benefited from it.

The main focus is to develop a system that monitors the energy consumed by all appliances to predict the electricity consumed in a month by a household. Internet of Things (IoT) makes it easy for us to connect hardware and software components of the system using Cloud and Database. A system has been proposed in which the hardware component such as current sensor sends the electricity consumed by all appliances in the household to the cloud database. An android app is made which gets the stored energy consumption data for the household from day one of the month till the current date from the database. It runs a test using the consumption data for predicting the electricity bill of a month using a machine learning regression model algorithm and calculates the expected bill amount to be paid at the end. The application also provides the user with many other features like visualizations to compare the energy consumption in the house day-wise using a bar graph and controlling the appliances to turn them on/off within room distance.

3.1 System Architecture and Block Diagram

The ACS712 current sensor measures the units of electricity flowing through an AC appliance. This data sensed by ACS712 is collected in the Things peak channel through either HTTP or web socket protocol via Node MCU. The Android application sends a GET/POST request to retrieve the data from the database for further processing. The received data is then visualized using a bar plot corresponding to the dates and it is given as input to the regression model that predicts the electricity units with high accuracy and precision. Then, the app calculates the monthly bill of the household based on energy consumption from the beginning of the month to the current date. It takes the energy consumed by appliances into account, which helps to monitor and reduce the daily usage to conserve electricity.

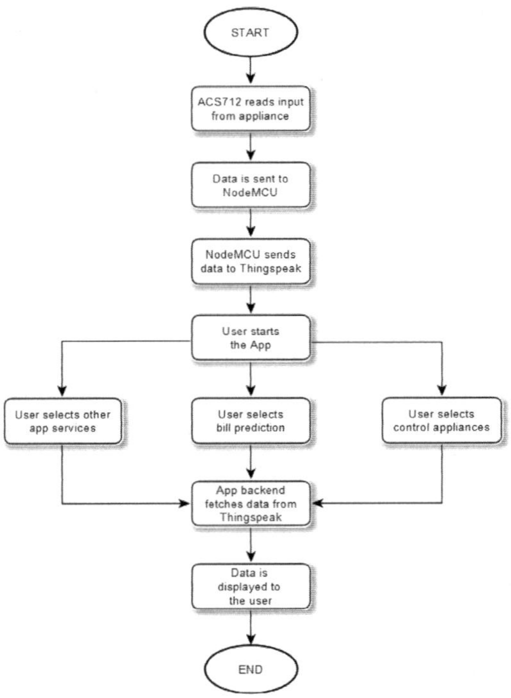

Figure 1. Flowchart for the energy consumption monitoring system.

Figure 1 explains the working flow of the entire energy monitoring system. ACS712 sends measured current to Node MCU, which sends the data to the cloud and then, the app retrieves this data to give corresponding output as requested by the user. The user has three choices: to view the bill prediction, to control appliances and to check the energy consumption. According to the requirement of the user, the data is fetched from the cloud and displayed to the user.

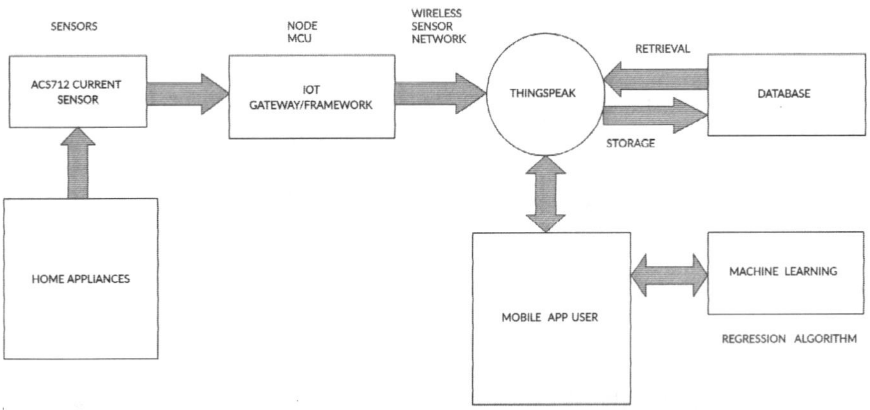

Figure 2. System Architecture of the Energy monitoring system for Smart Home Appliances.

Figure 2 shows the System Architecture of the Energy monitoring system. Current sensors measure the amount of current flow in every appliance such as LED bulb, fan, tube-light, etc., and sends the data to the cloud (Thingspeak) via IOT gateway i.e., Node MCU. Things peak maintains the record of electricity consumed by all household appliances each day in the database. The data is retrieved when the user uses the app and enter his/her choice to either get the bill prediction or control energy consumption. The system makes use of a machine learning regression model to predict the expected monthly bill accurately.

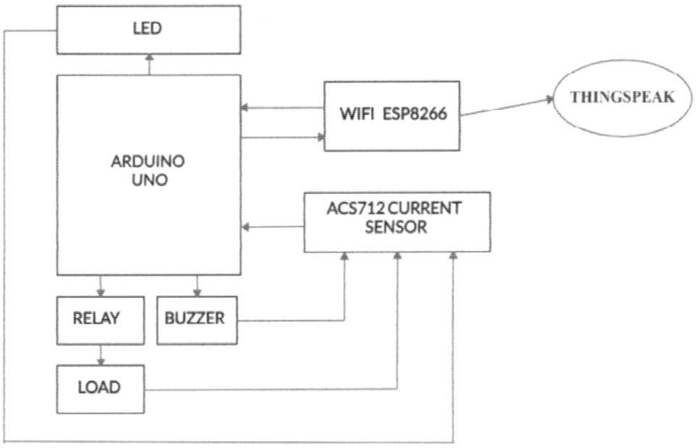

Figure 3. Block Diagram for Energy monitoring system for Smart Home Appliances.

Figure 3 shows the block diagram for the energy monitoring system. Hardware Components in the circuit include a buzzer, motor and LED which act as the electrical appliances in the prototype implementation. Node MCU connects Arduino to the internet and endures connection between cloud and hardware circuit. The voltage regulator is used to convert 12DC to 5DC voltage and the relay module is used to convert DC to AC current to make the load work. Load is any device which needs electricity to run. The current sensor measures the current flow in each appliance. From Arduino, current flows through the buzzer, motor, LED and the ACS712 current sensors find the current passing through these devices respectively and send the value to the Arduino board. Node MCU receives this value from Arduino and sends it to Thingspeak Cloud.

3.2 Hardware Components

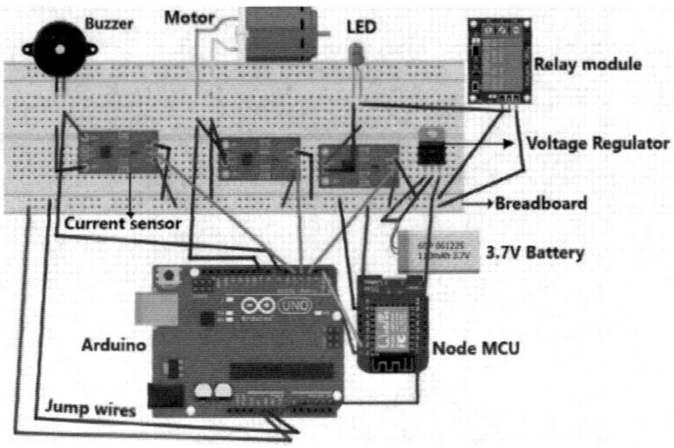

Figure 4. Hardware Circuit Connection for the Energy monitoring system.

Figure 4 shows the model hardware circuit of a simple energy monitoring system that functions the same as the proposed system for Smart Home. Hardware Components include buzzer, motor and LED in the circuit which acts as the electrical appliances in the prototype implementation. Node MCU connects

Arduino to the internet and endures connection between cloud and hardware circuit. The voltage regulator is used to convert 12DC to 5DC voltage and the relay module is used to convert DC to AC current to make the motor work. The current sensor measures the amount of current flow in each appliance. Jump wires are used to connect all the components to make a complete circuit.

Components used in the circuit:

- Arduino Uno
- ESP12/NodeMCU
- ACS712-30Amp Current sensor
- AC Appliances
- Jumping Wires
- Voltage regulator
- Relay module
- Battery and Breadboard

The Arduino UNO which is an open-source microcontroller board with input/output pins plays an important part in making the circuit by interfacing with all other components like NodeMCU, Buzzer, Current Sensor, etc. It needs to be powered from the computer through a USB cable for functioning. Sometimes, a voltage regulator is required for regulating the current entering the Arduino board. Arduino has the input/output pins which establish a connection with other components of the circuit through Jumping Wires. Arduino code needs to be run in the computer to start processing the data from sensors in the circuit. A reset button present on the board will restart the Arduino code running on the computer. Thus, this component plays a vital role in the development of this prototype. Node MCU is connected directly to Arduino in the circuit. From Arduino, current goes to the buzzer, motor, LED and the ACS712 current sensors read the current passing through these devices respectively and sends the value to the Arduino board. Node MCU receives this value from Arduino and sends it to Thingspeak Cloud. The ACS712 current sensor can be used to detect both AC and DC using the Indirect sensing method based on the Hall effect which explains the potential difference across a conductor when the magnetic field is applied perpendicular to the direction of current flow.

3.3 Software Components

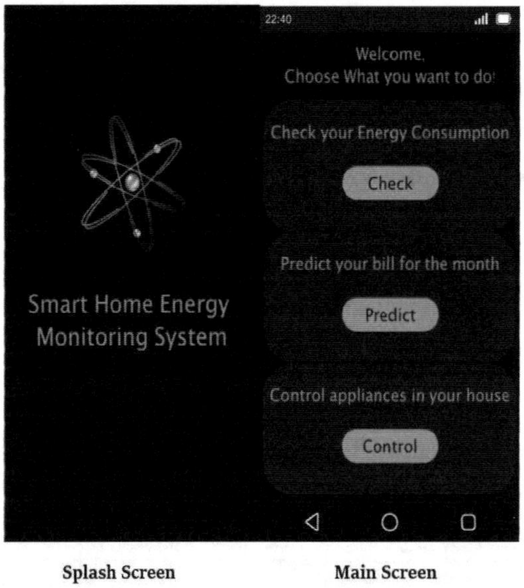

Figure 5. Splash Screen and Main Page Screen of the Android App.

IOP Conf. Series: Materials Science and Engineering 1085 (2021) 012026 doi:10.1088/1757-899X/1085/1/012026

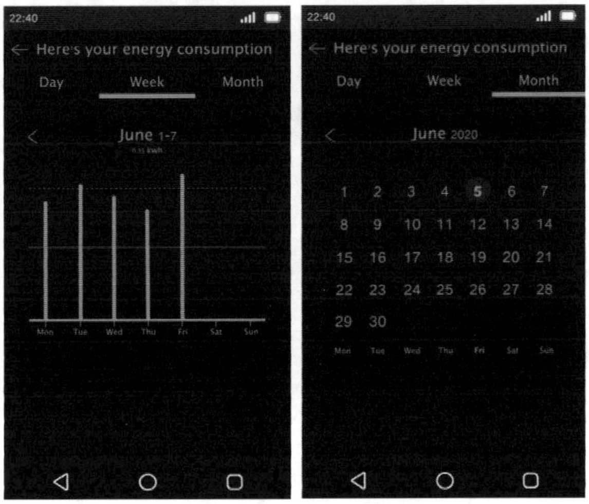

Daily, Weekly and monthly consumption of Energy

Figure 6. Daily, Weekly, and Monthly Energy Consumption Page of the Android App.

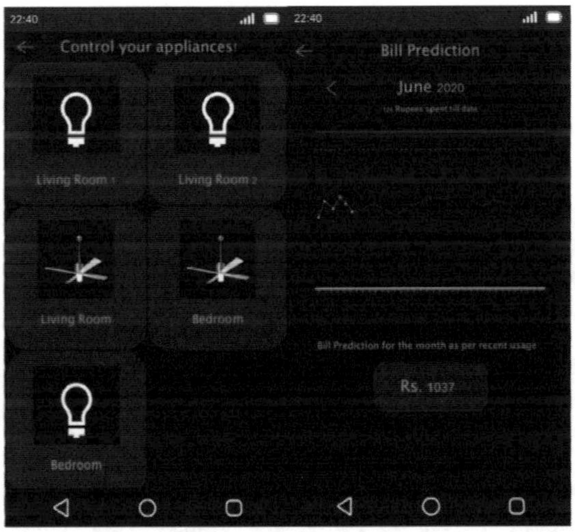

Controlling Appliances in the house hold, and Monthly bill prediction based on amount of electricity used till date

Figure 7. Bill Prediction and Smart Home Control Page of the Android App.

Figure 5, Figure 6, and Figure 7 show the user interface of an android application developed for the prototype. Figure 5 shows the splash screen and the main page of the application. From the main page, the user can choose the services of the app and experience it. Figure 6 shows the energy consumption service in the application and has an example bar graph to illustrate the interface of the application and depicts how the end users can monitor energy consumption on each day and total energy consumption in a particular week and month. Likewise, Figure 7 shows the bill prediction and the features of controlling the appliances in the app. The graph in Figure 7 (right) illustrates an example line graph and expected predicted amount of bill on each day of the month which depends on the usage of electricity till the current date. When the user opens the application, the splash screen appears for few seconds and directs him to the main page where he can choose to check energy consumption, make bill prediction or control the electrical appliances in his home. The Energy consumption page shows the visualization of energy consumed by his household from the beginning of the month till the current day plotted against days in a bar graph. This helps the user to understand his energy consumption properly and to keep track of the expense. The Bill prediction page gets the energy consumption data and previous month dataset to run the machine learning regression model to predict the expected monthly bill more accurately and precisely. The control page gives the user a Smart Home experience and he can turn on/off the appliances in his house connected to the system.

3.4 Cloud database and Regression model

The Thingspeak cloud database gets the energy consumption data from NodeMCU and maintains the record of electricity used by all household appliances each day. It also maintains a database with records of the previous month-end bill, total energy consumed in kilowatt, and number of days in that month. A Multiple linear regression model is trained with the energy consumption for n days with the number of days(n) as input and equivalent bill amount for the month as output. This type of regression model uses two or more independent variables to predict the output dependent variable. This technique helps to determine the variation of the model and the relative contribution of each independent variable in determining the output feature. The formula used by this regression algorithm is

$$y=b0+b1*x+b2*x2+b3*x3+..+bn*xn \quad (1)$$

where y is the dependent variable, x1,x2,x3,..xn are independent variable and b1,b2,b3...bn are constants.

Then the trained model is used to predict the total bill amount since day one of the month and the current day of the month. The dataset instances are split into (X_train, y_train) and (X_test, y_test) which are used for training and testing the machine learning model respectively. On running the testing phase, the model can be evaluated by using Root mean squared error, Mean Squared Error, Coefficient of Determination(R^2), Adjusted R^2, or Mean Absolute Error. The Thingspeak cloud database also maintains the record of all required user details during the login of an application corresponding to their energy consumption details. Hence using the Internet of Things (IoT), the hardware circuit, software application, and cloud database are integrated and linked together for the functioning of the entire prototype.

The data from the sensors are stored in the cloud database on daily basis. The multiple linear regression model will be trained based on the previous month's reports and predict the equivalent bill amount for the month daily by considering the energy consumption data from the beginning of the month until the current day consumption. In this way, sufficient data is formulated to train the machine learning model.

A User Registration channel in Thingspeak holds all the value of user details when they sign up in the mobile application. Here phone number is verified and is used as a unique primary key to recognize the user. A Bill prediction channel in Thingspeak that holds dataset with columns of date, kilowatts, expected bill, and actual bill. This column is the input in the machine learning model training phase. In the backend, Thingspeak uses MATLAB to plot this bar graph and update it every day from the starting of the month.

4. Novelty of the system

The system helps the user to predict the monthly bill amount to keep consumption of energy under control and save money. The visualization of the electricity usage and expected monthly bill against the days helps the user to understand and to keep track of the rate of energy consumption properly and find out if they are exceeding their threshold or not. Through the android application, the user can control all the electrical appliances connected to the system wherever he is inside the house. All these features are the main advantages of the proposed Energy Monitoring System for Smart Home.

The proposed method is cost-effective since it makes use of an application installed in a mobile device to generate the bill and predict the amount of energy consumed. The unique feature of this system is that it provides three different services in one android application i.e., showing the amount of energy consumed in the app, generating the bill and also providing the feature to control the energy consumption through the app. This system aims to build up sustainable energy IOT based system operating at low cost, maximum efficiency and accuracy in the context of smart homes. It uses a multiple linear regression algorithm to predict the amount of energy consumed with higher accuracy and precision. Thus, the proposed system incorporates the advantageous features of the existing systems mentioned in Section 2 along with their limitations and stands unique from them.

5. Conclusion

After understanding the growth and increased demand for electricity, if energy consumption is not controlled and is wasted it will bring a great impact on our lives soon. To sustain it for the future and to effectively control it and make people aware, an attempt has been made to develop a viable system of 'Energy Consumption Monitoring in Smart Home System' that solves the consumption problem faced in Households. The proposed system is used to calculate the energy consumption of the household and even make the energy monitoring to be handy. The mobile application provides an interface to the end-user to check daily electricity consumption retrieving data from Thingspeak cloud database and the regression model predicts the bill amount by the end of the month based on the consumption of electricity on each day till the current day and gives an estimated amount so that the users can be aware of how much they are spending on electricity. Hence it reduces the wastage of energy and brings awareness among its users. Thereby, it can be concluded that the proposed system is an efficient energy consumption proposition made with the hardware components Arduino, ACS12 current sensor, electrical appliances and Thingspeak as the cloud database from which an approach for larger systems can be drawn i.e., collecting sensor data and monitoring usage with a mobile application and prediction of the monthly bill using a regression model to conserve energy and to minimize electricity consumption.

6. Future work

There is a need for an app that is self-optimized i.e., without the actual need of the user, the app can itself control the energy consumption which can be done based on a threshold value set by the user. According to daily consumption of electricity/energy in the house, the user can set a limit/threshold value, and based on the same limit set by the user, the app can reduce the energy consumption by controlling the current flow and reducing the intensity of bulb or tube-light, speed of a fan, etc. This will thus eliminate the need of the user to control/monitor the consumption manually when the user is not using the application. This is one of the additional features that can be brought in the future research.

7. References

[1] Chupong C and Plangklang B 2017 Electricity bill forecasting application by home energy monitoring system *Int. Electrical Engineering Congress* (Pattaya: IEEE) pp 1-4

[2] Maheepala D C, Nayanajith R M N, Somarathna M W R P, Bandara R A A M and Hemapala K T M U 2018 Designing an Energy Monitoring, Analysing and Solution Providing System for Energy Auditing *4th Int. Conf. on Advances in Electrical, Electronics, Information, Communication and Bio-Informatics* (Chennai: IEEE) pp 1-5

[3] Mohammed A Z, Sabah M A, Mohammed F, Mohammed F A Ahmed and Abdelhay Ali 2015 Design and implementation of building energy monitoring and management system based on wireless sensor networks *10th Int. Conf. on Computer Engineering & Systems* (Cairo: IEEE) pp 230-233

[4] Chooruang K and Meekul K 2018 Design of an IoT Energy Monitoring System *16th Int. Conf. on ICT and Knowledge Engineering* (Bangkok: IEEE) pp 1-4

[5] Luan H and Leng J 2016 Design of energy monitoring system based on IOT *Chinese Control and Decision Conf. (CCDC)* (Yinchuan: IEEE) pp 6785-88

[6] Jinsoo H, Chang-Sic C, Wan-Ki P, Ilwoo L and Sang-Ha K 2014 Smart home management system including renewable energy based on Zigbee and PLC *IEEE Transactions on Consumer Electronics* **60** 198-202

[7] Al-Ali A R, Mohammed R, Imran, AliKarar M and Gupta R 2017 A Smart Home Energy Management System Using IoT and Big Data Analytics Approach *IEEE Transactions on Consumer Electronics* **63** 426-434

[8] Woong Hee K, Sunyoung L and Jongwoon H 2011 Real-time Energy Monitoring and Controlling System based on ZigBee Sensor Networks *2011 Int. Symp. on Intelligent Systems Techniques for Ad hoc and Wireless Sensor Networks (IST-AWSN), Procedia Computer Science* **5** 794-97

[9] Govindarajan R, Meikandasivam S and Vijayakumar D 2019 Cloud Computing Based Smart Energy Monitoring System *Int. Journal of Scientific & Technology Research* **8** 886-90

[10] Devadhanishini A Y, Malasri R K, Nandinipriya N, Subashini V and Padma G 2019 Smart Power Monitoring System Using IoT *5th Int. Conf. on Advanced Computing & Communication Systems (ICACCS)* (Coimbatore: IEEE) pp 813-16

[11] Jain R, Gupta S, Mahajan Cand Chauhan A 2019 IOT based Smart Energy Meter Monitoring and Controlling System *Int. Journal Of Research In Electronics And Computer Engineering IJRECE* **7** 1600-04

[12] Ganesh K S, Nilesh U B, Urvesh Y G and Suraj A J 2020 IOT Based Energy Monitoring System for Energy Conservation *Int. Research Journal of Engineering and Technology (IRJET)* **7** 758-63

[13] Collins K, Mallick M, Volpe and Morsi W G 2012 Smart energy monitoring and management system for industrial applications *IEEE Electrical Power and Energy Conf.* (London: IEEE) pp 92-97

[14] Sohaib S, Sarwar I, Muhammad Haseed I, and Mahmood A 2016 A low cost smart energy monitoring and control system for smart building *5th IET Int. Conf. on Renewable Power Generation (RPG)* (London: IET) pp 1-5

Optimize Continuous Integration and Continuous Deployment in Azure DevOps for a controlled Microsoft .NET environment using different techniques and practices

Dileepkumar S R[1]**, Juby Mathew**[2]

[1] Research Scholar, Lincoln University College (LUC),Malaysia
dileedil@gmail.com

[2] Faculty of Computer Science and Multimedia,Lincoln University College,Malaysia
jubymp@gmail.com

Abstract. Delivering a software is not a simple process, even if it is a completely new software created from the scratch, the enhancement of an existing software or fixing the bugs of an existing application. As per the new industrial standard most of the companies follow the DevOps methodologies to deliver the software products, which leverage the flexibilities of the software delivery with expected quality on time with minimal errors. One of the most important factors of the DevOps methodology is Continuous Integration and Continuous Deployment. As the entire software industry is moving towards Cloud Computing one of the most powerful, cost effective and easily maintainable environment to host the software applications. There are various Cloud computing providers available in the industry now. Most of them have its own advantage and dis-advantages. There are many researches going on various topics on these areas and the importance of these topics are getting more relevant in the industry now. There are lot of gaps on the Continuous Integration and Continuous Deployment concepts and more improvement options have to be identified and certified throughout the various researches. To understand the importance of this platform, we decided to concentrate the research on the Continuous Integration and Continuous Deployment in Azure DevOps for a controlled Microsoft .NET environment using different techniques and practices. Here our intention is to identify the various factors which helps the industry to do a better Continuous Integration and Continuous Deployment in the Azure DevOps for a controlled Microsoft .NET environment using different techniques and practices. Among the "n" number of factors identified on this subject, only couple of factors are considered as part of this research.

1. Introduction

.NET Framework is a platform developed by Microsoft, which helps to develop and run any type of applications/software. It is a programming model that provides run-time execution environment that manages various applications that target .NET Framework. .NET Framework provides memory management and other system services through Common Language Runtime (CLR). It also helps programmers enable to take advantage of robust, reliable code for all major areas of application development with the support of extensive class library. The .NET Framework also supports multiple programming languages. While creating service based applications that may be a soap based webservices, restful API, microservices or gRPC services in .NET or .NET core framework, the hosting of these applications could be fully controlled by Microsoft environment which is related to Web Application Servers (IIS, Kestrel, etc) and also as a service in Windows Service. Applications can be hosted independently outside of the .NET environment and fully integrated and operated in .NET framework. It can also be in cloud or on-premise environment. The area being concentrated is more on Azure cloud environment.[1]

DevOps is a set of practice that integrate software development and IT operation teams by working together across the entire software development life cycle. DevOps contains various tools that increases an organization's ability to deliver applications and services at a faster pace with the support of shorten development life cycle and continuous integration and deployment. It follows Agile methodology. Microsoft introduced Azure DevOps, Software as a service (SaaS) platform that follows the Agile methodology. Azure DevOps helps organizations to manage end-to-end software development and deployment process effectively and easily with the help of leading tools on the market.[2] Azure DevOps is good option for composing the DevOps tool chain.

A CI/CD (Continuous Integration Continuous Deployment) Pipeline is the new culture or practice which industry follows to eliminate the gap between development and operations teams by implementing automating the build process, quality control/testing and finally deployment process of the applications. It provides the smaller code changes options.CI/CD pipeline implementation is considered as the backbone of the DevOps environment in the modern industry, based on the set of principles and culture it follows. The collection of practices in CI/CD pipeline leverages the Development teams to deliver code changes more frequently and reliably. In software engineering, CI/CD is enforcing the organizations to introduce automation in building, testing and deployment of applications and bridges the gaps between operation team and development team and also their activities during the SDLC life cycle. The basic workflow of the CI/CD pipeline is represented in Fig.1.

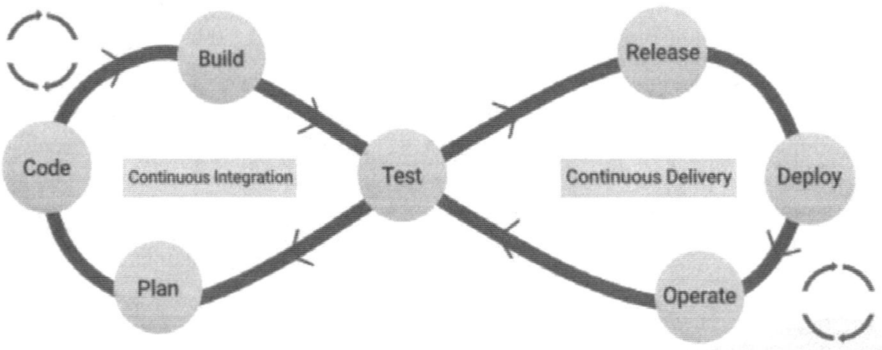

Fig 1. CI/CD basic workflow representation

2.Methodoloy

The review work of this paper accommodates the following phases for a systematic review of the literature. Fig.2 presents a specific set of steps followed in the management of this review paper.

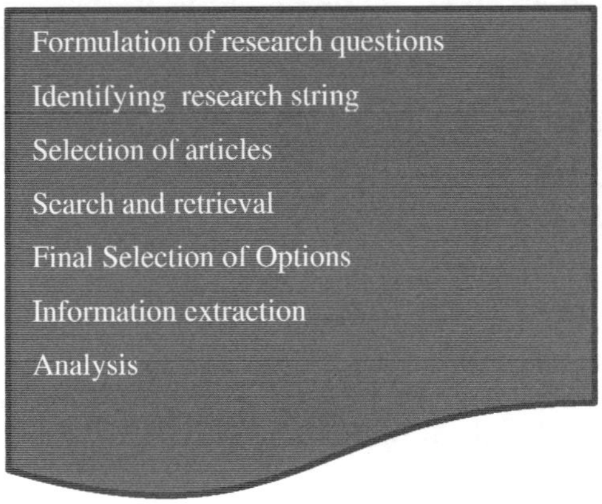

Fig 2. Structure of the review process

2.1 *Problem Statement*

Identifies the major impact of the above-mentioned drawbacks, corns, issues, pitfalls, difficulties and challenges in the respective Cost Based, Administration and Configuration, Architectural, Development and Technical and Quality Assurance, Manual Testing, Testing Automation and Testing Coverage areas while implementing the Continuous Integration and Continuous Development in Azure DevOps for a controlled Microsoft .NET environment.

2.2 Hypotheses

Suggesting the best techniques, policies and tools which will help to resolve the couple of above-mentioned drawbacks, corns, issues, pitfalls, difficulties and challenges in the couple of respective Cost Based, Administration and Configuration, Architectural, Development and Technical and Quality Assurance, Manual Testing, Testing Automation and Testing Coverage areas while implementing the Continuous Integration and Continuous Development in Azure DevOps for a controlled Microsoft .NET environment and optimize the Continuous Integration and Continuous Development in Azure DevOps for a controlled Microsoft .NET environment. The main research questions are based on the below mentioned respective areas. Cost Based, Administration and Configuration, Architectural, Development and Technical and Quality Assurance, Manual Testing and Testing Automation and Testing Coverage.

2.3 Research questions and Research String

The main purpose of this research is to comprehend the options to be considered to optimize Continuous Integration and Continuous Deployment in Azure DevOps for a controlled Microsoft .NET environment and how they are implemented and evaluated and in what ways or aspects they could be experimented with. To this end, we defined the following research questions:

Q1 What are the most relevant studies about CI/CD in Azure DevOps?
Q2 Which are the possible techiniques to be used in CI/CD in Azure DevOps?
Q3 What are the problems and challenges faced in CI/CD in Azure DevOps?
Q4 How CI/CD in Azure DevOps are evaluated?
Q5 Which are the different types of applications to be considered?
Q6 Which directions are most promising for future research?

Based on the above mentioned questions, our research focus points are based on the below mentioned policies.

1) Find out the best tools and infrastructure (based on the cost, performance and easiness of use) to implement the CI/CD pipeline
2) How to improve the coding standard and avoid poor coding related issues.
3) How to make sure all the environments (Development, QA, UAT, Performance and Production) are completely independent and loosely coupled on CI/CD pipeline.
4) How implement and configure a well-connected alert system for the Application exceptions and CI/CD exceptions.
5) How to implement a creative and user-friendly dashboard with respective data matrix.

Continuous integration (CI) and continuous delivery (CD) is a culture, collection of operating principles, and set of practices that support application development teams to deliver code changes recurrently and constantly. In an, organization the software engineering team members can integrate their work with cumulative frequency. Continuous delivery (CD) is mainly handling the packaging and deployment that CI covers on the build and test. The overall CI/CD workflow is represented in Fig.3.

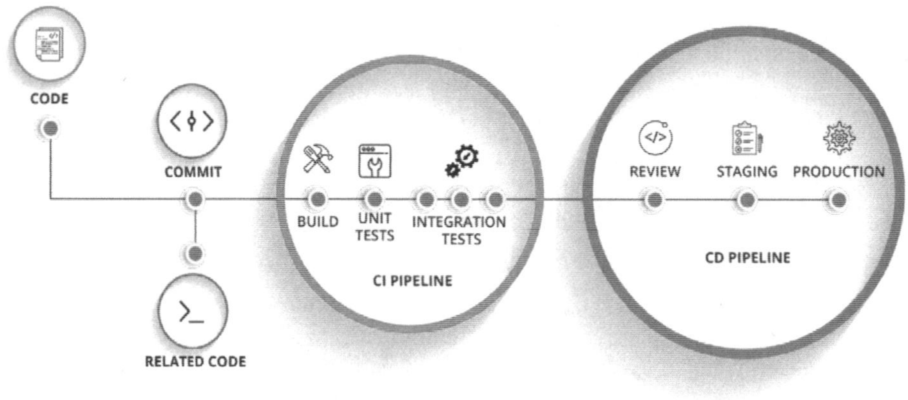

Fig 3. overall CI/CD workflow

The representation of an existing Continuous Integration and Continuous Deployment in Azure DevOps for a controlled Microsoft .NET environment can be found in Fig.4.

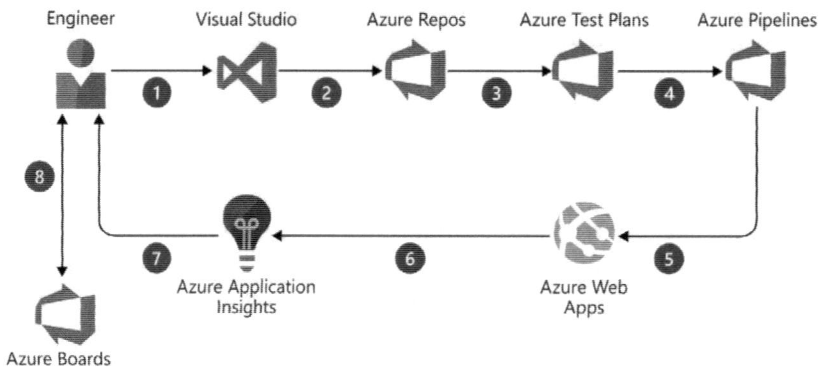

Fig 4. CI/CD in Azure DevOps for a controlled .NET environment

Methods to follow as part of the Research

The need to analyse the existing CI/CD pipelines if any to find out the optimize solutions are mentioned as following. Analyse the existing Infrastructure used for CI/CD pipeline and it's cost involved. Analyse the existing coding standard approaches and its drawbacks based on the existing code review comments repositories. Analyse the various environment established and how the CI/CD pipeline configured and how it coupled each other's. Extract the current matrix and dashboard mechanism and find out the improvement gaps. Analyse the alert mechanism and find out the drawbacks and gaps. Analyse alert data, alert source and prioritise alerts by multiple factors. Analyse test results obtained from unit test, integration test and UAT results in CI system by applying test analytics. The Recommendation Weightage of the options are represented in Fig.5.

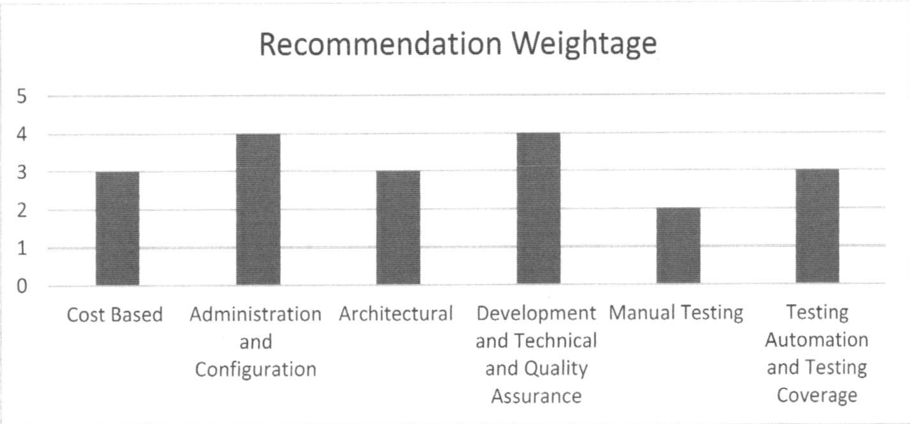

Fig 5. Recommendation Weightage

Most of the IT projects in the industry are developed or developing using any of the Microsoft products. .NET framework and the Azure Cloud platform are wildly using in the industry now a days. .NET is the development platform which helps to create any type of applications such as user interface, services, APIs and microservices, etc. Azure platform is mainly using for hosting the applications, backends and the respective deployments. The percentage of expecting optimization benefit chart is represented in Fig.6.

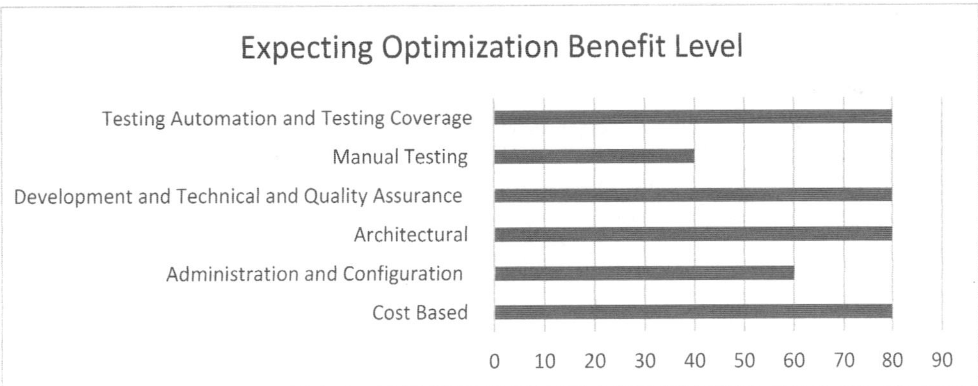

Fig 6. Expecting Optimization Benefits

Sampling, Data Collection Methods and Data Analysis Methods

Data sampling is a technique for the statistical analysis, which is used to select, manipulate and analyse a representative subset of data points to verify and identify trends and patterns in the larger data set being examined. The Data collections is planning to gather from the various application development team who use the Azure Cloud and Microsoft .NET platform to develop the software applications. Tools like Azure Application Insight, Rivery and Azure CI/CD dashboard are used to analyse the huge data to extract the statistical analysis data, used for the research.

3. Literature review

As this Continuous Integration and Continuous Deployment is new technique and introduced recently in the software industries the researches on these subjects are on going processes. As per the information collected from the various sectors the scope of the improvements in this area is wildly open. There are lot of existing drawbacks, corns, issues, pitfalls, difficulties and challenges that are facing in the current Continuous Integration and Continuous Deployment pipeline approaches.[3] There are various options that can be delivered as part of the various researches with problem solving statements.

As part of the research study, Optimize Continuous Integration and Continuous Deployment in Azure DevOps for a controlled Microsoft .NET environment using different techniques and practices, the research is focused on Azure DevOps for a controlled Microsoft .NET environment. The outcomes may can apply to any of the Cloud DevOps methodologies and other non .NET Microsoft environment and also a completely non-Microsoft environment with slight changes in the entire approaches and outcomes.In the cost based approaches, as per the studies of Forrester Research, Inc, 38 percent of infrastructure decision-makers of the organizations that implemented DevOps and their continuous deployment and release automation efforts in 2018, saw revenue growth of 10 percent or more from the prior year.[4] But the same time 25 percent of those that had not adopted DevOps reported only limited comparable growth. The output of the Continuous integration and continuous delivery implementations are the agility that can give them the competitive advantage to win, with higher confidence and faster feedback.[5]

The Administration and Configurations practices are mainly focusing on the tools which can be used to leverage the CI/CD implementations. CI/CD tools can be mainly categorized as Repository and Version Control tools, Build tools, Automation tools, Test Automation tools, and Monitoring tools. Managing the changes and Version Management are defined in the Repository and Version Control Tools. Git has become more popular and widely used version controlling system with CI/CD pipeline due to the open source and distributed architecture support [6], [7], [8], It also supports TFS too. The outstanding available tools for the CI/CD are Harness Jenkins, TeamCity, Bamboo, Buddy, GitLab CI, CircleCI and TravisCI, etc.

As per the research paper [9], an application's Architecture can be a key barrier. To better understand the characteristics that make an application amenable to CD, they compared the architectural characteristics of the applications that were easily moved to CD with those that were difficult to move to CD [10], [11]. Characteristics, in the form of Architecturally Significant Requirements (ASRs) [12], was identified [10]. Of those, Deploy ability and Modify ability often shape architectural decisions.

Development and Technical and Quality Assurance techniques is an another practice can be followed. In the early nineties, Booch initially advocated [13] and introduced Continuous Integration (CI) which is a software development practice as part of the extreme programming practices [14], foreseeing frequent builds integrations per day, then incorporated with an automated build machinery often hosted on a separate server. CI servers favors an early discover of integration errors without leaving the burden to developers, by the automated verification process. Over the years, researchers have investigated the CI adoption[11], and studied specific build phases, namely compilation [12] and testing. Besides the usefulness of automated compilation and testing, software projects can greatly benefit the execution of Automated Static Code Analysis Tools (ASCATs) within CI. For example, build breakages due to ASCATs checks can be used to force developers to follow coding standards/guidelines or to make sure source code is kept conform with the intended architecture as the project evolves. Automation testing tools also helps the testing team to run the regression testing faster and accurate. Sonarqube is leading tools for the .NET framework.

The most least considerable option is the Manual Testing. The CI/CD pipeline always pushes for continuous development, testing, and deployment, it is often inundated with manual testing. Most important issues with manual testing are the delay per iteration, which starts accruing and pushing the deployment back. QA team has to spend more time to do the application testing introducing painfully slow releases that defeat the main purpose of CI/CD and the manual testing is unable to keep up with the dynamic requirements. It is always necessary to run multiple tests depending on the objective of test cases and plans. Someone must identify test cases to conduct these tests, so the manual testing makes process slow. Concussively, the test cycles call for separate test environments that teams will have to build manually, upgrade, and tear down. The effort that goes into mimicking the end-user environment may require multiple permutations and combinations that the team will have to identify, build, and update.

Manual testing is a failing battle in CI/CD so need to elaborate clear advantages of CI/CD pipeline automation testing. Automation testing and Testing Coverage techniques is faster and more reactive feedback loops that continuously test the codes and share feedback within a duration of a few minutes, so it witnesses rapid acceleration. The detection of test procedures, room for parallel testing capabilities and automated cross-browser concurrent testing are useful features which will improve test coverage and reduce the testing time. The testing teams can set up test environments by using the automatic provisioning in just a few clicks! Testing team not even worried about the latest versions of operating systems and browsers because automation tool takes care of those. So, the teams valuable time for recreating the various environments can be easily reduced. Automation testing offers high scalability than manual testing, so development team can always meet the quality and security requirements. While manual testing can be reserved for specific testing types and modules, such as exploratory testing, it is best to automate testing for seamless integration of QA in the CI/CD pipeline. The table below (Table 1) describes the various techniques or areas used to optimize Continuous Integration and Continuous Development in Azure DevOps for a controlled Microsoft .NET environment.

Table 1. Comparison of Techniques/Area

Techniques /Area	Involvement Criteria	Advantages	Disadvantages
Cost Based	Analysis of existing Infrastructure used for CI/CD pipeline.	Realiable Cost Effective Options.	Resistance forms current custodians. Complete changes in the tools involved. More Training Requirements.
Administration and Configuration	Analysis of existing Administration and Configuration Options.	Esay maintenance and Least configuration efforts	Resistance forms current custodians. More Training Requirements. Less optimization expectation.
Architectural	Overall syetem and implementation architectural.	The most effective optimization options.	Resistance forms current custodians. Technical chanllenges. Resolution Duration. Cost Involvement.

Techniques /Area	**Involvement Criteria**	**Advantages**	**Disadvantages**
Development and Technical and Quality Assurance	SDLC Life cycle	The most effective optimization options. Effortless review process. Easy to incorporate with any type of source controls.	Re iterating SDLC life cycle. Complete changes in the tools involved. Initial Cost Involvement.
Manual Testing	Quality Assurance Team	Highly robest functional error free application	More Resource involvement. More Training Requirements. Time consuming process. Not helps on application implemenations.
Testing Automation and Testing Coverage	Quality Assurance Team and Development Team	Easy turn around for regression testing. Effective for frequent changes.	Time consuming process for the initial stages. Initial Cost Involvement.

4. Discussion

4.1 Techniques and Challenges

CI/CD pipeline will provide users with personalized and targeted suggestions [38]. Therefore, they address the problem of the regular implementations constrains in the digital era. Different tools have recently been developed, which can incorporate the process to find optimized CI/CD pipeline in a controlled .NET environment.

4.2 Evaluation matrices

The most prominent evaluation metrics in the research literature measure the accuracy of the system's predictions. Accuracy is measured based on the various techniques and options which are considered during the reserch.

4.3 Expected Contribution

Understand patterns identified drawbacks, corns, issues, pitfalls, difficulties and challenges from the respective area and improve the efficiency, accuracy, easiness and cost-effective implementation of the Continuous Integration and Continuous Development in Azure DevOps for a controlled .NET environment. Provide the helpful matrix to make the effective and cost effective industrial standard recommendations in the early stage of the implementation. How to design an effective alert or runtime communication mechanism with dashboard data representation. Best recommendation for the code analysis and code review process and the techniques and practices which help to do the proper test coverage for the manual and automated test cases.

5. Conclusion

Came up with various suggestions and techniques together with best practices and policies needed to follow to setup an Optimize Continuous Integration and Continuous Deployment in Azure DevOps for a controlled Microsoft .NET environment using different techniques and practices. In this paper, we presented possibilities of optimizing the Continuous Integration and Continuous Deployment in Azure DevOps for a controlled Microsoft .NET environment using different techniques and practices. We presented different options, methodologies, techniques and practice appropriate for this study. Looking through articles, available tools, CI/CD practices, and available methods, we can prove that lot of possibilities can be considered in optimizing the Continuous Integration and Continuous Deployment in Azure DevOps for a controlled Microsoft .NET environment using different techniques and practices. As part of this study our proposed solutions are mainly focusing on Cost Based, Administration and Configuration, Architectural, Development and Technical and Quality Assurance, Manual Testing and Testing Automation and Testing Coverage. Each option has it's own capabilities to leverages the CI/CD pipeline implementation in a Azure DevOps for a controlled Microsoft .NET environment, which will help to reduce the cost, effort, duration, process and manual interventions through out the SDLC life cycle of an application development and it's maintenance. The benefits will vary based on the options, techniques, practices and tools which are considering during the process. In order to succeed in the Continuous Integration and Continuous Deployment in Azure DevOps for a controlled Microsoft .NET environment using different techniques and practices, it is important to focus on the changing of the existing business, operational culture and to get the buy-in support from the executive team. Identifying the right approaches, which include the techniques, practice and tools that fit for the business needs. Leveraging our research, future work will be focused on any one/more of mentioned specific techniques, practice or tools to optimize the Continuous Integration and Continuous Deployment in Azure DevOps for a controlled Microsoft .NET environment effectively.

References:

[1] Mahesh Kuruba, Prashant Shenava and Jenny James 2018 Real-time DevOps Analytics in Practice *Preprint QuASoQ Workshop*

[2] Krusche S and Alperowitz L 2014 *In Proc. of ICSE'14. IEEE on Introduction of Continuous Delivery in Multi-Customer Project Courses*

[3] Continuous delivery vs. continuous deployment, "Practicing Continuous Integration and Continuous Delivery on Aure," Available online: (accessed on 15 June 2020).

[4] Rahman A, Helms E, Williams L, and Parnin C 2015 *In Proc. Agile Conf. (AGILE) on Synthesizing continuous deployment practices used in software development* pp 1-10

[5] Shahin M, Ali Babar M and Zhu L 2017 *Continuous Integration, Delivery and Deployment: A Systematic Review on Approaches, Tools, Challenges and Practices* IEEE Access vol 5 pp 3909-3943

[6] "Adopting Continuous Delivery" in Continuous Delivery Whitepaper. [Online] Available: Levi9 IT Services, 2016 [Accessed: Sept, 2017].

[7] Feitelson D G, Frachtenburg E, and Beck K L 2013 July – August *Development and Deployment at Facebook in IEEE Internet Computing* vol. 17 pp 8- 17

[8] Chen L 2015 *Towards Architecting for Continuous Delivery* in Software Architecture (WICSA) 12th Working IEEE/IFIP Conf. pp 131-134

[9] Chen L 2018 *Microservices: Architecting for Continuous Delivery and DevOps* in IEEE International Conf. on Software Architecture (Seattle: USA, IEEE).

[10] Chen L, Ali Babar M, and Nuseibeh B 2013 *Characterizing Architecturally Significant*

Requirements Software IEEE vol. 30 pp 38-45

[11] Ayewah Nand Pugh W 2010 *The google findbugs fixit in Proceedings of the 19th international Symp. on Software testing and analysis* ACM pp 241–252

[12] Beller M, Gousios G, and Zaidman A 2016 *Oops, my tests brokethe build: An analysis of travis CI builds with GitHub* (*PeerJPrePrints* vol. 4 [Online] Available: http://dx.doi.org/10.7287/peerj.preprints.1984v1)

[13] Hilton M, Tunnell T, Huang K, Marinov D, and Dig D 2016 *Usage,costs, and benefits of continuous integration in open-source projects in Proceedings of the 31st IEEE/ACM International Conf.* on Automated Software Engineering (ASE) pp 426–437

[14] Kim S, Park S, Yun J, and Lee Y 2008 *Automated continuous integrationof component-based software: An industrial experience in Proc. Int. Conf.* Automated Software Engineering (ASE) IEEE pp 423–426.

[15] Laukkanen E I, Paasivaara M, and Arvonen T 2015 *Stakeholder percep-tions of the adoption of continuous integration - A case study in AgileConf. (AGILE)* pp 11–20

Heart disease prediction system using Correlation Based Feature Selection with Multilayer Perceptron approach.

Ambily Merlin Kuruvilla[1] **and Dr.N.VBalaji**[2]
[1] Research Scholar, Department of CS,CA and IT, Karpagam Academy of Higher Education
[2]Dean, Faculty of Arts, Science and Humanitics, Karpagam Academy of Higher Education
E-mail: E-mail:ambilykuruvilla@gmail.com

Abstract. Cardiac disease prediction helps physicians to make accurate recommendations on the treatment of the patients. The use of machine learning (ML) is one of the solution for recognising heart disease-related symptoms. The goal of this study is to suggest a methodology for identifying the most relevant features of cardiac disease characteristics by applying a feature selection technique. The data set used in this study was Framingham heart disease dataset (FHS). It was collected from KAGGLE Machine Learning repository. There are 16 attributes and a mark in the dataset that has been validated by four ML classifiers. There are two feature selection methods, Correlation Based Feature selection (CBFS) and Principle Component Analysis (PCA) was used for the comparison in the study. By using CBFS Method five highly correlated features are selected for the study, and by using PCA thirteen features are selected. The experimental result shows that Correlation Based Feature Selection with Multilayer perceptron (CBFS with MLP) obtained the highest accuracy for this dataset.

1. Introduction

The research concentrates on the two feature selection methods for data reduction before building the predictive models by classification algorithms. These reduced features are then passed into the classification algorithms to design the models for the heart disease prediction. These models are used for the comparison of accuracy of the classifier. Principle Component Analysis and Correlation Based feature selection methods are used for finding out the reduced features. The selected features are inputted to four different classifiers such as Navie Bayes, ADABOOST, MLP and SMO. The accuracy of each model is compared with the other.

2. Background Study

Devansh Shah studied various attributes related to heart disease[1]. The study was conducted with Naïve Bayes, decision tree, K-nearest neighbor, and random forest algorithms[1]. The experimental result proves that K-nearest neighbor algorithm exhibits the highest accuracy.

Hamidreza Ashrafi Esfahani[2] formulated a model to predict cardiovascular disease. The model includes decision trees, Neural Networks, Rough set, Naïve Bayes and SVM for implementation. On comparing the results achieved, it was revealed that the hybrid model of Rough Set, Naïve Bayes and Neural Network obtained the highest accuracy. An ensemble strategy was implemented that allowed for the output to be combined that would result in better accuracy. The performance of the classifiers was compared with the parameters such as its precision, sensitivity, accuracy and F-Measure.

3. Proposed Methodology

Framingham Heart Study (FHS) from Kaggle Machine Learning repository is used for the study. Two feature selection methods along with four classification algorithms are used for the study. CBFS and PCA are the methods used for the dimensionality reduction. MLP, Navie Bayes, Sequential Minimum Optimiser (SMO) and ADABOOST algorithms are used for classification. The reduced feature set from both the feature selection methods are inputted to different classifiers. Eight different Machine Learning Models were created for Heart Disease Prediction. Accuracy of these Models are compared with each other.

4. Description of the Data Set

The Framingham Heart Study (FHS) dataset was collected from Kaggle. The dataset consists of 4241 records. It contain sixteen features including AGE, PREVALENT HYP, SYSBP, DIABP, GLUCOSE, SEX, EDUCATION, CURRENT SMOKER, CIGSPERDAY, BPMEDS, PREVALENT STROKE,BMI, HEART RATE, DIABETES, TOTCHOL and PREDICTOR VARIABLE.

5. Classification Algorithms

In Machine Learning various forms of classification techniques are available. Classification techniques used for this study was described below.

6. Multilayer Perceptron (MLP)

MLP is a subset of Artificial Neural Network. MLP comprises one or more than one hidden layers aside from one input and one output plate. The Perceptron is made of an input layer and a totally linked output layer. MLPs have the same levels of input and output, but could have several levels concealed within them.

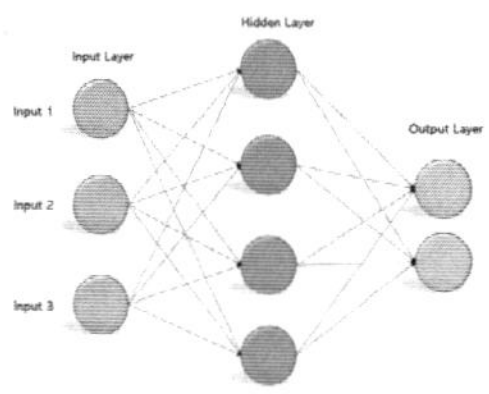

Figure 1. Different layers in MLP.

7. Adaboost

In machine learning, AdaBoost(Adaptive Boosting) is a supervised learning algorithm. It is used for combining several weak classifiers together to generate a strong classifier.

8. Naive Bayes

Naive Bayes is a Machine Learning algorithm based on Probability theory in statistics. The term naive suggests that the elements that go through the software are autonomous of each other,That is, the value of one characteristic, does not explicitly influence or alter the value of any of the other characteristics used in the algorithm. The Bayes theorem tells us how we can compute the conditional probability. The equation for conditional probability is,

P(Ai/Bi) =(P(Bi/Ai) * P(Ai)) /P(Bi)

P(Ai/Bi) defines the probability of an event Ai occurs corresponding to the event Bi has occurred.

P (Bi / Ai) is the conditional probability and it defines the probability of occurrence of event Bi corresponding to the occurrence of the event Ai.

P(Ai) and P(Bi) defines the probability of the events Ai and Bi occurs.

9. Sequential Minimal Optimization (SMO)

The sequential minimal optimization is more effective to solve the SVM problem compared to traditional Quadratic Programming algorithms such as the interior-point method. The SMO algorithm can be viewed as a method of decomposition by which a problem of optimization of multiple variables is decomposed into a set of sub problems, each optimizing an objective feature of a limited number of variables, usually only one, whereas all other variables are treated as constants which remain unchanged in the sub problem.

10. Feature Selection

During feature selection the most relevant features are extracted from the data set. Redundancy can be avoided using this method. Since irrelevant features are excluded from the input data, feature selection can increase the accuracy of prediction. In this study Correlation Based Feature Selection (CBFS) and Principle Component Analysis(PCA) is used for feature selection. After feature selection the reduced data set is applied to four different classification Algorithm.

11. Correlation Based Feature Selection(CBFS)

Correlation values are calculated by CBFS. The five highly correlated features are selected for the analysis. These features are given as the inputs for the classifiers.

Table 1. Features selected for the analysis by CBFS along with the correlation values.

S/N	Selected Features	Correlated Values
1	AGE	0.2254
2	PREVALENTHYP	0.2164
3	AGE	0.2254
4	PREVALENTHYP	0.2164
5	PREVALENTHYP	0.2164

12. Principal Component Analysis (PCA)

Thirteen features were selected by PCA during feature selection. The features selected by the PCA algorithm are AGE, PREVALENTHYP, SYSBP, DIABP, DIABETES, SEX, BPMEDS, TOTCHOL, PREVALENTSTROKE, CIGSPERDAY, EDUCATION, BMI, CURRENT SMOKER.

13. Result and Discussion

In the study two feature selection methods are used for comparison - Principle Component Analysis (PCA) and Correlation Based Feature Selection (CBFS). After dimensionality reduction the reduced dataset is applied to four different classification Algorithm such as

Multilayer Perceptron (MLP), AdaBoost, Navie Bayes and SMO. Five most correlated features were selected and applied to Classifiers in CBFS Method. Thirteen features were selected by Principle Component Analysis (PCA). The result is shown in Table4. From the results it is found out that Correlation Based Feature Selection (CBFS) along with MLP algorithm shows maximum accuracy.

Table 2. Comparison of Predictive accuracy of Models.

S/N	Alogrithm	CBFS	PCA	Before FS
1	MLP	84.9057	83.9151	84.1509
2	ADABOOST	84.8113	84.8821	84.8821
3	Navie Bayes	81.1792	80.0472	80.0472
4	SMO	84.8113	84.8113	84.8113

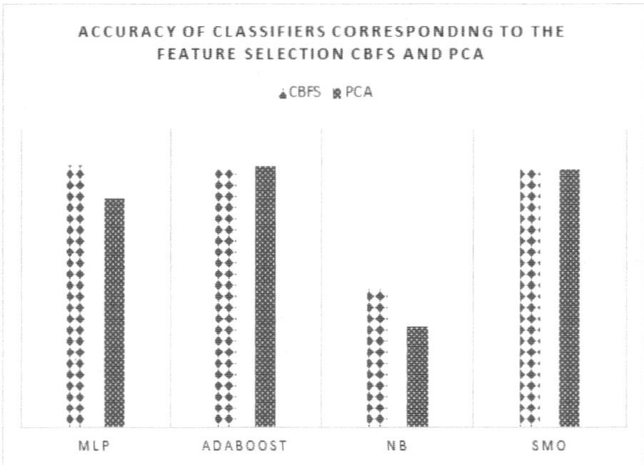

Figure 2. Comparison of Accuracy of classifiers corresponding to CBFS and PCA Feature Selection Methods.

Eight Models are generated by combining two feature selection algorithms and four classifiers. They are CBFS-MLP, CBFS-ADABOOST, CBFS-NB, CBFS-SMO,PCA-MLP, PCA-ADABOOST, PCA-NB and PCA-SMO.The accuracy of various Models are shown in Table.1 The results from table proves that Correlation Based Feature Selection along with Multilayer Perceptron (CBFS-MLP) Model perform better than the other Models with the accuracy of 84.9057 Percentage.

The result from Table 2 proves that CBFS-MLP combination shows better performance than MLP classifier. Also when we are comparing CBFS-MLP and PCA-MLP models accuracy measures proves that CBFS-MLP combination shows more performance.

Table 3. Comparison of Predictive Accuracy of MLP.

MLP	Precision	Recall	F Measure	ROC
Before FS	0.790	0.842	0.798	0.671
After CBFS	0.805	0.849	0.791	0.649
After PCA	0.784	0.795	0.668	3.62

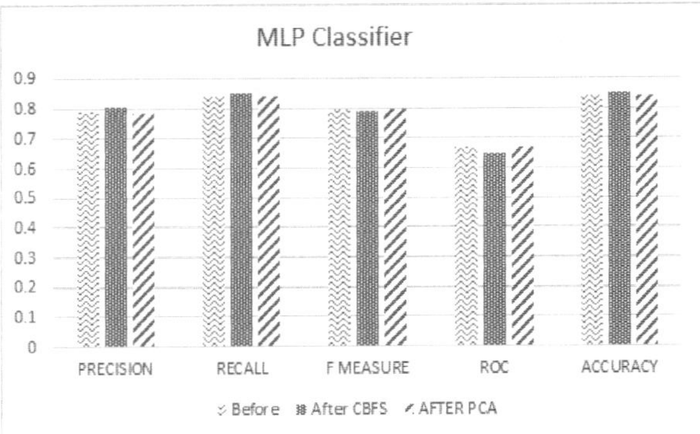

Figure 3. Comparison of Accuracy of MLP Classifier.

14. Conclusion and Findings

During the study the performance of two different feature selection methods CBFS and PCA are evaluated. Eight different classifier models are developed by combining the feature selection and classification algorithms.The performance of each model was evaluated.Performance measures such as Accuracy, Precision, Recall, F Measure and ROC are evaluated for finding out the best classifier. From the result it is proven that the model CBFS with MLP Classifier shows the maximum performance for FHS dataset.

References

[1] Devansh Shah, Samir Patel and Santosh Kumar Bharti 2020 Heart Disease Prediction using Machine Learning Techniques Springer.
[2] Hamidreza and Ashrafi Esfahani 2017 Cardiovascular disease detection using a new ensemble classifier IEEE International Conference on Knowledge-Based Engineering and Innovation (KBEI).
[3] V.V.Ramalingam 2018 Prediction of Heart Diseases Using Machine Learning International Journal of Engineering and Technology.
[4] Amin Ul Haq 2019 A Hybrid Intelligence System Frame Work for Prediction of Heart Disease Using ML, Mobile Information Systems.
[5] Shadman Nashif Heart Disease Detection by Using Machine Learning Algorithm and a Real time Cardiovascular Health Monitoring System World journal of Engineering and Technology,vol.06 no.04,2018,article id 88650.
[6] Poornima Singh 2018 Effective heart disease prediction system using data mining techniques International journal of Nano medicine.
[7] M. Gunay and T. Ensarı 2019 Predictive churn analysis with machine learning methods IEEE 26th Signal Processing and Communications Applications Conference (SIU), Izmir, 2018, pp. 1- 4.
[8] R. Suguna, M. Shyamala Devi and Rincy Merlin Mathew 2019 Customer Churn Predictive Analysis by Component Minimization using Machine Learning International Journal of Innovative Technology and Exploring Engineering.
[9] Shyamala Devi, Munisamy, Suguna Ramadass and Aparna Joshi 2020 Cultivar Prediction of Target Consumer

Class using Feature Selection with Machine Learning Classification Springer's book series "Learning and Analytics in Intelligent Systems Springer, LAIS vol. 3, pp. 604-612, 2019.
[10] Suguna Ramadass and Shyamala Devi 2019 Prediction of Customer Attrition using Feature Extraction Techniques and its Performance Assessment through dissimilar Classifiers Springer's book series Learning and Analytics in Intelligent Systems, Springer.
[11] R.Suguna, M. Shyamala Devi, Rupali Amit Bagate and Aparna Shashikant Joshi 2019 Assessment of Feature Selection for Student Academic Performance through Machine Learning Classification Journal of Statistics and Management Systems, Taylor Francis.
[12] M. Shyamala Devi, Rincy Merlin Mathew and R. Suguna 2019 Feature Snatching and Performance Analysis for Connoting the Admittance Likelihood of student using Principal Component Analysis International Journal of Recent Technology and Engineering.
[13] Debjani Panda, Ratula Ray, Azian Azamimi Abdullah and Satya Ranjan Dash 2019 Role of Feature Selection in Prediction of Heart Disease International Conference on Biomedical Engineering (ICoBE) IOP Publishing
[14] Bandari Sai Santosh, Dharma Sahith Reddy, M Sai Vardhan and Dr. Shaik Subhani 2019 Heart Disease Prediction with PCA and SRP International Journal of Engineering and Advanced Technology .
[15] Mothe Sunil Goud 2019 Heart Disease Prediction and Performance Assessment through Attribute Element Diminution using Machine Learning International Journal of Innovative Technology and Exploring Engineering.

IOP Conf. Series: Materials Science and Engineering 1085 (2021) 012029 doi:10.1088/1757-899X/1085/1/012029

A comparative study on machine learning algorithms for employee attrition prediction

P M Usha[1] and Dr N V Balaji[2]

[1] Research Scholar, Department of CA, CS & IT, Karpagam Academy of Higher Education, Coimbatore. E mail Id – pmusha.72@gmail.com
[2]Dean, Faculty of Arts, Science and Humanities, Karpagam Academy of Higher Education, Coimbatore. E mail Id – balajinv@karpagam.com

Abstract: The fourth industrial revolution introduces a wide range of technologies for an effective functioning of organizations with optimal usage of all resources including human resource. Machine learning is one of the driving technologies implemented in fourth Industrial Revolution. Machine learning can be utilised for developing models that can predict the retention or attrition of employees. The study is using machine learning algorithms like classification and clustering for preparing the prediction models. A comparison of these algorithms is done based on its performance. The performance is measured using parameters like accuracy, precision, recall, F Measure and time taken to build the model. The study is also finding the correlation between variables used in the work to the decision of staying back in the organization. The study is using the open-source tool Weka and also python for doing the same.

1. Introduction

The performance of an organization depends on the availability of adequate resources at its disposal. Human resources are considered as the most coveted resources of any organization. A replacement for a human resource is not as easy as replacement of any other resource. An organization can perform well only with right number of people with right skills to perform the right task. Stating the same reason, the attrition of employees is considered as a great menace for organizations.

The fourth industrial revolution introduces a wide range of technologies where human intelligence is emulated in performing various tasks. Machine learning is one of the driving technologies implemented in fourth Industrial Revolution. Machine Learning can develop technology to help smart industries [1]. It leads to an effective functioning of organizations with optimal usage of all resources including human resource. Machine learning can be utilised for developing models that can predict the retention or attrition of employees. Reducing the attrition rate is very important because losing an employee can be very costly for the organization as it is difficult to get an exact replacement. Selection and training of newly recruited employee can also cost heavy for the organization [2]. The study is using machine learning algorithms like classification and clustering for preparing the prediction models [3]. The algorithms used in this study are Naïve Bayes, J48, Decision Table and KMeans. A comparison of these algorithms is done based on its performance. The performance is measured using parameters like accuracy, precision, recall, F Measure and time taken to build the model. The dataset used in this study is collected in real time using a well-structured questionnaire from the employees of an organization working in IT sector. The study is also finding the correlation between variables used in the work to the decision of staying back in the organization. The study is using the open-source tool Weka [4] and also python for doing the same. Weka has got readymade tools for carrying out certain classification and clustering algorithms [5].

The organization of the paper is as follows. Next section explains methodology and background of work. Section 3 explains the results and discussion followed by a conclusion section.

2. Methodology

The researcher is using a real time data set for conducting the study. The data is collected using a well-structured questionnaire, which includes variables that influence the decision of an employee to leave the organization. Data is collected from an organization which is functioning in IT sector. The demographic features of employees like age, qualification, gender, experience, marital status, number of companies worked, qualification, number of dependents, monthly income and also the satisfaction level of employees in various aspects of organization like compensation, freedom to work, recognition given by superiors, promotion policy, growth opportunities, clarity of role, relationship with superiors, relationship with colleagues, training, work life balance, office timing and the like were collected using a well-structured questionnaire which was distributed to the employees using google form. The collected data was saved as a comma separated file (CSV).

The study is using python to understand the correlation between the variables used in the study to the target variable planstocontinue which denotes whether the employee has plans to continue in the organization for a longer period. This helps in identifying the variables that influences the decision of an employee leaving the organization the most.

Employee attrition prediction models are build using Weka and a comparison is done between the models based on accuracy, prediction, F Measure and time taken for building the model.

2.1 Correlation of variables in the data set

The dataset was given as input to a program written in python using the Anaconda Navigator and Jupyter Notebook to calculate the correlation between the listed variables and the target variable planstocontinue. Anaconda Navigator is a Python and R Distributor. It has editing tool, python interpreter and python packages. Jupyter notebook is opened from Anaconda and is used for creating and editing of python code. The correlation between variables are found by calculating the correlation coefficient using corr()function in python. An output of 1 indicates perfect positive correlation and if the output is -1 it indicates perfect negative correlation between the variables selected. If the output is near to 1 it indicates good correlation which denotes that the increase in one variable will result in increase of the other and vice versa. The correlation of the target variable and other variables were calculated [6].

2.2 Comparison of machine learning algorithms for building attrition prediction model

Organizations end up in a dangerous state of affairs when productive employees resign and leave the organization. If a machine learning model could be developed to predict the attrition of employees, necessary precautions could be taken to avoid that situation [7]. In this study, a comparison of the performance of various algorithms in constructing a model for the collected data set is executed. The models where developed using Weka which is an open-source software. Weka contains tools which can do data pre-processing, clustering, classification and the like [4]. The algorithms used for comparison in this study are Naïve Bayes, Decision table, Random Forest, J48 and K-Means. The parameters used to compare the performance of the algorithms are accuracy, precision, recall and F1 score. The following parameters are calculated using TP, TN, FP and FN. TP stands for True Positive which denotes the number of observations which are positive and also predicted as positive whereas FP stands for False Positive where the observation is negative, but is predicted as positive. TN stands for True Negative where the observation is negative and predicted as negative and FN is where the observation is positive where it was predicted as negative [7].

Accuracy stands for the ratio of correct predictions = (True Positive + True Negative) / (True Positive + True Negative+ False Positive + False Negative) (1)
Precision= The ratio of correct positive predictions to all positive predictions= TP / (TP+FP) (2)
Recall= The ratio of correct positive predictions to actual number of positive case=TP / (TP + FN) (3)
F1score = (2*Recall*Precision)/ (Recall + Precision) (4)

2.2.1 Naïve Bayes. Naïve Bayes is a classification algorithm which is dependent on Bayes Theorem. In Naïve Bayes classification we assume that each event is independent of the other [8]. Bayes theorem is based on the mathematical formula,

$$P(Y/X) = P(X/Y) * P(Y)/ P(X)$$

where X represents attribute, Y represents class, P(Y/X) represents the probability of occurrence of event Y given that event X has occurred, P(X/Y) represents the probability of occurrence of event X given that event Y has occurred, P(Y) is the probability of occurrence of event Y and P(X)denotes the probability of occurrence of X.

2.2.2 Decision Table. Decision table is a machine learning algorithm coming under the category of classification algorithms. Weka provides a tool for doing classification using decision table. A decision table is a visual representation of decisions taken under different sets of conditions. In decision table rows can represent objects, columns can represent attributes of objects and a special column can mention the decision.

2.2.3 J48. J48 is a decision tree algorithm. It is a classification algorithm. Weka has a readymade tool to perform J48 classification. In decision tree, branches are formed based on the values of the node. This helps in forming an appropriate classification. The attribute that has highest information gain and can classify the instances clearly is identified for the formation of the tree.

2.2.4 Random Forest. Random Forest is a classification algorithm which creates many decision trees. Each tree predicts from samples and the best answer is selected by way of voting. Different subsets of the features in the dataset are used for training separate decision trees. The decisions of each tree are aggregated to form the end decision. This process is called as bagging.

2.2.5 Simplekmeans. K-Means algorithm comes under the category of clustering algorithm. In k-means the centroids are randomly selected and the distance from each data point to the cluster centres is calculated. The datapoint is assigned to the cluster with minimum distance to the centre. Based on the new assignments, the centroids are again calculated. The iteration continues until there are no more changes. The algorithm will give k clusters as output.

3. Results and discussions

3.1 Correlation

After executing the corr() function in python with plans to continue as the target variable, the following results were obtained. The variable showing higher correlations are given below.

Table 1: List of fields having higher correlation with plans to continue.

Name of the field	Correlation coefficient	Name of the field	Correlation coefficient
Compensation	0.463996	Wlb	0.540620
Feedback	0.474139	Promotionpolicy	0.571525
Superiorrelation	0.455532	Officetiming	0.475550
Recognition	0.552022	Teamwork	0.508224
Growthopp	0.525373	Jobsecurity	0.579873
Training	0.459784	proud	0.653296
Opptouseskills	0.457808		

The above data shows that some variables are having a positive correlation to the decision of continuing in the organization for a longer period. For example, the variables like proud to work in organization(proud), job security(jobsecurity), promotion policy(promotionpolicy), work life balance(wlb), recognition and growth opportunity(growthopp) are having high positive correlation to the plan to continue in the organization. These results can be visually represented using a graph using the countplot() function in python. The following figures give a visual representation of the correlations present in the dataset.

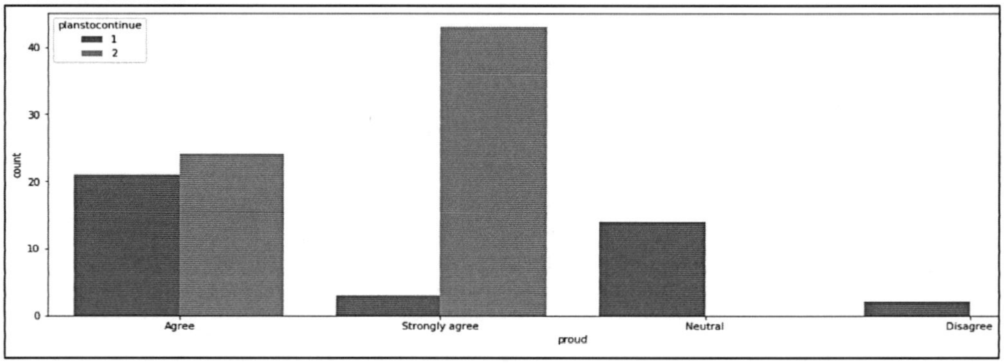

Figure 1: The chart representing the relation between the variables proud and plans to continue.

In the above figure, 1 represents "No" with a meaning that there are no plans to continue and 2 represents "Yes" which denotes that there are plans to continue. The chart above shows that majority of people who strongly agree and agree that they are proud to work in the organization have plans to continue in the organization. People who are neutral and disagree have no plans to continue in the organization.

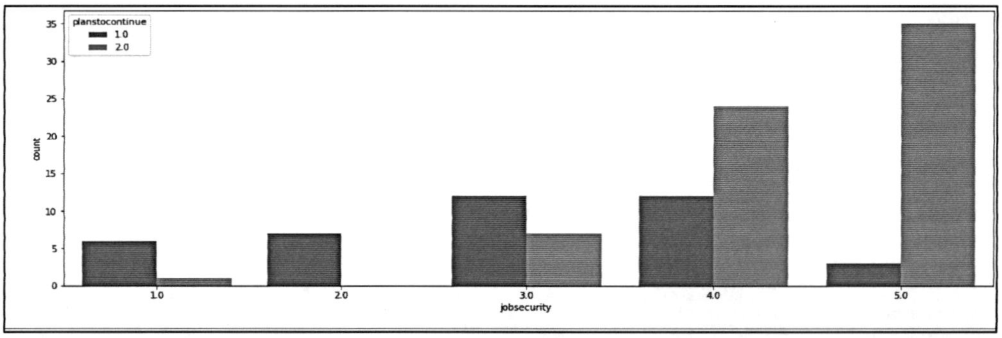

Figure 2: The chart representing the relation between the variables job security and plans to continue.

In the above figure, 1 represents "No" meaning there are no plans to continue and 2 represents "Yes" which denotes that there are plans to continue. In the x-axis 5 denotes highly satisfied, 4 denotes satisfied, 3 is neutral, 2 is dissatisfied and 1 is highly dissatisfied. The chart above shows that majority of people who are highly satisfied and satisfied with job security have plans to continue in the organization. People who are dissatisfied and highly dissatisfied have no plans to continue in the organization.

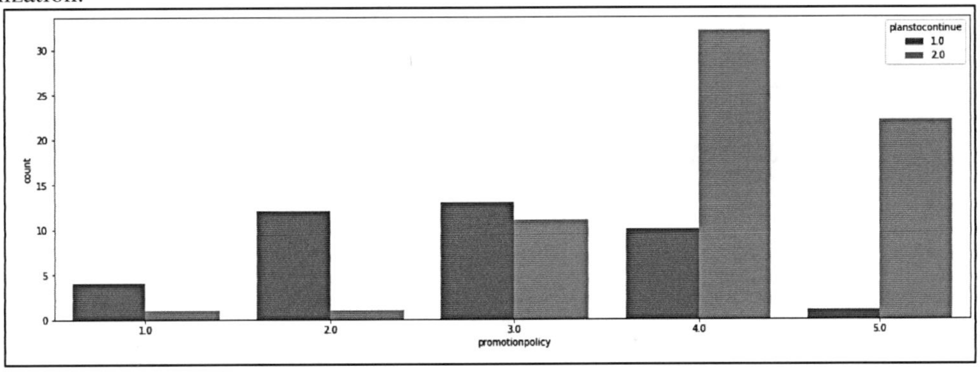

Figure 3: The chart representing the relation between the variables promotion policy and plans to continue.

In the above figure, 1 represents "No" meaning there are no plans to continue and 2 represents "Yes" which denotes that there are plans to continue. In the x-axis 5 denotes highly satisfied, 4 denotes satisfied, 3 is neutral, 2 is dissatisfied and 1 is highly dissatisfied. The chart above shows that majority of people who are highly satisfied and satisfied with the promotion policy have plans to continue in the organization. People who are dissatisfied and highly dissatisfied have no plans to continue in the organization.

The study has found the correlation between the target variable "planstocontinue" with other variables. It was found that the variables proud, jobsecurity, promotionpolicy, wlb, recognition and growthopp were having more correlation with the target variable planstocontinue. This illustrates that the satisfaction level of employees regarding the factors like pride to work in the organization, job security, promotion policy of the organization, work life balance, recognition given for work by superiors and the growth opportunities in the organization have great impact on the employee's decision to stay back in the organization.

3.2 A mathematical comparison of the results after implementation of selected algorithms
The study is also comparing the performance of various algorithms in predicting the decision of an employee to continue in the organization.

3.2.1 Naïve Bayes. The Naïve Bayes model gave the following result after computation.
TP = 58, TN=34, FP = 6, FN = 9
Accuracy = (TP+TN)/(TP+TN+FP+FN) = (58+34)/107=0.8598
Precision = TP/(TP+FP) = 58/ (58+6) =0.906
Recall = TP/ (TP+FN) =58/ (58+9) = 58/67=0.8657
F-Measure = (2*Recall*Precision)/(Recall+Precision) =88.5
Time taken to build model: 0 seconds

3.2.2 Decision table. The results of implementation of decision table classifier are as follows.
TP = 58, TN=23, FP = 17, FN = 9
Accuracy = (TP+TN)/(TP+TN+FP+FN) = (58+23)/107=0.757009
Precision = TP/(TP+FP) = 58/(58+17)=0.773

Recall = TP/ (TP+FN) =58/ (58+9) = 58/67=0.8657
F-Measure = (2*Recall*Precision)/(Recall+Precision)=0.8167
Time taken to build model: 0.05 seconds

3.2.3 J48. The results after the execution of J48 are as follows.
TP = 63, TN=28, FP = 12, FN = 4
Accuracy = (TP+TN)/(TP+TN+FP+FN) = (63+28)/107=0.8505
Precision = TP/(TP+FP) = 63/ (63+12) =0.84
Recall = TP/ (TP+FN) =63/ (63+4) = 63/67=0.9403
F-Measure = (2*Recall*Precision)/(Recall+Precision) =0.8873
Time taken to build model: 0.03 seconds

3.2.4 Random Forest. The random forest resulted in the performance as mentioned below.
TP = 63, TN=28, FP = 12, FN = 4
Accuracy = (TP+TN)/(TP+TN+FP+FN) = (63+28)/107=0.8505
Precision = TP/(TP+FP) = 63/ (63+12) =0.84
Recall = TP/ (TP+FN) =63/ (63+4) = 63/67=0.9403
F-Measure = (2*Recall*Precision)/(Recall+Precision) =0.8873
Time taken to build model: 0.06 seconds

3.2.5 Simplekmeans. TP = 45, TN=32, FP = 8, FN = 22
Accuracy = (TP+TN)/(TP+TN+FP+FN) = (45+32)/107=0.7196
Precision = TP/(TP+FP) = 45/(45+8)=0.8491
Recall = TP/ (TP+FN) =45/ (45+22) = 0.6716
F-Measure = (2*Recall*Precision)/(Recall+Precision) =0.75
Time taken to build model: 0.01 seconds

Table 2: Table showing the comparison of algorithms used in the study

Algorithm	Accuracy	Precision	Recall	F1score	Time in seconds
Naïve Bayes	85.98	90.6	86.57	88.5	0
Decision Tree	75.7	77.3	86.57	81.67	0.05
J48	85.05	84.0	94.03	88.73	0.03
Random Forest	85.05	84.0	94.03	88.73	0.06
KMeans	71.96	84.91	67.16	75	0.01

The above table shows that Naïve Bayes is performing better than other algorithms considering accuracy as well as time followed by J48 and Random Forest. Considering the time for building the model, J48 is exhibiting a better performance. Clustering gives lesser accuracy compared to the classification algorithms. Considering F1 score, J48 and Random Forest are showing the best output. F1 score is considered as a better parameter for performance comparison in case of datasets which are imbalanced. The dataset used here is a balanced dataset.

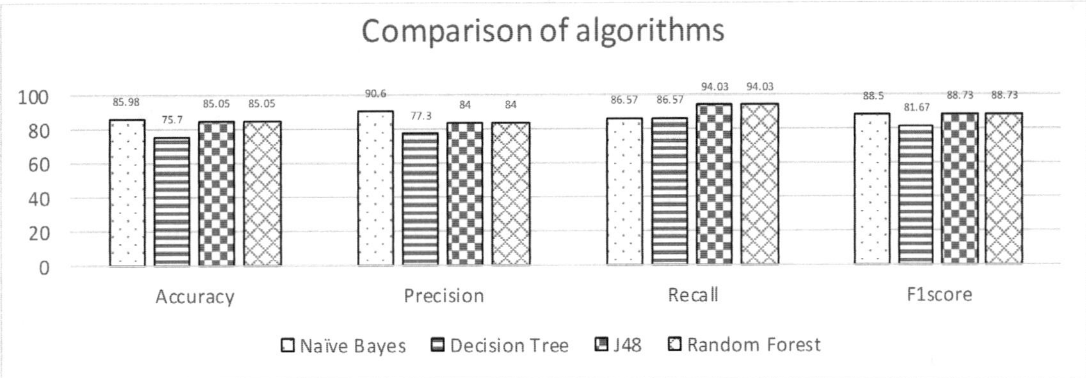

Figure 4 showing the comparison of various algorithms

4. Conclusion

The retention of employees is very much crucial as far as organizations are concerned. The employee's attrition can be reduced to an extend if we are able to identify the factors leading to attrition and also the factors leading to retention [9]. Machine learning can be identified as a very good tool for developing models for predicting attrition as well as retention [10]. The study has used python to find correlation of variables and has also compared the performance of various classification and clustering algorithms using the software Weka which is open-source software. The correlation coefficient calculated using python is giving the correlation between the variables like satisfaction level of employees regarding salary, training, jobsecurity and the like with the target variable planstocontinue. Hence the study has revealed that the factors like pride to work in the organization, job security, promotion policy of the organization, work life balance, recognition given for work by superiors and the growth opportunities in the organization have great impact on the employee's decision to stay back in the organization which in turn will help the organization to keep things in order. The efficiency of attrition prediction models is compared and it is found that for the dataset used in the study, maximum efficiency is achieved by Naïve Bayes method.

References

[1] Pedro Larrañaga, David Atienza, Javier Diaz-Rozo 2018 "*Industrial Applications of machine learning*", CRC Press, London

[2] Jain P K, Jain M & Pamula R 2020 Explaining and predicting employees' attrition: a machine learning approach. *SN Appl. Sci. 2*, 757

[3] https://tdwi.org/articles/2019/04/05/adv-all-predictive-analytics-and-employee- retention.aspx

[4] https://www.cs.waikato.ac.nz/ml/weka/

[5] Monika Lambsonge, Girish Talmale, Poonam Prasad February-2020 Review paper on: WEKA Tool With different Data Mining Technique, *International Journal of Emerging Technologies and Innovative Research* , Vol.7, Issue 2, page no.891-897

[6] Kothari C R 2019 "Research Methodology Methods and Techniques", New Age International (P) Limited

[7] Alduayj Sarah & Rajpoot, Kashif 2018, Predicting Employee Attrition using Machine Learning, 2018 13th Int. Conf. on Innovations in Information Technology (IIT)

[8] Nafizatus Salmi and Zuherman Rustam 2019, Naïve Bayes Classifier Models for Predicting the Colon Cancer, IOP Conf. Series: Materials Science and Engineering

[9] Sri Harsha B, A. JithendraVaraprasad, L.V N Pavan Sai Sujith March 2020, "Early prediction of employee attrition", *INTERNATIONAL JOURNAL OF SCIENTIFIC & TECHNOLOGY*

[10] Mohbey K K , (2020) Employee's Attrition Prediction Using Machine Learning Approaches, *Machine Learning and Deep Learning in Real Time Applications,* pp 121-128

IOP Conf. Series: Materials Science and Engineering 1085 (2021) 012030 doi:10.1088/1757-899X/1085/1/012030

Empirical Statistical Analysis and Cluster Studies on Socio-Economic Status (SES) Dataset

V. Balasankar [1], Dr. Suresh Suresh Varma Penumatsa [2], Dr. T. Pandu Ranga Vital[3,*]

[1]Research Scholar, Department of CSE, Adikavi Nannaya University, Rajahmundry, Andhra Pradesh, India. Email- balasankar.v@gmail.com

[2]Professor, Department of CSE, Adikavi Nannaya University, Rajahmundry, Andhra Pradesh, India. Email- vermaps@yahoo.com

[3]Associate Professor, Department of CSE, Aditya Institute of Technology and Management, Tekkali, Srikakulam, Andhra Pradesh, India. Phone: +91 9491936489 Email- vital2927@gmail.com

Abstract. Socio-economic status (SES) levels and conditions are extremely influential variables in the study of a particular area of society or any society. Social factors, for instance, the position of caste, religion, marital status, education levels, give good assessment results for us about a person's goals and the method of achieving their objectives. Generally economic status of any family is needy upon the social factors, for instance, the size of the family, educators in family and levels, and the level of the friendly environment in the family. SES with machine learning (ML) especially cluster analysis is important to identify important features or dimensions of the SES dataset, evaluate the rakings of dimensions and dimensional reductions. In this research, we collected 1742 samples (household information) as per socio-economic ratios and area (rural and urban) wise ratios with good questionnaires between 2018 and 2019 from Rajamahandravaram, East Godavari District, AP, India. We conduct the statistical analysis and cluster analysis for identifying the important factors of SES levels and their problem analysis. In cluster analysis, we apply k-means, hierarchal clustering (HC), and hierarchal with principal component analysis (PCA). The good projection results related to HC and PCA-HC specifies passements of SES class values.

1. Introduction

The word 'socio' originates from the word 'social' and denotes to the people, and the ways (level) they fit into the community or residential area in which they live. It reflects how well they are instructed, educated, and working a job and so on. Economic denotes to the money related situation of the individuals inside society and incorporates, earning capabilities, own house and the resources, and so on. As such, it is a sociological and economical total assessment measures of an individual's work insight and a person's or family's economic and social situation according to other people, based on occupation, education, income, and other factors. The assessment of the family's SES is measured using the income of the household, education in house, earning members in the house, other sources of income, and assets of that house. Mainly, SES levels are three that are low, middle, and high SES. Some researchers are described as the SES indicator low, middle, upper-middle, and high as per indicator variables education, income, occupation, and wealth [1] [11]. SES levels are differentiated in rural and urban areas. Some of the research works are confined to the linkages between urban and rural provincial establishment as it is most parts accepted that these two zones are autonomous to one another with the distinct limit in financial, political, and social aspects. Regardless, whether urban and rural provincial zones, one can't fulfill all the requirements of the individuals without depending upon others. It is thusly relying upon one another to address different issues and angles which are unavoidable. The mutuality relationship between urban and rural settlement is important. In such a manner, the collaboration is perceived through elements of components in these settlements by factors for the flow of people (migration), merchandise and enterprises flow, thoughts and data flow, culture, and assets flowing from one place to another. So, communication like the progression of individuals, merchandise and enterprises, ideas, and assets; to help livelihoods, that is required instead of two separate and isolated socio-economic entities [2][13]. The improvement of positive mental states is seen as a continuous interaction between the child's ' intrinsic inequalities, the family qualities, and the more and wider environment. The family's financial status seems to assume a characteristic part in deciding the positive mental and psychological conditions of the youth, with this hypothesis [3][4]. Poverty and neediness as the socio-economic development whereby the assets accessible to society are utilized to fulfill the needs of the meager few while the many don't have even their fundamental needs met [5].

SES with machine learning (ML) research is vital to predicting and preventions related to SES factors like poverty, education, administration, and so on, in that particular area. We can analyze SES assessments with unsupervised ML models like k-Means, hieratical, and hybrid cluster analysis with PCA (principal component analysis) for effective and vital results. In this research, we analyze the SES dataset of Rajamahandravaram using unsupervised cluster analysis like k-means and hieratical clustering with PCA analysis.

The main Objectives are based on the above problem that the present study was carried out with the following objectives.

1. To find out the impact of SES on the people's lifestyles of Rajamahandravaram using statistical analysis.
2. To examine the SES levels using cluster analysis with visualizations.
3. To explore the impact of SES on the resilience of education factors with cluster visualizations.
4. To find out the correlation among SES feature attributes.
5. To find out the ranking of attributes and analyze the projections using the HC-PCA algorithm.

The organization of the paper as follows.

In section 2, we describe in detail various research works related to SES and ML. In section 3, we describe the proposed model, and k-means, HC, and PCA algorisms and their mechanisms. In section 4, we analyze SES attributes statistical values, conduct the cluster analysis, and projects the high ranging attributes using HC-PCA algorithms.

2. Literature Survey

In this section, we reviewed numerous reputed journal papers and various works on SES from reputed articles, and some of the paper works of different authors are presented. Conger et al. (2009) [6] reviewed SES related to individual development and family processes. In this review, they discussed the integrationist model for family life and SES, incorporates social selection and social causation positions. Mainly, they focused on concepts that are 1. New Millennium Economic Climate, 2. Social class and SES measurements, 3. SES underlying assumption studies and family life 4. SES predictor variables quality of relations, sustainability, and stability 5. Empirical Findings like Couple Relationships and SES, 6. Romantic Relationships and FSM (Family Stress Model), 7. Child Development and Parenting relationships, and SES, 8. FSM and Children's Development, 9. The IM (Investment Model) and Development children, 10. Predicting from the Individual to SES, 11. The Interactionist Perspective. They concluded with some suggestions that are 1. The SES QAS

(Quantitative Assessment), 2. Research Methodology Improvements and 3. Current Theoretical Frameworks (CTF) Extensions and Elaborations. Schäfer et al. (2012) [7] researched age-related health issues related to SES that there are 3 psychosocial socioeconomic disadvantages, maltreatment, and social isolation". For this, they experimented with 1037 members from New Zealand. In this research, they found that study members suffered from 3 age-related health issues at age of 32 that are high inflammation levels, metabolic risks, and major depression. Lampos et al. (2016) [8] studied SES classification using Facebook user's language and behavior. At first, we detail a 3-way characterization task, where clients are delegated having an upper, center, or then again lower financial status. A nonlinear, generative learning approach utilizing a composite Gaussian Process bit gives essentially better grouping precision (75%) than a serious direct other option. By transforming this assignment into a double order – upper versus medium and lower class – the proposed classifier arrives at the exactness of 82%. Every day, the feeling and assessment of various individuals over the world are reflected as short messages utilizing microblogging stages Ghazouani et al. (2019) [9] reviewed and researched Twitter user's classification according to SES levels. In this, they mainly focused on the topics like "Tweet Gathering and Analysis, Classification Methods of SES, Data Corpuses and Resources, Inference Methods for SES Evaluation on Twitter, Regression Methods, Demographic Attributes, Temporal Information, Social Network Relations, Spatial Information, SES Features and Indicators, and Evaluation of Socioeconomic Status".

SES is a significant social and economic perspective broadly concerned. Evaluating individual SES can help related associations in settling on an assortment of policy choices. Customary methodology experiences a very significant expense in gathering enormous scope SES-related review information. With the pervasiveness of advanced cells, cell phone information has become a novel information hotspot for anticipating singular SES with minimal effort. In any case, the assignment of foreseeing singular SES on cell phone information likewise proposes some new difficulties, including inadequate individual records, scant express connections, and restricted named tests, indifferent in earlier work confined to territorial or household unit situated SES expectation [10]. The social determinants of well-being writing regularly SES as a critical factor in representing ladies' height. a setup marker of human government assistance at the populace level—utilizing customary relapse. Notwithstanding, this writing comes up short on a precise distinguishing proof of the prescient influence of SES just as the conceivable non-direct connections between the proportions of SES (material abundance, education, and occupation) in anticipating variety in ladies' tallness [11]. Zhao et al. (2020) [12] evaluated the predictive power of SES. In this research, they used demographic health surveys from 66 middle and low countries 1,273,644 women individual samples from 1994 to 2016. They assessed this data with 7 ML algorithms and concluded with the best ML performer NN. The NN was performed for SES prediction with 31.52 variances or MSE value.

3. Methodology and Models

3.1 Proposal Model

The figure1 shows the proposed model of the Socio-economic system with statistical and cluster studied models. In this, we collected the information from each house of rural and urban areas of the Rajahmundry constitution, district of East Godavari, A.P, India. We have gathered all the information with a good questionnaire and store the necessary information in the secondary storage section. After that, we extract the information with features and classes into a data set as *.csv format. Some of the information is plotted on the Rajamahandravaram (Rajahmundry) map with longitude and latitude measures. The targeted class attribute contains four classes that are rich, upper-middle-class, middle class, and poor. For this investigation, we extract the feature attributes as per household information that are personal-data, Socio-status, Economical-status, Living-status, Health-wealthy status, and so on. In this, we have to use 1742 household records with 49 attributes and construct *.csv format file and stored into the secondary storage section as well as plot the data points on the Rajamahandravaram map with full details for the convenience of visualization. We have to conduct the statistical analysis of the SES data set with statistical parametric equations like mean, median, max, and min and generate the reports for analysts. Cluster analysis is important to analyse the data set for the identification of important factors related to the target class values, dimensional reduction, and ranking of the attributes. In clustered studies, we choose the k-means and HC clusters using PCA. Before that, we conducted the pre-processing, the data set using the min-max method for better analysis.

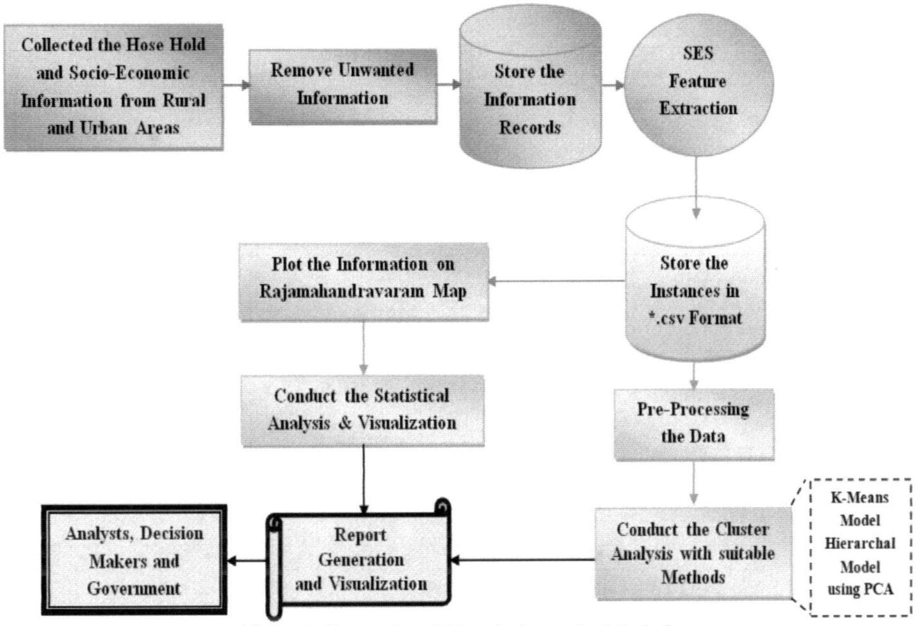

Figure 1: Proposal model for cluster analysis include

3.2 Data Set Description

Rajahmundry renamed as Rajamahandravaram is one of the major consistencies of East Godavari district in Andhra Pradesh, India. We gather information about each house from this constitution area of rural and urban. Nearly, we collected the 1742 samples as per socio-economic ratios and area wise ratios with good questionnaires between 2018 and 2019. Some of the data is plotted on the Rajamahandravaram Map using longitude and latitude values. The figure 2 shows location details and detailed information about plotted houses clicking on that point of more details button. For this experiment, we used 48 feature attributes and one class that is the status (rich, poor, middle, and upper-middle classes) described in table 1.

Table 1: Data Set attributes description

Dataset	Data Type	Description
Area(R-0/U-1)	Discrete (Integer)	House hold from Rural (0) or Urban (1)
Family Size	Continues (Integer)	Total members in House, range is 1 to 16
Male Size	Continues (Integer)	Total male members in House, range is 1 to 8
Female Size	Continues (Integer)	Total female members in House, range is 0 to 8
below 18	Continues (Integer)	Total members less than18 age in House, range is 0 to 5
above 18	Continues (Integer)	Total member >=18 age in House, range is 0 to 12
married people	Continues (Integer)	Total married members in House, range is 0 to 8
No. of children	Continues (Integer)	Number of children in House, range is 0 to 2
No. of literates	Continues (Integer)	Number of literates in House - range is 0 to 12
High Qualification	Discrete (Integer)	Qualification of house members range 0 to 5 0-very low 3-moderate 5- very high
No. of Workers	Continues (Integer)	Number of workers in House - range is 0 to 8
Child work below 15	Continues (Integer)	Number of Child workers in House - range is 0 to 1
Occupation (0-5)	Discrete (Integer)	Occupation of house hold members range is 0 to 7 0-very low or none 4-moderate 7- very high

Major Work	Discrete (Integer)	Work Category 0 to 5 0-very low 3-moerate 5- very high
Ration Cards	Discrete (Integer)	Ration Cards of house 1-white and 2- pink
Health cards	Discrete (Integer)	Health cards range is 0 to 1, 0-no 1-yes
No. of Diseased people	Continues (Integer)	Number of Diseased members in House - range is 0 to 2
No. of Handicapped	Continues (Integer)	Number of Handicapped members in House – range is 0 to 1
Bikes	Continues (Integer)	No. of Bikes in House range 0 to 3, 0 for none 3 for '3 or more'
Cars	Continues (Integer)	No. of Cars in House range 0 to 3, 0 for none 3 for '3 or more'
others	Continues (Integer)	No. of other Vehicles in House range 0 to 3, 0 for none 3 for '3 or more'
Own house (1) or rental (0)	Discrete (Integer)	Status of House 0 for 'Rental' 1 for 'Own'
land(cents)	Continues (Real)	Agriculture Land in cents range is 0 to 2000.0
Gold	Continues (Real)	Gold in grams in House range is 0 to 1500.0
Annual Income	Continues (Real)	Annual income of house range is 27000.0 to 15000000.0
Income from Govt	Continues (Real)	Income from Govt. of house range is 0 to 1480000.0
income from pension	Continues (Real)	Income from Pension of house range is 0 to 40000.0
Income from private	Continues (Real)	Income from private or own of house range is 0 to 15000000.0
Social status (0/1/2/3/4)	Discrete (Integer)	Social Status 1 for 'ST' 2 for 'SC' 3 for 'OBC' 4 for 'OC' 0 for 'none'
Nearest Hospital in Km	Continues (Real)	Hospital distance range 1.0 to 6.0
Nearest Primary School in Km	Continues (Real)	Primary School distance in KMs. range 1.0 to 5.0
Nearest High School in Km	Continues (Real)	High Schools distance in KMs range 1.0 to 10.0
Nearest College in Km	Continues (Real)	College distance in KMs range 2 to 14.0
Nearest University	Continues (Real)	University distance in KMs range 30.0 to 45.0
Addicted persons to smoke drinks	Discrete (Integer)	Habits of drink and smoke in house 0-None 1-Partial 2-Addicted 3- Extreme
Building model	Discrete (Integer)	Building model range is 0 to 5 0-low level 3-moderate 5-high level
Water sources	Discrete (Integer)	Water Facilities 0- none or poor 1-moderate 2-good facility 3-very good
Toilets facilities 1/0	Discrete (Integer)	Toilet facilities 0 for 'no' and 1 for 'yes'
electricity1/0	Discrete (Integer)	electricity facilities 0 for 'no' and 1 for 'yes'
TV 1/0	Discrete (Integer)	TV facilities 0 for 'no' and 1 for 'yes'
Fridge 1/0	Discrete (Integer)	Fridge facilities 0 for 'no' and 1 for 'yes'
Air Condition1/0	Discrete (Integer)	Air Condition facilities 0 for 'no' and 1 for 'yes'
Heater1/0	Discrete (Integer)	Heater facilities 0 for 'no' and 1 for 'yes'
Computer 1/0	Discrete (Integer)	Computer facilities 0 for 'no' and 1 for 'yes'
Fuel for cooking 1/0	Discrete (Integer)	Fuel for cooking 0 for 'Non-Gas' and 1 for 'Gas'
Income status in past 5 years (2/0/1)	Discrete (Integer)	0-Decremental Income 1-Remain same 2-Increment
Internet 1/0	Discrete (Integer)	Internet facility 0-No and 1-Yes
Migrated family or not	Discrete (Integer)	Migrated from other places 0-No and 1 - Yes
SES Levels (1-4)	Discrete (String)	Levels 1-Poor 2-middle 3-upper middle 4-Rich

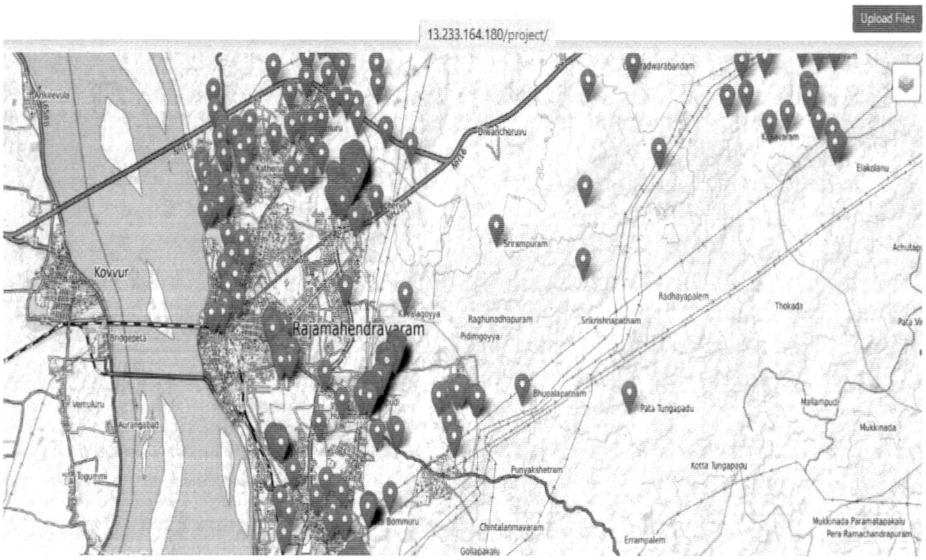

Figure 2: Plotting data points according latitude and longitude on Rajamahandravaram map

3.3 Plotting Data Points

The figure shows the Rajamahandravaram map where it contains the collected house hold data points that plotted on map with location points (blue coloured bubble points) with latitudes and longitudes. We established this map with attach with our data set using latitudes and longitudes values. Moreover, in this we provide statistical analysis results of the data set, individual data, bar charts and pi charts of the attributes, searching and upload data record facilities, defined queries projections and so on. The local cite information with IP address that is 13.233.164.180.

3.4 k-Means algorithm

K-Means is an apportioning clustering technique which displaces objects by moving starting with one cluster then onto the next beginning from an initial partitioning [14][15]. The motivation of the cluster examination is to segment n observations into K groups in which every perception has a place with the cluster with the closest mean. It is one of the simplest unsupervised ML that solve the problems with clustering. The K-Means is a transformative and evolutionary method that picks up its name from its technique for activity [16] [17]. The k-means algorithms as follows

step 1: The model chooses K data points as the centres of initial cluster using mean values

step 2: Each data point in the data is allocated to the nearer cluster group using Euclidean distance between each data point and each centre of the cluster.

step 3: Each cluster is determined by cluster average data points.

Step 4: Steps 2 and 3 repeats until the clusters converge.

3.5 Hieratical Clustering

Agglomerative HC has been the predominant approach to deal with embedded group of clusters. The main aim is to focus toward useful models and methods that both efficient and effective [18]. It is regularly useful to recognize strategy, including a minimization standard and the objective structure of a 2-way tree speaking to the fractional request on subsets of the power set; instead of an execution, which identifies with the detail of the calculation utilized. Similarly, as with numerous other multivariate strategies, the items to be classified have mathematical estimations on cluster factors or characteristics. A mathematical system of this sort isn't the one in particular which can be utilized to plan grouping calculations. Reasonable elective types of capacity of a rectangular cluster of qualities are not conflicting with review the issue in mathematical terms [19][20].

3.6 Principal Component Analysis (PCA) Algorithm

PCA is a symmetrical linear transformation that moves the information to new coordinate system to such an extent that the best variance by any projection of the information comes to lie on the primary arrange (first principal component (PC)), the second most noteworthy difference lies on the subsequent coordinate (second PC), and etc. PCA used to lessen measurements of information absent a lot of loss of data. It is utilized in AI in frequently.

1. Get the mean vector, in this calculation every data point value is subtracted by the mean value and divided by the Standard deviation (SD). Below equation shows mean vector value.

$$Z = \frac{\text{Factor Value - Mean Value}}{\text{Standard Deviation (SD)}}$$

In this, every variable value is changed in normalized value (same scale value).

2. Construction of Covariance Matrix using assembled samples from mean datapoint value matrix.

This progression is to perceive how the components of the informational collection are transforming from the mean with respect to each other. At times, the components are very related or associated so they contain abundance or excess information [21]. Thusly, in order to recognize these relationships, we compute the covariance model network. The covariance grid is a $n \times n$ symmetric lattice with n measurements that has as entrances the covariance identified with each and every likely pair of the underlying components.

The covariance n x n Matrix is

$$\begin{bmatrix} Cov(x_0,x_0) & Cov(x_0,x_1) & .... & Cov(x_0,x_n) \\ Cov(x_1,x_0) & Cov(x_1,x_1) & .... & Cov(x_1,x_n) \\ .... & .... & .... & .... \\ Cov(x_n,x_0) & Cov(x_n,x_1) & .... & Cov(x_n,x_n) \end{bmatrix}_{n \times n}$$

3. Register the Eigen values and Eigenvectors.
4. Register the premise vectors.
5. Correspond to each feature data point as a LC (linear combination) of principle vectors.

4. Experimental Setup and Result Analysis

In this section, we discuss the data set attributes empirical statistical analysis. Furthermore, we apply unsupervised ML models k-means and HC cluster on SES Dataset and analyse their abilities.

4.1 Statistical Analysis

We collected the data from rural and urban areas of the Rajahmundry constitution, East Godavari District, A.P., India. For this, collected sampling data is as per ratios of socio-economic status. The rural area samples are 946, and urban area samples are 796 (Total 1742). As per the statistical analysis of the household dataset, some of the houses contain the average 4 to 5 where the mean value is 4.381 and Std. Dev is 1.467. Some of the houses have only one member (min value is 1), and some of the houses contain 16 (max value). Each house contains at least one male person (min value male persons in a house is 1) and a maximum of 8 male persons as well as on average 2 to 3 persons per one house. On the other hand, the female persons' min value is 0, and the max value is 8, and mean and SD values are 1.975 and 0.776 respectively, which means every house contains on an average one to two females. As per statistics some good conditions that very fewer child workers, average young generation 2 to 3 people in every house and average 1 to 2 workers in each house. Another good thing, the number of diseased people and the number of disabled persons is very less percentage that the mean values are 0.066 and 0.024 respectively. Table 2 shows the detailed statistical analysis of each attribute of the data set.

Table 2: Detailed statistical analysis of Rajahmundry SES data set

Attributes Min, Max, Mean and Standard Deviation Statistics					Yes or No Attributes Statistics		
Attribute	**Min**	**Max**	**Mean**	**SD**	**Attribute**	**No**	**Yes**
Family Size	1	16	4.381	1.467	**Health cards**	689	1053
Male	1	8	2.406	0.773	**Own House**	466	1276
Female	0	8	1.975	0.776	**Toilets facilities**	168	1574
below 18	0	5	1.592	0.882	**electricity**	0	1742
above 18	0	12	2.802	1.183	**TV**	442	1300
married people	0	8	2.037	0.375	**Fridge**	1134	608
No. of children	0	2	0.131	0.362	**Air Condition**	1481	261
No. of literates	0	12	0.738	0.826	**Heater**	1523	219
No. of Workers	0	8	1.693	0.693	**Computer**	1421	321
Child. work	0	1	0.006	0.079	**Other Type of Attributes Statistics**		
No. of Diseased people	0	2	0.066	0.268	**Attribute**	**Type**	**Value**
No. of Handicapped	0	1	0.024	0.152	**Ration Cards**	**white**	660
land(cents)	0	2000	140.576	202.602		**Pink**	1082
Gold(grams)	0	1500	34.788	51.792	**Fuel for cooking**	**Gas**	1575
Annual Income	27000	15000000	331504.6	467944.2		**other**	167
Income from Govt.	0	1480000	24055.68	117628.8	**Social status**	**ST**	31
income from pension	0	40000	408.726	1351.704		**SC**	190
Income from private	0	15000000	308868	467484.8		**BC**	896
Hospital in Km	1	6	3.637	0.836		**OC**	625
Pri mary School in Km	1	5	2.846	1.901	**Addicted persons to smoke and drinking in House**	**None**	736
High School in Km	1	10	3.832	1.899		**Partial**	826
College in Km	2	14	6.866	2.057		**Addicted**	158
University in Km	30	45	35.065	4.185		**Extreme**	22

Some other Types of Attribute Statistics					
Attribute	**Type**	**Value**	**Attribute**	**Type**	**Value**
Literacy and Educators Houses	**None or Below 10th**	296	**Occupation Major Work Occupation in House**	**No Work or Very less**	6
	10th Standard	428		**Seasonal Workers**	463
	Inter Level or ITI	386		**Average or Daily wagers**	497
	Degree Level	272		**Permanent Low salary**	345
	Technical Degree or Other	249		**Permanent Middle Salary**	364
	P.G. level	101		**Permanent High Salary**	62
	Professional or Ph.D. Level	10		**Business or Organizers**	5
Having Bikes in House	**None**	939	**Having Cars in House**	**None**	1607
	One	671		**One**	128
	Two	118		**Two**	5
	More Than Two	14		**More Than Two**	2
Having Other Traveling Recourses	**None**	1615	**Target Class (SES levels)**	**Rich**	73
	One	121		**Middle class**	794
	Two	5		**Upper Middle class**	526
	More Than Two	1		**Poor**	349

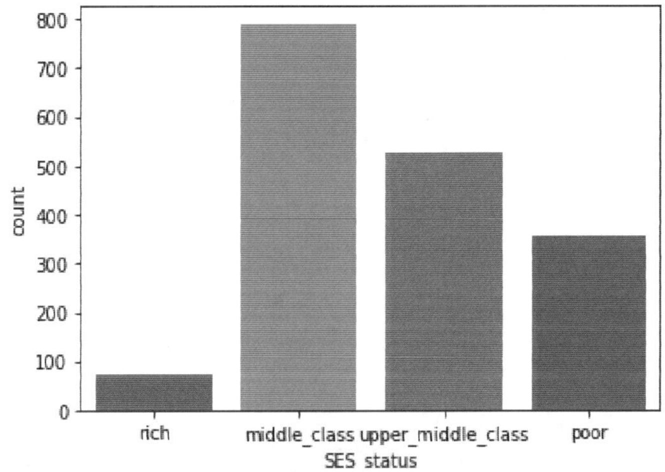

Figure 3: Counting SES Levels in Rajamahandravaram area

An important thing for the economic status that fully depends on annual income for each house and its resources that are from the public, private, asserts, and work, and so on. As per statistics, the annual income min value is 27000/- and the max value is 80, 000, 00/-. The main income resources from private, government, or pension schemes. The detailed analysis is shown in table 1. The educational and health resources are also available within the distance of every house. The figure shows the SES level samples that are rich, middle-class, upper-middle-class, and poor as per the ratios of Rajamahandravaram. The Rich and Poor positions are in the 4th and 3rd, and the middle class and upper-middle-class occupy the first and second positions in the area. Figure 3 shows the counting number of SES status levels were low to high as per ratios

4.2 Matrix (Heat Map): Figure 4 shows the Correlation heat map of each attribute of the Rajamahandravaram SES dataset. The correlated values are specified with the colour that the indicator mentioned in the figure. The values are measured between -1 and 1. If the cell colour is a dark red that its indication is neutrally correlated with each other. The dark blue indicates the zero correlation. The attribute electrical is correlated marginal with all other attributes. The relationship between attributes in the dataset is vital reasons to evaluate the fitness of the data set that one variable value is fully dependent on other variables or weak associations with others. Sometimes the variable relations are peculiar in that one unknown variable value depends on two or more variable values. The correlation values are useful for modelling and analysing the data in a better way. The correlation is defined using two attributes or variables of statistical value relationships. The correlation value is between -ve (negative) and +ve (positive) values. +ve value of correlation represents that change of attributes or variable movement in the same direction. -ve value of correlation represents that change or relationship of attributes or variable movement in the reverse direction. Neutral or zero correlation represents that both variables are divergent or unrelated. In some situations, two or more of the variables in the dataset are related very rightly named as multicollinearity that it impacts more and more on the performance of some algorithms like linear regression. In this situation, we can remove highly correlated attributes from the experimental dataset for improving model performance.

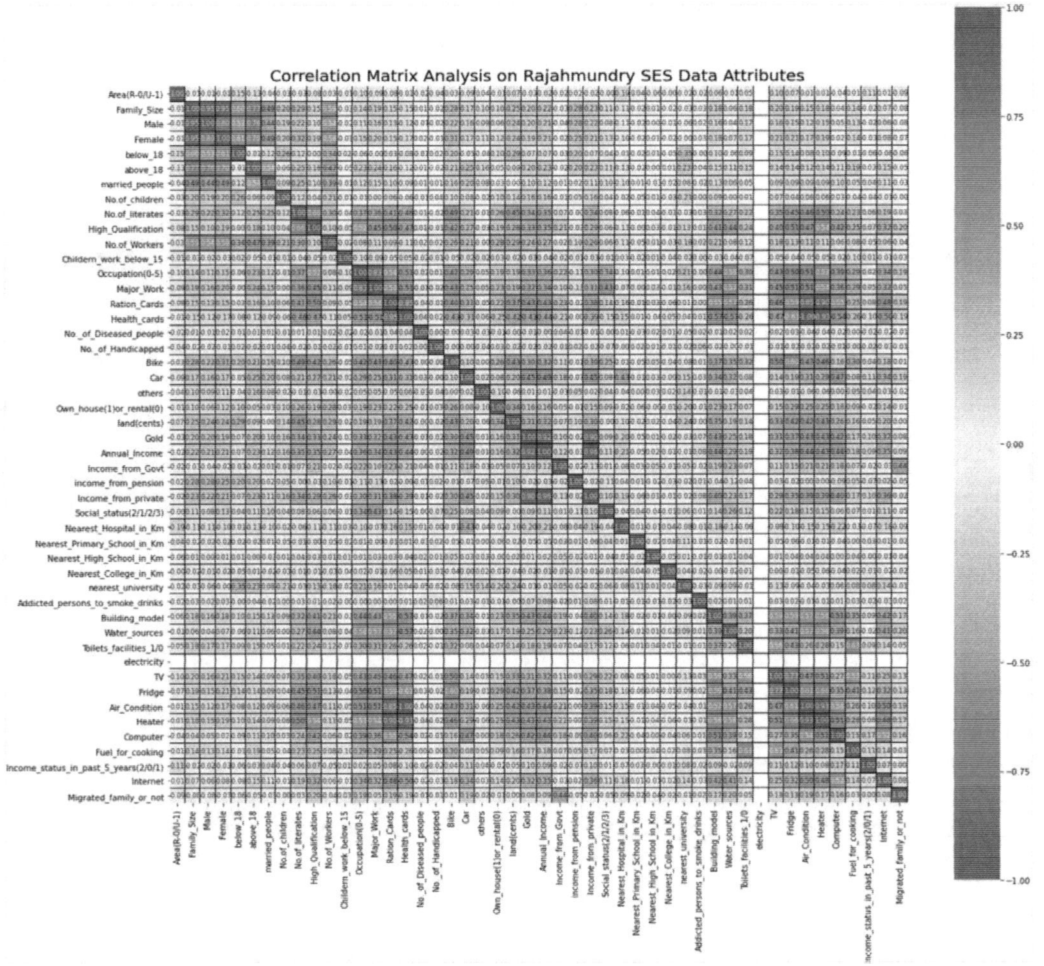

Figure 4: Correlation Matrix Analysis of Rajamahandravaram SES Dataset

4.3 k-Means Clustering

In the k-means cluster analysis, we give the value 4 for the number of clusters that clusters are cluster 0, cluster 1, cluster 2 and cluster 3. The total data set size is 1742(no. of houses information). The cluster 0 contains 485 instances, the cluster 1 is dealing with 480, the cluster 2 contains highly 547 instances and the cluster 3 contains 230 instances. The table 2 shows the k-means unsupervised ML algorithm centroids of each cluster. In this we describe some of the attributes centroids related to each cluster and full data set also. The total data set concened with the target class attribute SES level vlaue Middle-class, and cluster 0 and 3 are related to Upper-middle. The cluster 0 is related to class attribute value poor and cluster 1 is related to middle. The detailed cluster centroids details are shown in the table 3.

Table 3: K-Means Cluster Centroids of Rajahmundry SES data set

Attributes	Full Data (1742)	Cluster 0 (485)	Cluster1 (480)	Cluster2 (547)	Cluster3 (230)

Family Size	4.3812	4.4763	3.8438	4.4863	5.0522
Male	2.4059	2.4474	2.175	2.4516	2.6913
Female	1.9753	2.0289	1.6687	2.0347	2.3609
below 18	1.5924	1.4515	1.4063	1.6399	2.1652
above 18	2.802	3.0289	2.4354	2.8867	2.887
married people	2.0373	2.0804	1.9812	2.0329	2.0739
No. of children	0.1309	0.1134	0.1021	0.1225	0.2478
No. of literates	0.7382	1.1567	0.1854	0.5978	1.3435
High Qualification	0	3	0	0	0
No. of Workers	1.6935	1.7464	1.4708	1.7623	1.8826
Child works below 15	0.0063	0	0.0146	0.0055	0.0043
Occupation (0-5)	2	4	1	2	2
Major Work	1	5	1	2	2
Ration Cards	1	2	1	1	2
Health cards	1	0	1	1	0
No. of Diseased people	0.066	0.0577	0.0604	0.0878	0.0435
No. of Handicapped	0.0235	0.0227	0.0333	0.0201	0.013
Bike	1	1	0	1	1
Own house (1) or rental (0)	1	1	0	1	1
land(cents)	140.5758	210.2268	28.1104	102.9068	318
Gold	34.7878	60.4396	9.0156	21.9481	65.0174
Annual Income	331504.6	586360.8	91041.67	222734.9	554608.7
Income from Govt	24055.68	72288.66	395.8333	6133.455	14347.83
income from pension	408.7256	154.6392	420.8333	405.8501	926.087
Income from private	308868	514080.4	89654.17	213515.5	560400
Nearest Hospital in Km	3.6366	3.3897	3.7229	3.7623	3.6783
Nearest Primary School in Km	3.8462	3.8536	3.9958	3.7331	3.787
Nearest High School in Km	3.8318	3.8309	3.7125	3.8062	4.1435
Nearest College in Km	6.8657	6.7711	6.9354	6.9744	6.6609
nearest university	68.0654	69.066	68.4104	68	65.3913
Building model	1.0327	1.6866	0.25	0.9452	1.4957
Water sources	1	2	1	1	1
TV	1	1	0	1	1
Fridge	1	1	0	1	1
Air Condition	0	1	0	0	1
Heater	0	1	0	0	1
SES Levels	Middle	Upper-Middle	Poor	Middle	Upper-Middle

Figure 5: k-Means Cluster Analysis (clusters vs SES classes) of Rajamahandravaram SES Dataset

Figure 5 shows the cluster analysis between cluster number and target calsses. In this plot, the X-axis specifies the SES status and Y-axis indicates the cluster number. The clusters form different colors and describes related SES stus. The cluster 0 (colour blue) contains most of instaces from Upper-middle class very less instances from middle and rich classes. The cluster 1(red colour) is constructed with more poor and less middle-class instances. The cluster 2 colour is green that it is constructied with most of middle-class elements and very less elements of poor. The cluster 3 is combination of middle and upper-middle class elemets.

4.4 Hierarchal Cluster (HC) Attribute (Column) wise Examination on SES Dataset: The

Figure 6 shows the HC column-wise complete link clustering. In this, the attributes closure relations of SES data set are analysed one to other. The Euclidian distance is the measurement of the HC construction. The complete HC network is constructed with 7 clusters. Each cluster is represented with each colour shown in figure 6. The cluster1 is constructed with the pairs of attributes like occupation and Major Work with height 0.4, married People and income with height 3.93 status, female and below 18 with height 0.334. In the C1 cluster, highly corelated and related attributes are female and below-18-age.

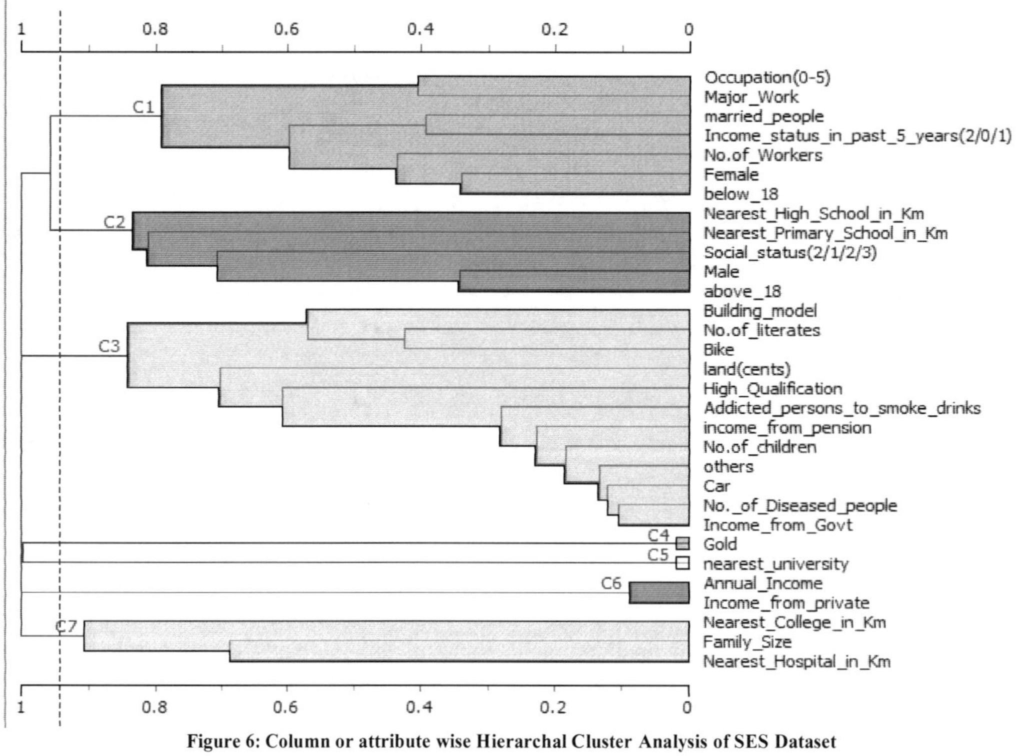

Figure 6: Column or attribute wise Hierarchal Cluster Analysis of SES Dataset

The cluster 2 is constructed with the pair of attributes like male and above-18-age with 0.33 height, this pair is connected to the social-status with height of 0.71. In this, the connected attributes in chain and increase the height values. In cluster 2 highly corelated attributes are male and above-18-age. The cluster 3 is constructed with 12 essential attributes. In this cluster, highly correlated attributes Diseased-people and income-from-Govt with height 0.109 and next chain connection attributes are car, others and so on. The cluster 3 and cluster 4 are constructed with single attributes that are gold and nearest-university. The cluster 6 is constructed with two attributes that are income-private and annual-income with height 0.88. The cluster 7 has 3 attributes that are nearest-hospital and family-size with 0.64 connected to nearest-college attribute with height 0.912. As per analysis, the cluster 6 contains highly correlated attributes than other clustered attributes and cluster 7 contains non-correlated or low-correlated attributes where distance is very high than other cluster attributes. The overall cluster analysis is with complete linkage, maximum depth is 8, the height ratio is 14%, and distance measurements are calculated with Euclidian distance formula.

.

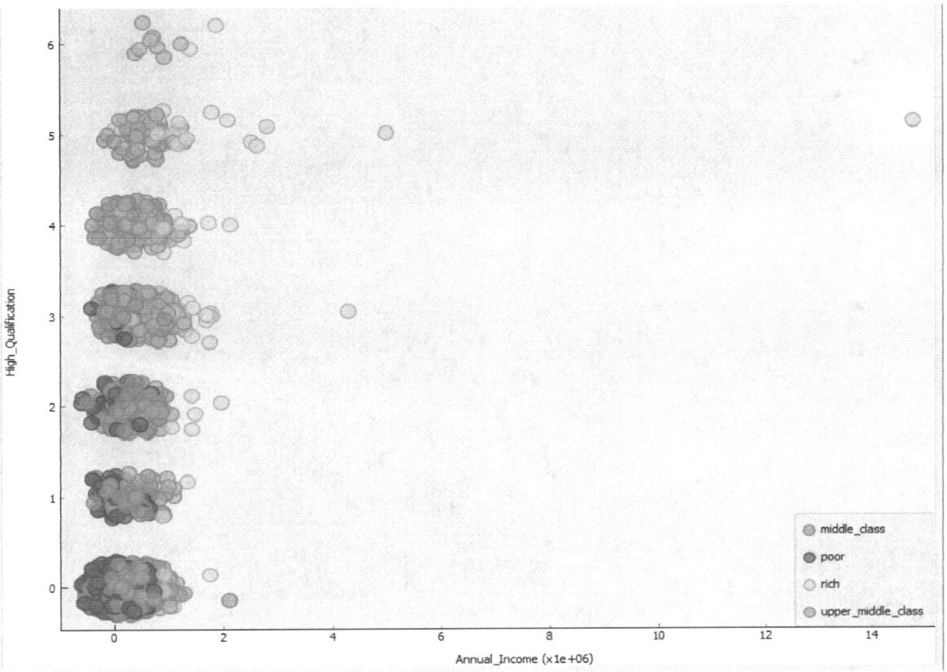

Figure 7: Hierarchal Cluster Attribute Analysis: Target class, Annual- Income and High-Qualification

Figure 7 describes the resulting HC Attribute Analysis on attributes Target class, Annual- Income, and High-Qualification. The X-axis represents the Annual income (AI x 1e+06), Y-axis represents the High Qualification that the range is 0 to 6, and the target class data points represented with colour bubbles (red-poor, green-rich, blue-middle, and orange-upper-middle). As per the analysis, income is related to higher education impacts on SES levels that poor class attribute data points decrease inversely proportional to the higher education. Most of the data points in higher education levels 3,4,5 and 6 contain middle, upper-middle, and rich classes only. The highest degree of Ph.D. (level 6) is pursued by rich and upper-middle classes only. In this analysis, we observe that poverty is related to education

4.5 HC Row wise Analysis on Rajahmundry SES Dataset:

Figure 8 describes the complete link HC with row-wise class attribute values like Rich 'class, Poor-class, Middle-class, and Upper-Middle-class classes. In this, the target class values (Rich, Poor, Middle, and Upper-Middle of SES dataset) are elaborated with 8 clusters in closure relations at the 24 distance. As per feature attribute values and target class values, the closure relation clusters are determined and analysed using Euclidean distance measures. The distance is at 3.38 in cluster 5 that the values are rich. C1 cluster related to rich-class, C2 cluster specifies the upper-middle-class values, C3, and C4 clusters represent the rich and middle-class independently. The C5 contains all 4 values that most of the data points are middle and upper-middle classes, as well as the C6 cluster, which covers middle and upper-middle-class values. C7 and C8 clusters dedicate to the rich class attributes.

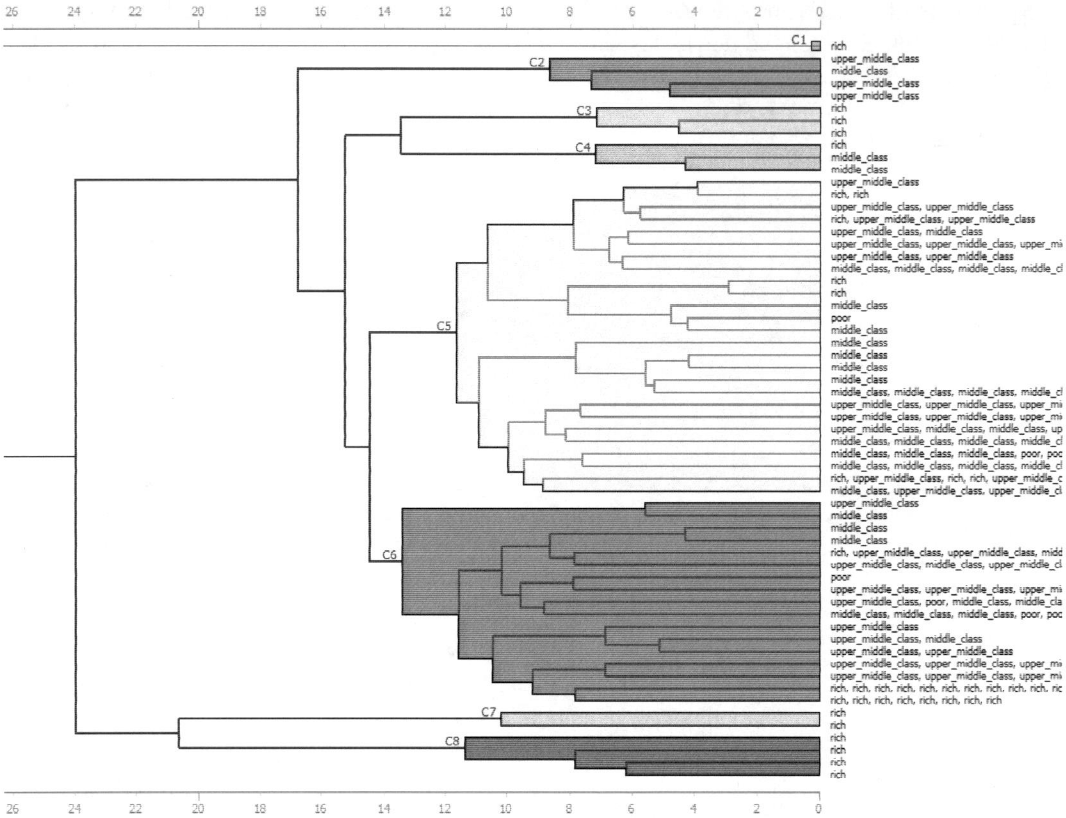

Figure 8: Row or Record wise Hierarchal Cluster Analysis of SES Dataset

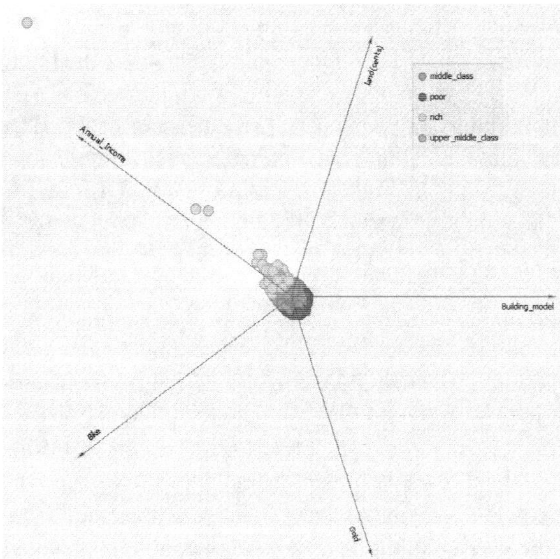

Figure 9: HC Projections Analysis without applying PCA

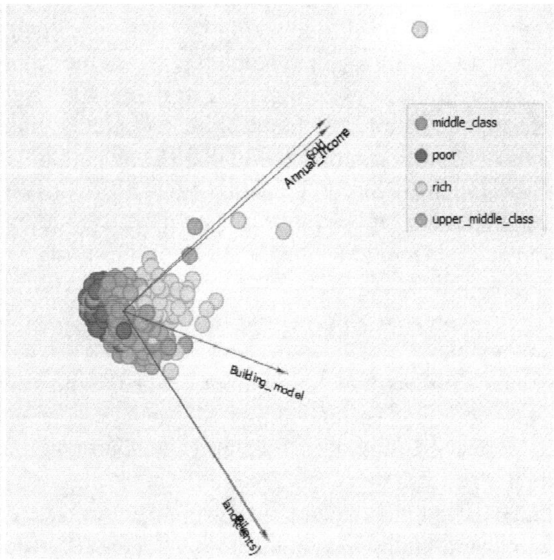

Figure 10: HC Projections Analysis with applying PCA

4.6 HC without PCA Projections Analysis:

Figure 9 depicts the HC projections without PCA according to attributes concerning Annual-Income, land in cents, having a bike, Gold, and Building models. The blue elements indicate the middle-class, the red elements specify the poor data points, the green-coloured elements describe rich, and the orange-coloured data points specify the upper-middle-class. Most of the rich data points (green coloured dots) have high Fraction of Annual-Income, land in cents, having a bike, Gold, and Building models. These components are related to the HC target class values like rich, poor, middle, and upper-middle classes. The projections are described in Figure 9 without implementing PCA. These five dimensions are the first 5 listed ranks in the experiment according to SES classes with an accuracy of 99.13%.

4.7 PCA-HC Projections Assessments on SES Dataset:

Figure 10 demonstrates the HC cluster projections with PCA according to attributes concerning Annual-Income, land in cents, having a bike, Gold, and Building models. The blue elements indicate the middle-class, the red elements specify the poor data points, the green-coloured elements describe rich, and the orange-coloured data points specify the upper-middle-class. Most of the rich data points (green coloured dots) have high Fraction of Annual-Income, land in cents, having a bike, Gold, and Building models. These components are related to the HC target class values like rich, poor, middle, and upper-middle classes. The projections are described in Figure 9 implementation of PCA-HC. These five dimensions are the first 5 listed ranks in the experiment according to SES classes with an accuracy of 99.93%.

5. Conclusion

Statistical analysis and Machine learning are very essential and crucial current days for analysing the Socio-Economic Status problems. In this research work, we collected and analysed Rajamahandravaram SES dataset using statistical models and identify the important features of the SES. As well as, we conducted the cluster analysis like k-means, HC, and HC-PCA on SES dataset and get good visualization results from this experiment.

References

[1] López-Goyburu, P., & García-Montero, L. G. 2018. The urban-rural interface as an area with characteristics of its own in urban planning: A review. Sustainable cities and society, 43, 157-165.

[2] Chapman, B. P., Fiscella, K., Kawachi, I., & Duberstein, P. R. 2010. Personality, socioeconomic status, and all-cause mortality in the United States. American journal of epidemiology, 171(1), 83-92.

[3] Wang, M. T., Henry, D. A., & Degol, J. L. 2020. A development-in-sociocultural-context perspective on the multiple pathways to youth's engagement in learning. In Advances in Motivation Science (Vol. 7, pp. 113-160). Elsevier.

[4] Perrotta, C., & Williamson, B. 2018. The social life of Learning Analytics: cluster analysis and the 'performance'of algorithmic education. Learning, Media and Technology, 43(1), 3-16.

[5] Mishra C. P. 2012. Nexus of poverty, energy balance and health. Indian journal of community medicine : official publication of Indian Association of Preventive & Social Medicine, 37(2), 71–78.

[6] Conger, R. D., Conger, K. J., & Martin, M. J. 2010. Socioeconomic Status, Family Processes, and Individual Development. Journal of marriage and the family, 72(3), 685–704.

[7] Schäfer, I., Hansen, H., Schön, G. et al. 2012 The influence of age, gender and socio-economic status on multimorbidity patterns in primary care. first results from the multicare cohort study. BMC Health Serv Res 12, 89.

[8] Lampos, V., Aletras, N., Geyti, J. K., Zou, B., & Cox, I. J. 2016, March). Inferring the socioeconomic status of social media users based on behaviour and language. In European Conference on Information Retrieval pp. 689-695. Springer, Cham.

[9] Ghazouani, D., Lancieri, L., Ounelli, H., & Jebari, C. 2019, September. Assessing socioeconomic status of Twitter users: A survey. In Proceedings of the International Conference on Recent

Advances in Natural Language Processing (RANLP 2019) (pp. 388-398).

[10] Danese A, Moffitt TE, Harrington H, et al. 2009 Adverse Childhood Experiences and Adult Risk Factors for Age-Related Disease: Depression, Inflammation, and Clustering of Metabolic Risk Markers. Arch Pediatr Adolesc Med. 2009;163(12):1135–1143.

[11] Daoud, A., Kim, R., & Subramanian, S. V. 2019. Predicting women's height from their socioeconomic status: A machine learning approach. Social Science & Medicine, 238, 112486.

[12] Zhao, T., Huang, H., Yao, X., & Fu, X. 2020. Predicting individual socioeconomic status from mobile phone data: a semi-supervised hypergraph-based factor graph approach. International Journal of Data Science and Analytics, 9(3), 361-372.

[13] Bose, E., & Radhakrishnan, K. 2018. Using unsupervised machine learning to identify subgroups among home health patients with heart failure using telehealth. CIN: Computers, Informatics, Nursing, 36(5), 242-248.

[14] Qi, J., Yu, Y., Wang, L., & Liu, J. 2016, October. K*-means: An effective and efficient k-means clustering algorithm. In 2016 IEEE International Conferences on Big Data and Cloud Computing (BDCloud), Social Computing and Networking (SocialCom), Sustainable Computing and Communications (SustainCom)(BDCloud-SocialCom-SustainCom) (pp. 242-249). IEEE.

[15] yakur, M. A., Khotimah, B. K., Rochman, E. M. S., & Satoto, B. D. 2018, April). Integration k-means clustering method and elbow method for identification of the best customer profile cluster. In IOP Conference Series: Materials Science and Engineering (Vol. 336, No. 1, p. 012017). IOP Publishing.

[16] Meng, Y., Liang, J., Cao, F., & He, Y. 2018. A new distance with derivative information for functional k-means clustering algorithm. Information Sciences, 463, 166-185.

[17] Jothi, R., Mohanty, S. K., & Ojha, A. 2019. DK-means: a deterministic k-means clustering algorithm for gene expression analysis. Pattern Analysis and Applications, 22(2), 649-667.

[18] Vital, T. P., Lakshmi, B. G., Rekha, H. S., & DhanaLakshmi, M. 2019. Student Performance Analysis with Using Statistical and Cluster Studies. In Soft Computing in Data Analytics (pp. 743-757). Springer, Singapore

[19] HC1. Murtagh F. 2011 Hierarchical Clustering. In: Lovric M. (eds) International Encyclopedia of Statistical Science. Springer, Berlin, Heidelberg.

[20] HC2. Murtagh, F., & Contreras, P. 2017. Algorithms for hierarchical clustering: an overview, II. Wiley Interdisciplinary Reviews: Data Mining and Knowledge Discovery, 7(6), e1219.

[21] Vital, T. P., Satyanarayana, L. V., & Murthy, G. S. N. (2020). Empirical Analysis on Uddanam CKD with Unsupervised MLs Including PCA. In Computational Intelligence in Pattern Recognition (pp. 571-587). Springer, Singapore.

Study of Piezoelectric transducers used for Energy Harvesting Applications

Nilimamayee Samal[1*], O Jeba Shiney[2] and Abhinandan Patra[3]

[1,2]School of Electrical, Electronics and Communication Engineering, Galgotias University, Noida India
[3]IBM India Pvt. Ltd.

Email id: nilima.samal1@gmail.com *

Abstract: To manage the soaring power demand, various types of Energy Harvesting Systems (EHS) are being developed. The energy harvesting from unutilized natural renewable sources using piezoelectric transducers is one of them. Day-by-day different analytical models are being reported with different piezoelectric transducers to improve the energy efficiency and output power of the energy harvesting systems. The latest Energy Harvesting Systems (EHS) technology includes the material with outstanding piezoelectric properties, compact fabrication style and distinctive design. They are capable of generating electric power of microwatt to milliwatt range from the unutilized energy sources such as environmental, industrial, vehicular sources and human motion etc. In this study, the reported research works on piezoelectric energy harvesting system (2007-2020) have been reviewed and discussed. The review basically focuses on the constitutional/compositional development of the piezoelectric transducers used in energy harvesting processes.

1. Introduction

In the era of 'Emergence-of-Energy', the waste management and the sustainable energy management techniques have become the highlighted topics of research. The recovery of energy in usable form from the waste and unused power has become the recent topics of discussion now-a-days. The waste and unused power exits in several forms in our surrounding and mostly generated from different environmental, industrial, structural, vehicular sources, human activities etc. These sources are schematically represented in fig-1. The useful electric power is generated from the unused physical energy by using different transducer materials, such as piezoelectric transducers, magneto-electric transducers, thermo-electric transducers, semi-conductor and capacitive transducers etc. The several ways to increase the efficiency of energy conversion process are improving energy density of electrical storage, reducing power-consumption of the sensors, using self-powered sensor for conversion of energy and transmitting power to the sensors through a central source. The most pragmatic way among these methods is to use an active or self-powered sensor for energy harvesting [1]. Secondly, when the mechanical stress is more prevalent over the other sources of unused power, the piezoelectric transducer comes into picture for energy harvesting.

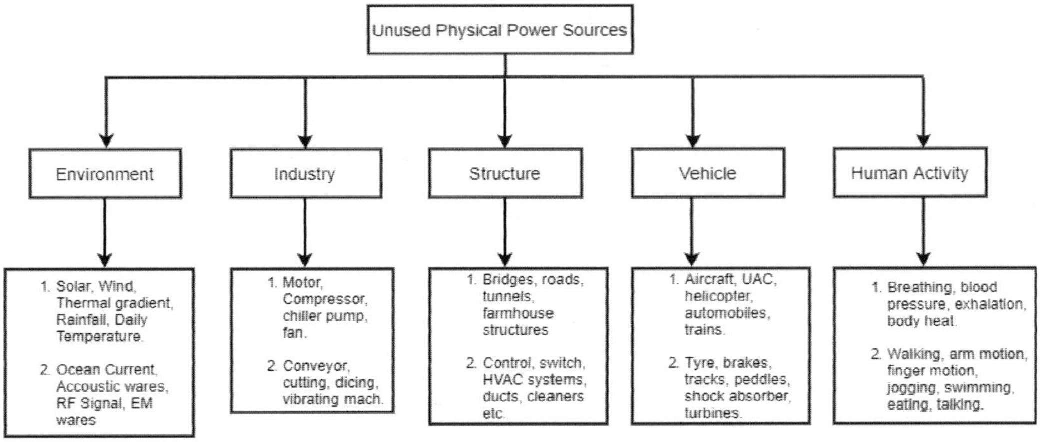

Figure 1. Unutilized Physical Energy Source

The block diagram representation of the basic principle of conversion of unused physical energy to used electric-energy is represented in fig-2. Prime component of this conversion process is a piezoelectric transducer. The electric energy produced by the piezoelectric material as a consequence of the applied mechanical vibration is stored after proper signal conditioning. The mechanical loss in the conversion process is due to unmatched mechanical impedance(z), damping factor, reflector etc. the mechanoelectrical transduction loss is due to coupling factor(k) and piezo-electric coefficient, i.e multiplication of piezo-electric strain constant (d^{eff}) and piezo-electric voltage constant (g^{eff}). The electric loss is due to unmatched electrical impedance. [2]

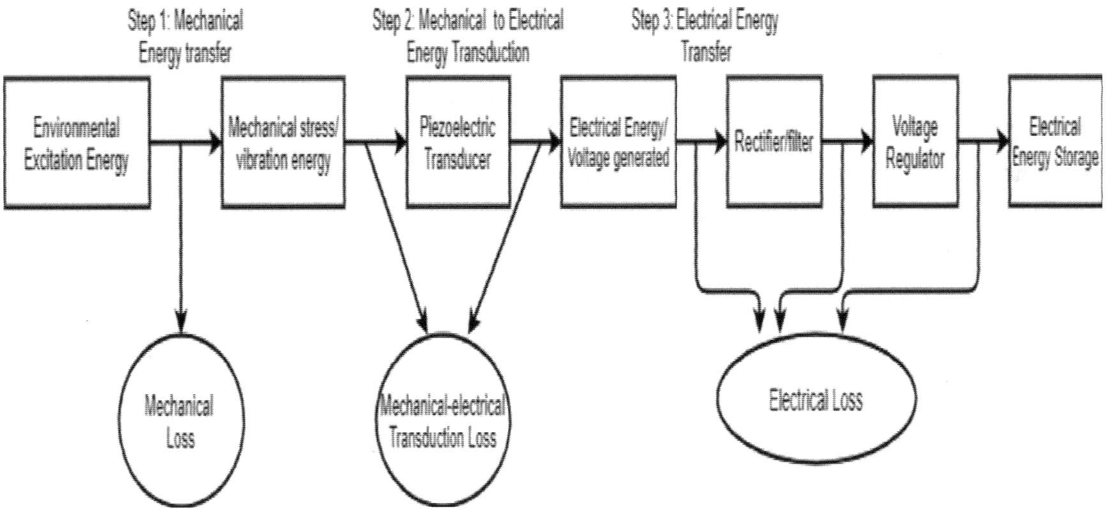

Figure 2. Conversion flow of physical energy to electric-energy using piezo-electric transducers

The advancements in piezo-electric energy harvester focusses on the augmentation of piezoelectric properties, viz. piezoelectric strain co-efficient (d^{eff}), piezoelectric voltage co-efficient (g^{eff}), coupling factor

(k) etc. The optimized response of the energy harvesters are also greatly influenced by the structural design of the transducer. In this article the piezo-electric transducers with different structures, configurations and compositions used for various energy harvesters have been reviewed and discussed.

2 Energy harvesting using piezoelectric ceramics and single crystals

The piezoelectric ceramics and single crystal are broadly classified into two categories: Hard Doped and soft doped. As the hard doped piezoelectric materials are able to withstand a high mechanical or electrical shock/stress, they are suitable for high power applications. The soft piezoelectric materials are characterized by its relatively large permittivity, high electo-mechanical coupling factor and large piezo-electric constant. They are used as sensors, receiver, actuator and low-power transducer. [3]

S.Shahab et al. have done an experimental case study on PZT-5H for vibration-energy harvesting and reported a comparison study on soft(PZT - 5H, PZT - 5A) and hard(PZT - 4, PZT - 8) piezoelectric ceramics [Table1]. In Table1, the parameters compared are piezoelectric strain constant (d_{31}), relative permittivity ($\varepsilon^{S}/\varepsilon$), electro-mechanical coupling factor (k^2), density (ρ) mechanical quality factor (*Q)* [4]. The vacuum permittivity is 8.85 Pf/m. Some other comparative studies on analytical and experimental parameters of piezo-electric ceramic and single crystal used for vibration EHS have also been reported [5-7].

Table 1. Properties of the soft & hard piezo-electric ceramic [4]

Properties	PZT - 5H	PZT - 5A	PZT - 4	PZT - 8
d_{31} [pm / V]	(-)274	(-)171	(-)123	(-)97
$\varepsilon^{S}/\varepsilon$	3400	1700	1300	1000
ρ [kg/m^3]	7500	7750	7500	7600
Q	65	75	500	1000
K^2	0.151	0.119	0.107	0.0924
$k^2 Q$	9.83	8.89	53.5	92.4

3 Energy harvesting using piezoelectric polymers

The piezoelectric polymers are preferred over the piezoelectric ceramics and single crystal because of their low acoustic impedance, flexibleness for fabricating them in required shape and size and their cost-effectiveness. The piezoelectric polymers are broadly classified into three categories (fg-3):

a) Bulk piezo-polymers
b) Voided-charged polymer
c) Piezo-electric polymer composite

The most dominant piezoelectric polymer used for energy harvesting is PVDF (Polyvinylidine fluoride). PVDF film of 6-25μm thickness goes through different stages/processes like polymerization, stretching and poling to become suitable for transducer applications [9].

J.B.Rocha et al. integrated electro-active ß-PVDF with silver and aluminum electrode on shoe-sole and connected the electronic circuit required for enhancement of energy-transfer and electric storage-efficiency in the interest of harvesting energy from the walk. They reported that the power generated from PVDF ranges from 10 to 100 mw. The variation in the generated power is due to the area, position, dimensions and the no of PVDF foil embedded (fig-4) [10]. A multilayer structure of co-polymer of Vinylidene fluoride & tri-fluoro ethylene P (VDF - TrFE) with aluminum electrode has also been used for energy harvester [11]. The PVDF based energy harvester via displacement amplification module has been demonstrated by Y.H.Shin et al [12].

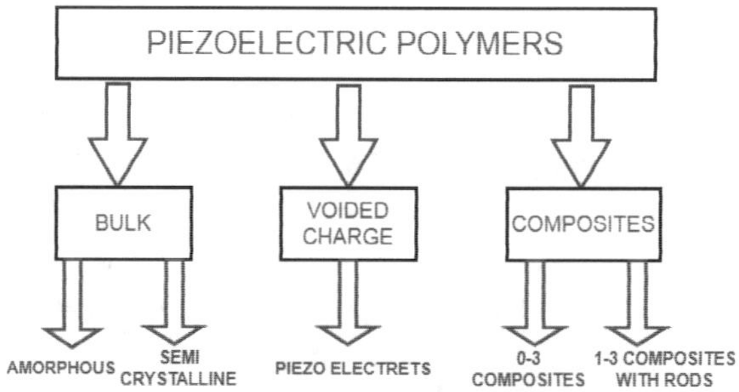

Figure 3. Classification of Piezoelectric Polymers

Figure 4. Electronic circuit connection to piezoelectric material embedded on shoe-sole [10]

The advances in this type of transducers depend upon the developments in the in-organic ceramics, organic polymers [13, 14], 2-D materials [15] & biologically derived systems [16], enhancements in piezoelectric properties [17] and structural designs [18]. Now-a-days, researchers are focusing on multi-direction energy harvesting using 3-D piezo-polymer micro-system. M.Han et al. demonstrated a 3-D piezo-polymer micro-system to harvest vibration energy. The device is made up of PVDF(3mm x 50µm x 9µm dimension and 1.2 mm vertical displacement) with Cr (10 nanometer) / Au (100 nanometer) / Ti (20 nanometer) as top electrode and Cr(10 nanometer) / Au (100 nanometer) as bottom electrode. The harmonic vibration excites selected vibration mode of the 3-D structure, as a result the electrical energy is harvested through the metallic electrode contacts [19].

4 Energy harvesting using Piezo-polymer composite:

Piezo-polymer composite comprises of piezoelectric ceramic and polymer. There are two types of configurations of a diphase composite, i.e. parallel and series configuration. MFC (Micro Fiber Composite) and NFC (Nano Fiber Composite) are more dominantly used in energy harvesting systems.

4.1 MFC (Micro Fiber Composite)

The MFCs are highly flexible, durable and mechanically tough. They have a high electro-mechanical coupling coefficient, good impedance matching and higher transducer sensitivity. So these are preferred over the conventional piezoelectric elements to be used in energy harvesting systems. [9]

A HPG (Hybrid Piezoelectric generator) for mechanical energy harvesting has been developed by Md.M.Alam and D.Mandal. They have used NCMF (Native Cellulose Micro Fiber) in active phase & PDMS (poly dimethyle siloxane) in inactive phase to form microfiber composite and used MWCNTs (Multi Wall Carbon Nanotubes) as conducting fillers. They reported, the NCMF-HPG delivers an O.C. o/p voltage of around 30 V and S.C. o/p current of around 500 nA (fig-5) and the power density is around 9.0 $\mu W/cm^2$ [20].

4.2 NFC (Nano Fiber Composite) or Nanohybrid

The Nano Fiber Composites (NFCs) are composed up of a piezoelectric fiber (Nano Scale range) and a polymer. The nano fibers have a small size and large surface area. They are not mechanically and chemically stable, when used alone. But when they combine with polymer and form a composite, their stability increases noticeably. Hence the nano fiber polymer composites are preferred over the microfiber composites [21].

Xi Chen et al. used PZT nano fiber (Diameter-60 nm & length- 500μm) with platinum electrode and polydimethylsiloxane (PDMS) to form a nano composite. They demonstrated Nano generator using PZT nano fibers for mechanical-energy-harvesting. The o/p power produced was 0.03 μW, when the load resistance was 6 MΩ [22]. S.H.Wankhade S.H. Wankhade et al. have fabricated a device using nanohybrid of PZT-PVDF for harvesting energy from different human motions/activities, such as fingers tapping, foot tapping & bending. It has been reported that the device is capable of producing an o/p voltage of 55volt and a power density of 36 microwatt per $c.m^2$ [23].

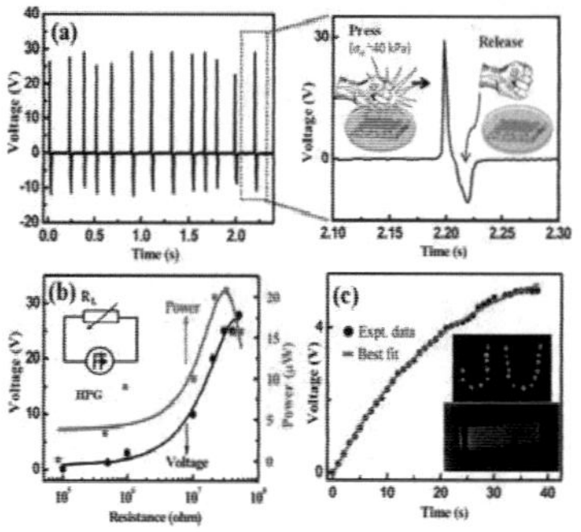

Figure 5. Open circuit o/p voltage(v/g) with respect to time, (b) v/g and inst. power o/p with respect to load resistance (R_L) (c) voltage-time plot of capacitor getting charged from HPG [20]

5 Energy harvesting using piezo-metal composite

Based on the design, the piezo-metal composites are classified as 'Cymbal' and 'Moonie'. Both these types of transducers comprise of a piezoelectric disc with metal capping on top and the bottom fig-6 [21]. The displacement of moonie transducer is because of the flexural movement of end-caps and the displacement of a cymbal is due to both flexural and rotational motion of end caps. The cymbal is preferred over moonie for energy harvesting because of its higher displacement, high stability, high contact surface and its lesser fabrication cost [22].

S.Gareha et al. have simulated a compression based PCT traffic model, which can be used to generate electric power. On the basis of their simulation study, it has been concluded that a total electric power of 170 kw per km can be generated using a number of PCTs (Piezoelectric Cymbal Transducer) for a high way with a traffic density of 600 vehicles per hour [23] and the generated power cab be utilized for feeding power to the road side light systems, emergency communication units etc. Bo Ren et al. used 0.71 $Pb(Mg_{1/3}Nb_{2/3})$- 0.29 $PbTiO_3$ based cymbal for a vibration energy harvesting system and reported that the power density of this modified transducer is around 3 times larger than the $Pb(Zr_xTi_{1-x})$ ceramic [24].

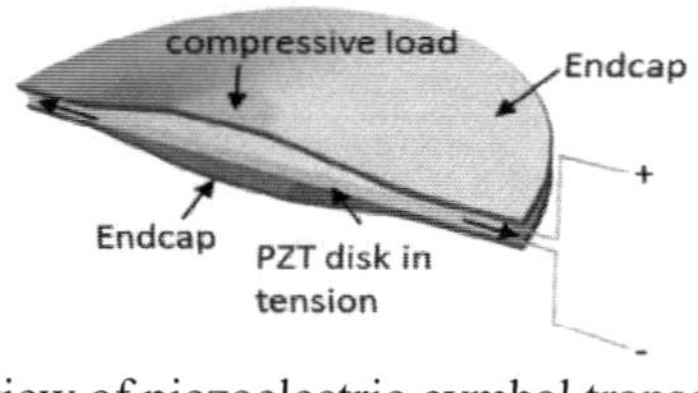

Figure 6. Half view of piezoelectric cymbal transducer (PCT) [21]

6 Summary

From the review of the reported research works from 2008 to 2020 in the area of piezoelectric energy harvesting, it is observed that the most abundantly used piezoelectric material are PZT and PVDF because of their outstanding piezo-electric properties and large conversion efficiency. Now-a-days small powered ICs are being incorporated in the controlling circuit to decrease the input power requirement [30-35]. Some of the important piezoelectric energy harvesting applications with the piezoelectric material used is listed in Table-2.

Table 2. Summary of important piezoelectric energy harvesting applications

Type of Piezoelectric transducer based on composition	Composition	Energy harvesting Application	Physical energy used for energy harvesting	Reference
Piezoelectric Ceramics	PZT	Shoe-sole insertion	Human motion	25, 26
Piezoelectric Polymer	PVDF	Shoe sole insertion	Human motion	11, 26
Piezoelectric Polymer	PVDF	Wind generator	Wind flow	27
Piezo-Film	PVDF	Rain fall energy harvester	Rain drop impact	28, 29
Microfiber Composite	NCMF-PDMS	Hybrid piezoelectric generator	Mechanical vibration	14
Nanofiber Composite	PZT nanofiber-PDMS	Nano generator	Mechanical vibration	22
Cymbal	$0.71Pb(Mg_{1/3}Nb_{2/3})$-0.29 $PbTiO_3$	Vibration energy harvester	Mechanical vibration	24
Piezoelectric Polymer	PVDF	Roadway energy harvester	Vehicular vibration	12
3D piezo polymer microsystem	PVD	Biomedical, Vibration energy harvester	Body movement	19
Nano hybrid	PZT-PVDF	Nano Generator	Human Motion	23

7 Conclusions

The present state-of-art in piezo-electric transducers for energy harvesting have been reviewed and discussed in this article. The piezoelectric transducers, such as piezoelectric ceramics, piezoelectric polymers, composites, cymbals etc. are being used in energy harvesting processes since decades. The majority of piezoelectric EHS reviewed are able to generate power in μw to mw range depending upon the ambience. Recently, the development of multidirectional energy harvesting systems, use of Nano-hybrid piezoelectric transducers etc. are very impressive. From the review it is realized that even though the PEH technology have been developed a lot, still there is a scope of increasing the stability and consistency of the system.

References

1. Priya S 2007 *Journal of Electroceramics Springer* **19** 167-184.
2. Kim H et al. 2007 *IEEE Transactions on Ultrasonics, Ferroelectrics and Frequency Control* **54(9)** 9628087.
3. N. Technology, Hard and Soft Materials (Micromechatronics Inc.) http://www.mmech.com/hard-and-soft-matl. Accessed 26 June 2020.
4. Sahab S et al. 2017 Energy Technology

5. Karami M A et al. 2011 *IEEE Transactions on Ultrasonics, Ferroelectrics, and Frequency Control* **58(7)** 12116852.
6. Badel A et al 2006 *IEEE Transactions on Ultrasonics, Ferroelectrics, and Frequency Control* **53(4)** 8852545.
7. Rakbamrung P et al 2010 *Sensors and Actuators A: Physical* **163(2)** 493-500.
8. Kim H.S et al. 2011 *International. Journal of Precision Engineering and Manufacturing* **12** 1129-1141.
9. Bauer S and Bauer F 2008 *Springer Series in Materials Science* **114** 157-177.
10. Rocha J G et al. 2010 *IEEE Transactions on Industrial Electronics* **57(3)** 11137562.
11. Zhang L et al. 2013 *IEEE Transactions on Ultrasonics, Ferroelectrics, and Frequency Control* **60(9)** 13712663.
12. Shin Y H et al. 2018 *Applied Energy, Elsevier* **216** 741-750.
13. Wang X et al. 2007 *National Library of Medicine* **316(5821)** 102-105.
14. Dagdeviren C et al. 2014 *PNAS* **111(5)** 1927-1932.
15. You Y M 2017 *Science, AAAS* **357(6348)** 306-309.
16. Zhu H et al. 2014 Nature Nanotechnology **10** 151-155.
17. Li F et al. 2018 *Nature Materials*, **17** 349-354.
18. Qi Y et al. 2011 *ACS Publication* **11(3)** 1331-1336.
19. Han M et al. 2019 *Nature Electronics* **2** 26-35.
20. Alam Md M and Mandal D 2016 *ACS Appl. Mater. Interfaces* **8(3)** 1555-1558.
21. Zhu J et al. 2018 *ACS Appl. Mater. Interfaces* **10(23)**19940-19947.
22. Meehan A et al. 2011 *IEEE Transaction on Power Electronics* **26(1)** 176.
23. Khaled F et al. 2015 *IEEE Transactions on Industrial Electronics* **62(6)** 3508-3515.
24. Hakova M 2018 *Analytica Chimica Acta, Elsevier* **1023** 44-52.
25. Ren B et al. 2010 *J. Appl. Phys.* **107(3)** 034501.
26. Turkmen A C and Celik C 2018 *Energy, Elsevier* **150** 556-564.
27. Viola F et al. 2014 *9th Intl. Conf. on Ecological Vehicles and Renewable Energies* (Monte-Carlo, Monaco) p. 1-7 14399307.
28. Ilyas M A and Swingler J 2015 *Energy, Elsevier* **90(1)** 796-806.
29. Acciari G et al. 2018 *IEEE Transactions on Industry Applications* **58(1)** 458-458.
30. Chen X et al. 2010 *Nano Lett., ACS* **10(3)** 2133-2137.
31. Julsgaard M et al. 2016 *Veille Scientifique, AGA* **151(1)** 110-119.
32. Kuang Y et al. 2017 *J. Phys. D: Appl. Phys, IOP Science* **50(34)** 1-11.
33. Kim H.W et al 2004 *Jpn. J. appl. Phys.* **43(9A)**, 6178.
34. Gareh S et al. 2019 *Intl. J. Renewable Energy Research* **9(4)** 2101.
35. Zhao J and You Z 2014 *Sensors, MDPI*, **14(7)** 12497-12510.

Real Time Resource Scheduling Framework for Smart City

Girish Talmale [1*] **,Urmila Shrawankar** [1]

Department of Computer Science and Engineering, G H Raisoni College of Engineering, Nagpur, India

* Email : girishtalmale@gmail.com

Abstract. Smart City plays a vital role in social development. Real time scheduling of different tangible and intangible resource in distributed environment in smart is challenging problem. The different disaster management tasks smart city applications such as floods, earthquake, pandemic, fire, disaster relief, and control of hazardous area, major accidents, control of terrorism and other attacks require real time emergency resources scheduling. The existing techniques uses for smart city resource management are suffering from low resource utilization and high response time, migration and preemption overhead. This work paper presents cluster based real time resource scheduling framework for emergency situation in smart city. The edge computing resources are group into cluster. The proposed framework increases resources utilization and reduces response time, migration, preemption overheads. The emergency tasks are scheduled to these computing cluster are allow to migrate within the cluster to ensure to reduce migration, switching overhead, response time and increase resource utilization and success rate.

1. Introduction

The smart city improves the leaving standard of life. The resource scheduling is used to manage and optimize the different smart city resources such as energy, transportation infrastructure, health, environmental and human etc. in smart city. There is requirement to schedule these resources in such way that improve resource utilization and reduce response time. Lot of research work has been done in the past on emergency response systems[1]. The resource scheduling for fire disaster management using dynamic model, resource allocation model under time constraint[2], resource management under multiple disaster places[3][4]. Real time resource scheduling in distributed environment is still open research problem[5]. Different resource management techniques for edge computing survey are represented by Tang [6]. The adaptive algorithm for smart city resource management is presented by Zhang[7]. Fog and mobile computing resource management and development in real time resource management in fog computing is done[8]. The different challenges face in smart city resource management is describe by Law[9]. Real time application in smart city and its requirement and scheduling techniques is represented by Gasco[10]. Optimal job shop scheduling techniques is mentioned by Liu M [11]. The workflow scheduling approach is used in edge computing data centre is described [12]. Energy efficient resource scheduling in cloud computing data centre is described [13]. Dynamic service migration mechanism in edge computing is developed[14]. Resource task scheduling with low cost for mobile edge computing is presented in some research work [15]. The smart city application tasks scheduling for edge computing is presented using multi-server techniques [16]. This paper proposed real time resource scheduling by grouping edge computing resources into cluster to ensure real time actions in emergency situations with high resource utilization, success rate and reduce migration, preemption overheads as shown in Figure.1.

The objectives of this research work are as follows:

1. To propose cluster based real time resource scheduling framework for distributed environment in smart city
2. To implement proposed scheduler using edge computing to ensure real time response and increase resource utilization
3. Analysis and Performance measure of proposed scheduler with benchmarking.

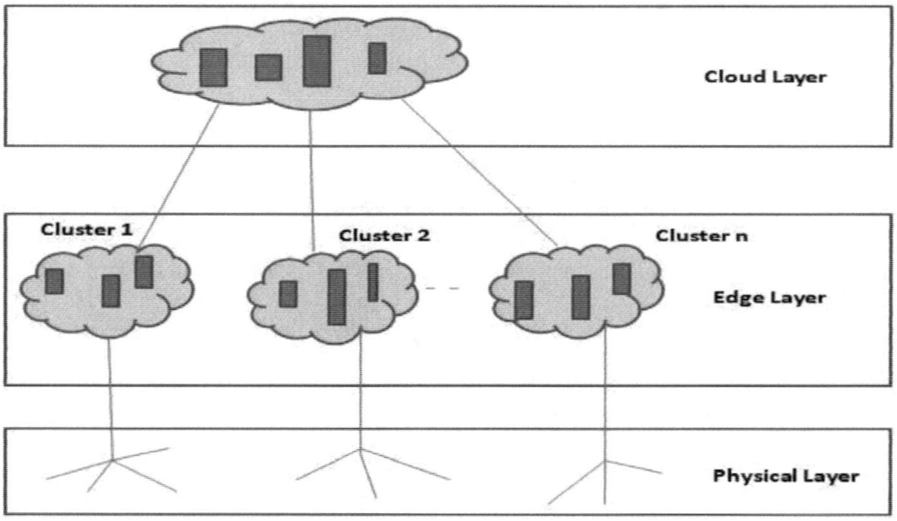

Figure.1. Cluster Based Resource Scheduling Framework

2. Methodology

The cluster based real time tasks scheduling model consist of set of processing nodes P and T as real time emergency smart city application tasks set.

$T = \{T1, T2, T3 --- Tn\}$ (1)

The task is characterized by Ti(ei,di,pi) where

ei: Tasks Ti worst case execution time
di: Tasks Ti deadline
pi: Tasks Ti period

The tasks utilizations represented as

U=ei/pi

The implicit deadline tasks model is used where deadline of task is equal to period of the tasks

di=pi .The set of computing resources are as mention in equation 2.

$P = \{P1, P2, P3 --- Pm\}$ (2)

The scheduling of tasks to processing nodes is global scheduling problem where the set of tasks is maintained in one global run queue and tasks scheduler dispatch tasks to any free processing nodes. The maintenance of long run queue is difficult as the number of tasks in common run queue increases as shown in Figure.1.

2.1. Cluster Formation

Cluster based resource scheduling group the processing nodes into cluster. The propose scheduler assign c number of processing nodes in each cluster and create m/c number of cluster as shown in equation 3.

$$C1, C2, C3 - - - - - C(m/c) \qquad (3)$$

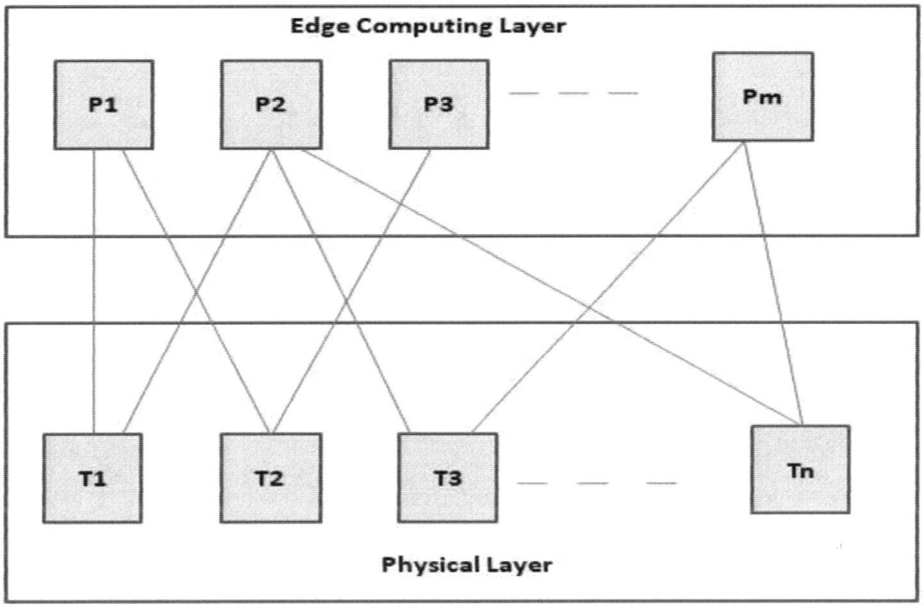

Figure 2. Global Tasks Scheduling

3. Implementation Details

Implementation of proposed algorithm presented as follows.

3.1. Cluster Based Resource Sharing Algorithm

The Cluster based resource scheduling consist of two steps:

Cluster Tasks Assignment: Task assignment assigns the tasks to cluster based on worst fit tasks assignment techniques.

Cluster Tasks Scheduling: It schedule the order of execution of tasks executed on the processing nodes as shown in Fig.3.

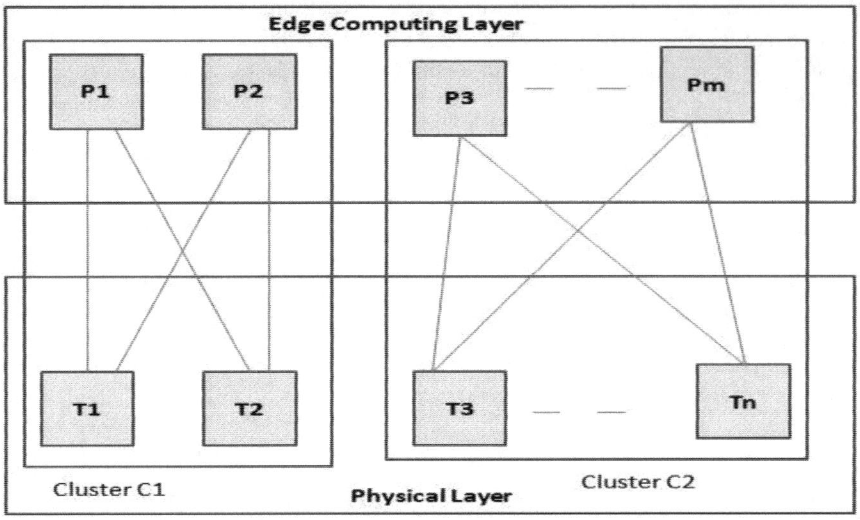

Figure 3. Cluster Tasks Scheduling

Algorithm: Cluster Based Real Time Resource Scheduling

$Initialize\ Tasks\ Set\ T = \{T1, T2, T3 - - - - - Tn\}$
$Initialize\ Processing\ Nodes\ P = \{P1, P2, P3 - - - Pm\}$
$Initialize\ Size\ of\ Cluster\ c = 2^i$ For i={1,2,3,----------k}
$Create\ Clusters\ Set\ C = \{C1, C2, C3 - - - C(\frac{c}{m})$
$for\ each\ cluster\ C1\ to\ C\left(\frac{c}{m}\right)\ do$
  $for\ each\ task\ \ From\ Ti\ from\ i = 1\ to\ n\ do$
    $if\ U(Ti) \leq c\ then$
      $c = c - U(Ti)$
    $else\ if\ U(Ti) > c$
      $Select\ next\ C$
    $else$
      $Tasks\ not\ Schedulable$
    $end\ If$
  $end\ for$
$end\ for$

4. Results and Discussion

The performance analysis of proposed cluster based real time resource scheduler is done using simulation for real time tsunami disaster management system on edge computing platform. Tsunami disaster management system consists of two part detection of tsunami and prevention measure after tsunami. The detection of tsunami includes the real time alert tasks for citizens, government agencies, rescue team, hospitals teams, wave tracking and identification tasks, ocean monitoring tasks. Tsunami disaster prevention includes tasks for emergency healthcare, tracking of living things in affected area, transpiration etc. The different tasks involved in detection and prevention of tsunami disaster management system are processes on edge computing cluster to provide real time response. The simulation result shows that the proposed cluster based real time scheduler increase resource utilization and reduces response time as compare to global scheduling as shown in Table1.

Table1. Comparison of Global Scheduling with Cluster Scheduling

Performance Parameters	Global Scheduling	Cluster Scheduling
Resource Utilization (%)	83	96
Migration Cost (%)	93	57
Preemption Cost (%)	81	67
Success Rate (%)	74	91
Response Time (%)	56	36

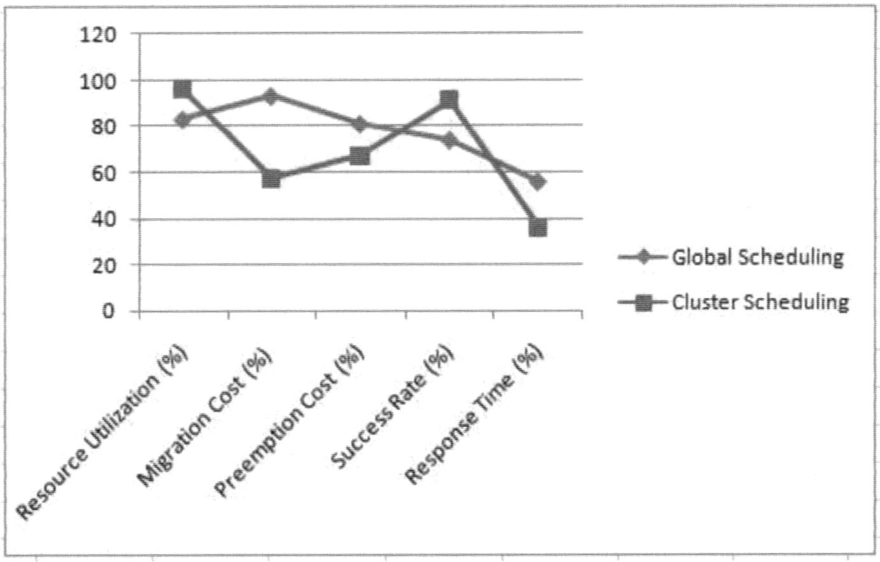

Figure 4. Performance Analysis of Cluster Scheduling

The proposed scheduler increase resource utilization and success rate by approximately 14% and 22%. The response time using proposed scheduler reduces by 55%. The migration and preemption overhead with proposed scheduler is reducing by approximately 63% and 20% respectively.

5. Conclusions

Smart city plays an important role in urban and social development. Emergency resource management is challenging tasks due to new requirement of smart city such as real time response, distributed recourses and self organization. The propose cluster based real time resource scheduling group the edge computing nodes into cluster and different disaster management tasks scheduled on these cluster computing nodes for real time response. Cluster scheduling ensures the efficient scheduling of emergency applications tasks. The proposed cluster based real time scheduler increase resource utilization and reduces response time. Cluster based real time scheduler reduces migration, preemption and response time. The tasks acceptance rate that is success rate is increased in proposed resource scheduling algorithm for smart city applications.

In future the proposed resource sharing framework applicable in other application areas such as defense, agriculture, etc for efficient resource scheduling.

References

[1] Li C, Dai Z, Liu X and Sun W 2020 *Sustainability* **12** 7: 2957
[2] Vlacheas P ,Giaffreda R. , Stavroulaki V. ,Kelaidonis D ,Foteinos V. , Poulios G. , Demestichas, P , Somov A. , Biswas,A.R. and Moessner K2013 *IEEE Communications Magazine* **51** 102-111
[3] Zheng X, Li M, Guo J 2020 *International Journal of Communication System*
[4] Sodhro A, Pirbhulal S, Luo Z, Hugo V 2019 *Journal of Cleaner Production* **220** 1167-79
[5] Chen Y, Tang S and Zhou L 2018 *Inform Sci* **433** 510-526.
[6] Tang H, Zhang X, Fu S, Liu X, Wu Q and Qi L 2020 *Transaction on Emerging Telecommunication Technology*
[7] Zhang Y and Gu X 2018 Cluster Computing. **22(S4)** 9445-9453
[8] Ahmed E, Chatzimisios P, Gupta B, Jararweh Y and Song H. 2018 *Transaction on Emerging Telecommunication Technology* **29(4)** e3307-09
[9] Law K and Lynch J 2019 *IT Professional*. **21(6)** 46-51.
[10] Gascó M and Hernandez M 2018 *Communications of the ACM* **61(4)** 50-57
[11] Liu M, Yao X and Li Y 2020 *Applied Soft Computing* **87** 105954
[12] Qureshi B 2019 *Future Generation Computing System* **94** 453-467.
[13] Tang X, Liao X, Zheng J and Yang X 2018 *Cluster Computing* **21(3)** 1581-1593
[14] Chen M, Li W,Fortino G, Hao Y, Hu L, and Humar I 2018 *ACM Transactions on Internet Technololgy* **1**(1) 15
[15] Zhang W, Zhang Z, Zeadally S and Chao H 2018 *IEEE Communications Letters* **23** 4-7
[16] Deng Y, Chen Z, Yao X,Hassan S and Wu J 2019 *IEEE Access* **7** 14410-14421

Evaluation of Speckle Noise Reduction and Feature Enhancement in Prolapsed Mitral Valve Leaflet Echocardiography

C Preethi[1,4,5], M Mohamed Sathik[2,5] and S Shajun Nisha[3,4,5]

[1] Ph.D Research Scholar, [2] Principal, [3] Assistant Professor & Head
[4] PG & Research Department of Computer Science
[5]Sadakathullah Appa College, Affiliation of Manonmaniam Sundaranar University, Tirunelveli, Tamil Nadu, India

Corresponding author's e-mail: cpreethi292@gmail.com

Abstract. Rheumatic heart disease has a substantial impact on morbidity and mortality for both men and women in developing countries. It is a complication of autoimmune phenomenon known as acute rheumatic fever in response to group A streptococcus bacteria. It often causes damage to valves and lacks its functionality. Initial manifestations of prolapsed valves are evident in echocardiography in the form of valve bulging, commissural fusion and restricted leaflet motion. However, these echocardiogram images are inevitably degraded by multiplicative noise known as speckle noise at the time of image acquisition and transmission. Hence despeckling is of vital importance for enhancing ultrasound image quality. In this paper, a comparative analysis of well established despeckling filters compiled for speckle suppression and feature enhancement has been performed. In addition to this, the performance of these filters is validated using different quantitative metrics. The experimental result shows that feature enhanced Fast Non-Local Means (FNLM) filter outperforms other state-of-art filtering techniques in the aspect of higher PSNR value and preserves structural similarity in terms of SSIM value near to unity. In conclusion, the preprocessing intends to suppress speckle noise and enhancing perceived visual quality that aids for further development of virtual cardiac model.

1. Introduction

The wide accessibility and advancement of digital technology assist the medical experts to make the diagnosis of the disease more precisely, for instance, the X-rays, Computed Tomography (CT), Magnetic Resonance Imaging (MRI), Ultrasound (US) and Positron Emission Tomography (PET) imaging. Among these diagnosis-aid technologies, ultrasound turns into a primary modality in recent years due to its noninvasive, faster and benign environment [1][2]. Cardiac ultrasound, also termed as echocardiography remains the best modality choice for the diagnosis of cardiovascular diseases such as Rheumatic heart disease (RHD). It is regarded as the leading cause of death representing 60 to 65% of all global deaths. Although any cardiac valve may be affected, mitral valve is the most commonly affected valve, resulting in mitral stenosis or mitral regurgitation [3]. Stenosis occurs when the mitral valve opening is narrowed whereas regurgitation occurs when the mitral leaflets does not seal properly, inducing the blood to flow backward [4]. However, diagnosis of mitral valve abnormalities in echocardiogram videos is still an open challenging task. Speckle noise may be encountered in these videos due to scattering of ultrasound signals from the small rapidly moving structures within the soft tissues. Furthermore granular pattern of speckle limit the post-processing techniques such as valve segmentation and motion tracking [5]. Speckle noise is modeled as a multiplicative noise based on the following equation (1).

$$p(x, y) = q(x, y) * n_{mul}(x, y) + n_{add}(x, y) \quad (1)$$

Where $p(x, y)$ is the speckle corrupted image, $q(x, y)$ is the original ultrasound image, $n_{mul}(x, y)$ and $n_{add}(x, y)$ are the multiplicative and additive components [6]. Since the additive component of speckle noise is very small it can be ignored as given in equation (2)

$$p\ (x, y) = q(x, y) * n_{mul}(x, y) \quad (2)$$

Over the years, plenty of filtering techniques with better noise suppression capabilities have been introduced. The main aim of this study is concerning a thorough review of various classical filters such as fast non-local means filter, Gaussian filter, Median filter, Crimmins and Lee filter for removing speckle noise followed by contrast-limited adaptive histogram equalization (CLAHE) method to yield a desired level of feature enhancement in cardiac ultrasound images.

The rest of the paper is summarized as follows. Section II addresses the brief summary of well known speckle denoising approaches. Section III describes the framework for speckle suppression and feature enhancement of cardiac echo videos. Section IV deals with experimental results and analysis. Section V concludes the work summary and highlights the future directions.

2. Background

In the past decades, lots of filtering techniques have been contributed for medical ultrasound image despeckling. In general, these filters are based on spatial and frequency domain. Mohamed Yaseen Jabarulla et al. [7] they proposed a study based on spatial filters such as kuan, Frost, Mean, Median and Speckle Reducing Anisotropic Diffusion (SRAD) filters. Their results showed that denoising effect and overall visual quality of hepatic cyst ultrasound images is enhanced using SRAD filter with a PSNR value of 31.11 dB.

In [8], Mrunal N. Annadate et al. recommend a Hybrid Median Filter (HMF) which is a modification of median filter for feature preservation on fetal ultrasound videos. Despite of using three step ranking, this HMF filter failed to tackle the problem of computational complexity.

P M Shankar [9] analyzed the behavior of denoising based on local statistics in B mode ultrasound images. They have proposed a maximum likelihood adaptive filter depends on the variation of window size. The performance level of this filter preserves the edges in echoic target, but declining fine details in low frequency area.

In [10] Sumit Kushwaha et.al presented an approach for quantitative analysis of four state-of-the-art filters, namely Median, Lee, Frost and SRAD filters on B-mode Ultrasound images of human kidney. They suggested quantitative comparisons on the basis of PSNR and SSIM with varying kernel levels and iterations. The Experimental results have been proven that SRAD and Median filters are the remarkable filters for despeckling.

The proposed idea of non linear anisotropic diffusion filtering based on partial differential equation (PDE) has been incorporated by Sumit Kushwaha [11]. For the experimental analysis purpose, eight clinical B-Mode Ultrasound images of human liver from different patients have considered. The denoising behavior of existing speckle suppression filters such as Malik Filter, LEE Filter, Frost Filter and Anisotropic Diffusion for Memory Based on Speckle Statistics Filter (ADMBSS) has been compared with a proposed approach in terms of PSNR. The authors claimed that the proposed method revealed satisfactory results in comparison with standalone filters.

Adil H. Khan et.al [12] proposed a novel schur based approach where the ultrasound image is segmented into a small sequence of overlapping segments to obtain orthogonal vectors. A non empty subset of orthogonal vectors with maximum Eigen values has been selected for the maximum filtration of speckle noise. The proposed approach has been examined on both simulated and real ultrasound images of kidney, foetal neck, musculo skeletal nerve and lymph node and found outperforming with benchmarking techniques such as homomorphic wavelet despeckling (HWDS), Wiener, Frost and Gamma. Overall performance of the proposed schur method can be verified as the prominent method mainly on the basis of parameters PSNR (peak signal-to-noise ratio), SNR (signal-to-noise ratio), Alpha and CNR (carrier-to-noise ratio) with optimal resolution. However, speckle noise is either inefficiently suppressed or higher degree of distortion is relatively included into despeckled images.Hence, previous studies provide the realization of speckle suppression and feature enhancement to preserve the quality of the echocardiographic images.

3. Overview of Speckle Reduction and Feature Enhancement Framework

Preprocessing improves the interpretability of information in digital images for human perception. The core objectives of preprocessing deployed in this research work are noise reduction and contrast-limited adaptive histogram equalization (CLAHE) as depicted in Figure 1. Typical echocardiography videos used in our analysis were obtained from publicly available dataset echocardia (www.echocardia.com) [13]. A total of 10 mitral valve echo videos of subjects was collected from four standard perspective views, parasternal long axis (PLAX) view, parasternal short axis (PSAX) view, apical 4-chamber (A4C) view and esophageal view. The echo videos of patients affirmed with pathology of either mitral stenosis or mitral regurgitation, which was in MP4 format are reviewed for this study.

The obtained echo videos were sampled by a default sampling rate (25 frames per second) to individual frames and from each video ten successive frames were incorporated in the analysis. Furthermore, as all the subject's protected health information (PHI) must be anonymize to secure the patient's privacy, optical character recognition (OCR) was used to de-identify the clinical metadata annotations from each frame dealt with patient identification, date of image acquisition and so on [14]. Text within the image was identified using this OCR technique based on non-uniform contrast background. Suspected clinical PHI in a bounding box was redacted by reassigning with lowest pixel value typically black which would result in anonymized echo images. After de-identification of sensitive data, corrupted images were obtained by addition of speckle noise with variance 0.001. Further, speckle noise has been significantly suppressed using different filters with 3x3 mask size. Finally, the denoising behavior of aforementioned filters is further improved by using a CLAHE method with clip limit value one.

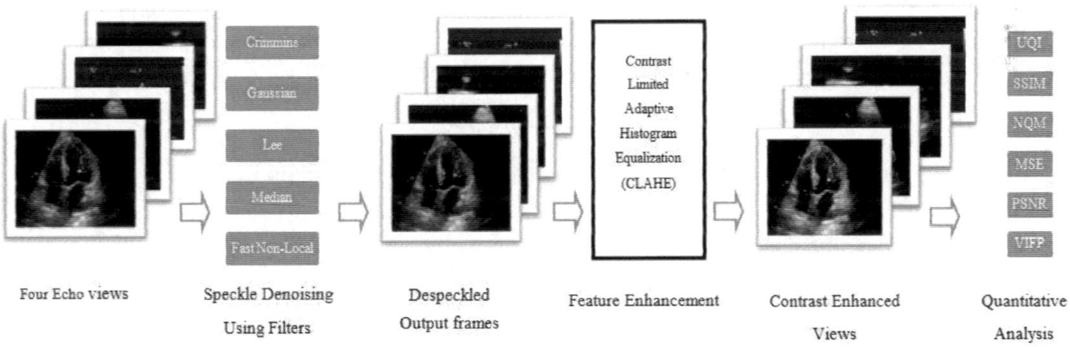

Figure 1. Framework for speckle denoising and feature enhancement

3.1. Speckle Reduction Filters

This section addresses speckle reduction filters and their mathematical implementations in detailed fashion.

3.1.1. Lee Filter

The Lee filter [15][16] adopts a linear multiplicative model based on minimum mean square error (MMSE). It is a patch based processing and every pixel in the noisy image is processed individually within the moving window or the kernel. It reduces the speckle noise by replacing the center pixel of a kernel with the average intensities of the window. This method can be implemented based on the following equation

$$D_{i,j} = \overline{Y} + W(C_p - \overline{Y}) \quad (3)$$

Where $D_{i,j}$ is the despeckled image, $\overline{Y}$ is the mean of the pixel values in the kernel, C_p is the center pixel of the kernel and weighting function W is given by

$$W = \frac{\rho^2}{\rho^2 + \sigma^2} \quad (4)$$

Where ρ^2 is the variance of pixel values in the kernel and σ^2 is the variance of the pixel values in the reference image.

3.1.2. Median Filter

Median Filter [17][18] which is used in our approach is a non linear filter based on order statistics. The idea of median filter is to replace the noisy value of the center pixel with a median of its pixel neighborhood. Prior to process the speckled image with this filter, at first all the pixels in the window must be sorted in numerical order. It is estimated using the mathematical equation as follows:

$$\hat{f}(u,v) = med(f(u-i, v-j)), i,j \in K \quad (5)$$

Where $f(u,v)$is the speckle corrupted original image, $\hat{f}(u,v)$is despeckled output image, K is two dimensional mask or kernel. Typically, size of a kernel is often selected as odd numbers and lowest window size needs to be chosen to reduce the time complexity of denoising.

3.1.3. Crimmins Speckle Filtering

The concept of geometric despeckle filter introduced by Crimmins [19] uses complementary hulling noise reduction technique. These filters are non-linear in the sense that it is either incrementing value of dark pixel or decrementing light pixel by comparing the intensity of the pixel with the 4 set of its neighbors (North to South, East to West, Northeast to Southwest and Northwest to Southeast). The denoising behavior of these filters smoothen the edges of ultrasound images as well as it is iteratively used over filtered image to reduce noise level.

3.1.4. Gaussian Filter

Gaussian filter [20] is a convolution based linear filter that replaces individual higher value pixels by a weighted average of pixels on the periphery. It suppresses speckle noise based on Gaussian distribution. The 2D Gaussian filtering with a mean value zero and standard deviation as one is given in equation (6)

$$G(x,y) = \frac{1}{2\pi^{\sigma^2}} e^{-\frac{(x^2+y^2)}{2\sigma^2}} \quad (6)$$

Where σ represents standard deviation and x and y are the coordinates of an image.

3.1.5. Fast Non-Local Means (NLM) filter

Non-Local Means is a patch based denoising method in which a target pixel is compared with similar patches and restored the particular target pixel by averaging the group of neighborhood pixels. This traditional method makes use of Gaussian Euclidean distance to find out cues of similar patches. The performance of this pixel wise approach deteriorates depends on the assumption that all the pixels can be linked to other pixels with self-similarities. The problem of computational complexity due to this pixel-wise method is minimized by using block-wise method, namely fast non-local means filtering [21]. Consider the image region over Ω as blocks $b_{i,j}$ and performing the same pixel-wise approach on each block. Let $NLM(b_{i,j})$ be the restored block on the image I and $w(i,j)$ be a weighted average of all pixels in block $b_{i,j}$. It can be formulated as in equation (7)

$$NLM(b_{i,j}) = \sum_{b_{i,j} \in \Omega} w(i,j) I(b_{i,j}) \quad (7)$$

When setting the argument *fast-mode* as true, uniform spatial weighting is employed to compute Euclidean distance between patches. Otherwise, Gaussian Euclidean distance is applied over the patches. Overall, it provides edge preservation and noise suppression that would be degraded by other denoising methods [22].

IOP Conf. Series: Materials Science and Engineering 1085 (2021) 012033 doi:10.1088/1757-899X/1085/1/012033

3.2. Contrast-Limited Adaptive Histogram Equalization (CLAHE)

Contrast-Limited Adaptive Histogram Equalization (CLAHE) method significantly enhancing low contrast of ultrasound frames at good scale. The method partitions the image into non- overlapping tiles that contain an M number of pixels using desired tile size. After that Histogram is computed for each tile, part of the histogram bin that exceeds the corresponding clip limit is clipped. Furthermore, the amount of clipped pixels is distributed by any one of the distribution functions, namely uniform, Rayleigh and Exponential [23]. Finally, transform function is computed for neighboring tiles and interpolation method is adopted for binding the transform function to obtain the enhanced image. In our methodology, uniform distribution in combination with Bilinear Interpolation [24][25] is applied over the pixels to obtain desired contrast levels as shown in Figure 3.

4. Experimental Results and Analysis

The framework for the echocardiography video speckle suppression using denoising filters combined with contrast-limited adaptive histogram equalization improves the overall visual quality of video frames. In this section, qualitative analysis is performed with prolapsed mitral valve echocardiography images. To evaluate the performance, the original echo images are corrupted with noise variance 0.001 and window size 3x3 is selected to filter noisy cardiac images .Visual comparison of all speckle suppression filters that constitutes four cardiac views is presented in Figure 2. Out of all filters, the fast non-local mean filter excels over the other filtered output in terms of visual quality. Figure 3 shows the output of CLAHE method based on a uniform distribution .The output generated using fast non-local mean filter provides an acceptable level of contrast enhancement with clip limit value 1.

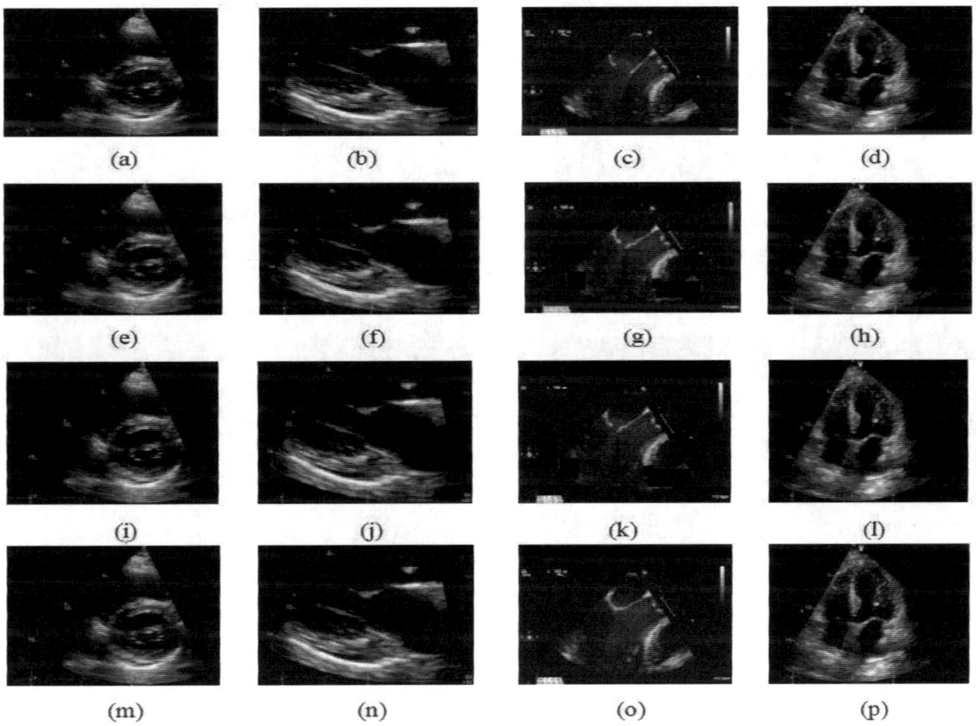

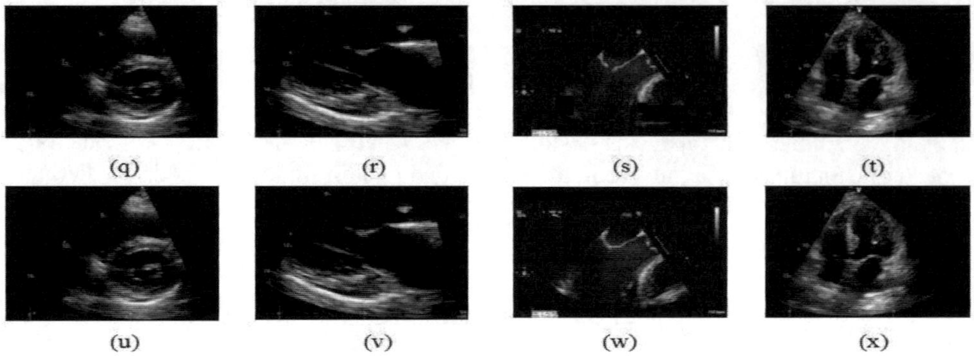

Figure 2. Comparison of various despeckling filters. (a)-(d) Four Perspective echo viewpoints of mitral valve frames with noise variance 0.001. *Left to right:* parasternal short axis, parasternal long axis, esophageal view, Apical four chamber view, (a)-(d) Noisy images,(e)-(h) Denoised images using Lee filter,(i)-(l) Denoised images using median filter,(m)-(p) Denoised results of Crimmins,(q)-(t) Denoised images using Gaussian,(u)-(x) Denoised results of Fast non-local means.

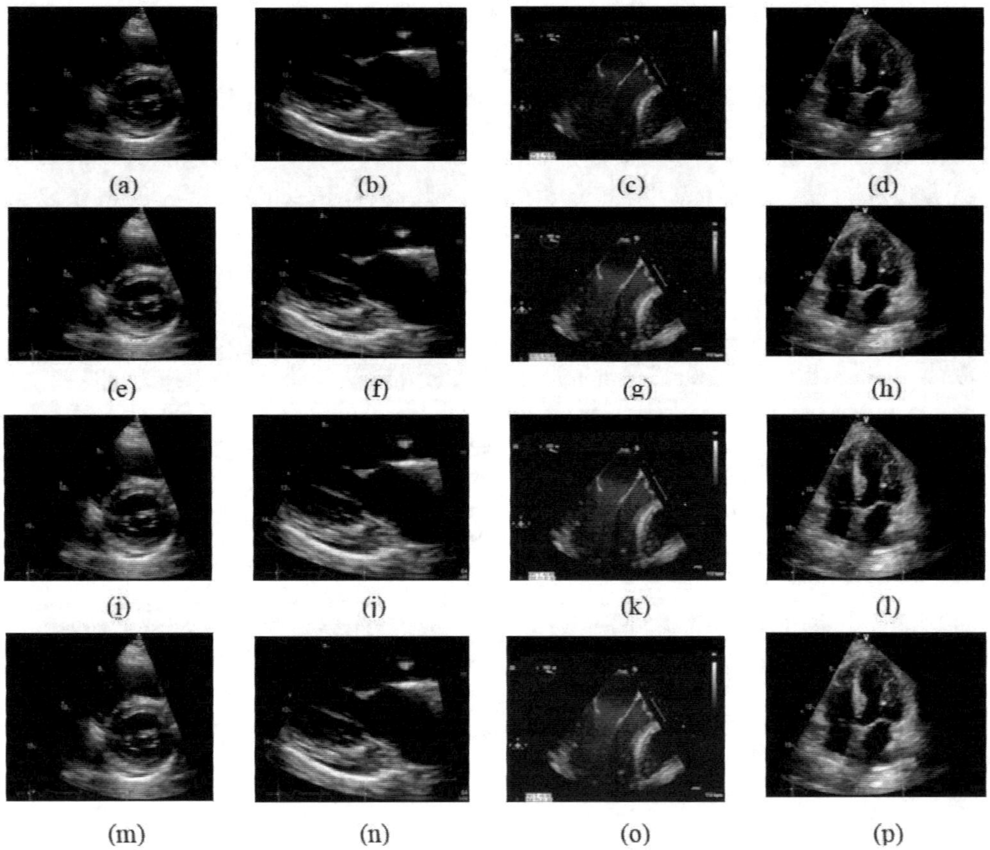

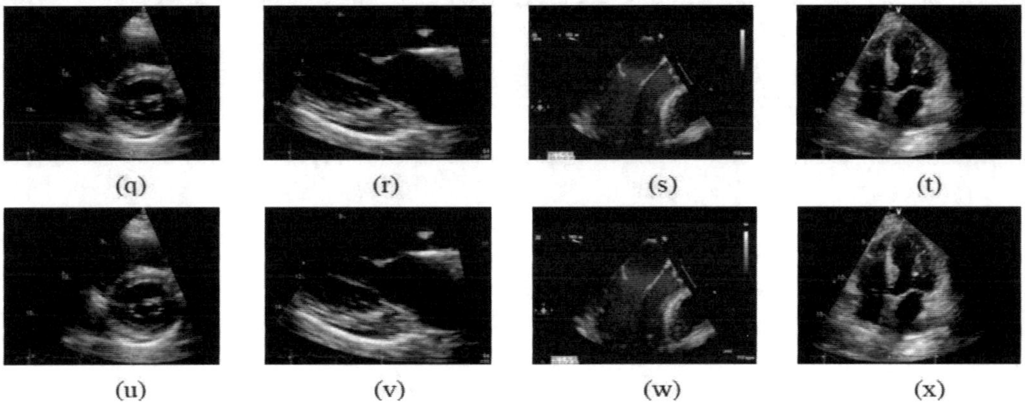

Figure 3. Speckle-filtered contrast enhanced echo frames using CLAHE algorithm on different filters. First row shows the original images from four perspective echo view points, second row shows the speckle-filtered contrast enhanced output with Lee and Median filter, Third row and fourth row depicts the output with Crimmins and Gaussian, fifth row outputs the enhanced features and edges with fast non-local mean filter.

Table 1, 2 and 3 presents resultant performance analysis of four echocardiography views based on metrics such as mean squared error (MSE), peak signal-to-noise ratio (PSNR), The Structural Similarity Index (SSIM), Universal Quality Index (UQI), Pixel-based Visual Information Fidelity (VIFP), Noise Quality Measure (*NQM*). From the observing overall performance of Table 4, it is evident that denoising effect achieved by fast non-local means filter is more than up to par with higher PSNR and low MSE values, though the behavior of Gaussian also retained similar higher values. Meanwhile Figure 4 depicts an average graphical variation of speckle suppression results as measured by different metrics.

Table 1. Cardiac view based performance analysis in terms of PSNR and SSIM.

	PSNR					SSIM				
	Crimmins	FastN1	Gaussian	Lee	Median	Crimmins	FastN1	Gaussian	Lee	Median
PLAX	0.64022	0.832306	0.845355	0.312752	0.58827	0.868037	0.990874	0.993364	0.845712	0.96324
PSAX	0.641035	0.898397	0.861226	0.377294	0.61369	0.934232	0.993382	0.995426	0.840189	0.971151
A4C	0.646836	0.852189	0.843328	0.345792	0.58143	0.859982	0.996824	0.994082	0.861449	0.960758
Eso phageal	0.597907	0.844974	0.869521	0.264345	0.704023	0.967936	0.999706	0.992091	0.856252	0.984164

Table 2. Cardiac view based performance analysis in terms of UQI and MSE.

	UQI					MSE				
	Crimmins	FastN1	Gaussian	Lee	Median	Crimmins	FastN1	Gaussian	Lee	Median
PLAX	0.723068	0.931058	0.973978	0.870946	0.934233	9.612673	2.558305	2.504329	62.28427	43.46513
PSAX	0.778664	0.944226	0.979499	0.885657	0.946049	7.349555	1.817618	1.759247	38.307152	28.34722
A4C	0.633661	0.989767	0.967559	0.889676	0.942545	11.24112	2.419919	2.842148	52.64713	44.79878
Eso phageal	0.977909	0.989523	0.915755	0.951962	0.98485	4.813599	1.534757	1.597735	35.171993	18.90657

Table 3. Cardiac view based performance analysis in terms of VIFP and NQM.

	VIFP					NQM				
	Crimmins	FastN1	Gaussian	Lee	Median	Crimmins	FastN1	Gaussian	Lee	Median
PLAX	0.64022	0.832306	0.845355	0.312752	0.58827	31.59618	42.98828	44.0803	24.8329795	38.34751
PSAX	0.641035	0.898397	0.861226	0.377294	0.61369	32.35361	42.92463	44.21692	23.4753697	38.43324
A4C	0.646836	0.852189	0.843328	0.345792	0.58143	32.17508	43.54755	44.38987	26.9631126	38.37024
Eso phageal	0.597907	0.844974	0.869521	0.264345	0.704023	31.02127	46.0564	42.19617	19.9239791	36.84979

Table 4. Overall Performance Analysis of Various Filters.

	PSNR	SSIM	UQI	MSE	VIFP	NQM
Crimmins	39.17681	0.907547	0.778326	8.254237	0.6315	31.78653
Fast Non-Local[a]	**44.90212**	**0.995197**	**0.963644**	**2.08265**	**0.856967**	**43.87922**
Gaussian	44.89687	0.993741	0.959198	2.175865	0.854858	43.72082
Lee	27.44004	0.89956	0.89956	47.1026363	0.325046	23.79886
Median	33.10179	0.951919	0.951919	33.87942	0.621853	38.0002

[a]Best performance metrics are highlighted as bold

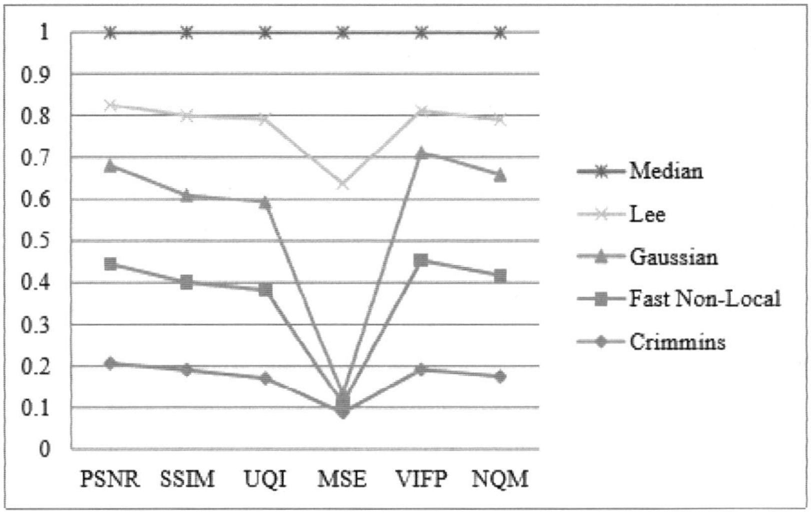

Figure 4. Average variations of resultant image quality as measured by different metrics

5. Conclusion and Future Directions

This paper has explored the complete pipeline for speckle suppression and feature enhancement of mitral valve echocardiography videos. The processing pipeline used various denoising filters and limits over amplification of noise using contrast adaptive feature enhancement (CLAHE) algorithms. The presented work proved that Fast non-local means filter outperforms the other filtering techniques in terms of resultant image visual quality and structural edge preservation. Besides, quantitative analysis is also evaluated in terms of different quality metrics. The experimental analysis suggested that CLAHE based on the uniform distribution in combination with bilinear interpolation highlight the features with denoising. Future work can be extended with various file formats and noise levels to enhance overall efficiency as well, which can be further directed towards the development of virtual cardiac model.

References

[1] Huang Q and Zeng Z 2017 A Review on Real-Time 3D Ultrasound Imaging Technology *BioMed Research International* **2017**

[2] Wang S, Huang T Z, Zhao X Le, Mei J J and Huang J 2018 Speckle noise removal in ultrasound images by first- and second-order total variation *Numerical Algorithms* **78** 513–33

[3] Sultan M S, Martins N, Costa E, Veiga D, Ferreira M J, Mattos S and Coimbra M T 2019 Virtual M-Mode for Echocardiography: A New Approach for the Segmentation of the Anterior Mitral Leaflet *IEEE Journal of Biomedical and Health Informatics* **23** 305–13

[4] Harb S C and Griffin B P 2017 Mitral Valve Disease: a Comprehensive Review *Current Cardiology Reports* **19**

[5] Perperidis A 2016 Postprocessing Approaches for the Improvement of Cardiac Ultrasound B-Mode Images: A Review *IEEE Transactions on Ultrasonics, Ferroelectrics, and Frequency Control* **63** 470–85

[6] Dass R 2018 Speckle Noise Reduction of Ultrasound Images Using BFO Cascaded with Wiener Filter and Discrete Wavelet Transform in Homomorphic Region *Procedia Computer Science* **132** 1543–51

[7] Jabarulla M Y and Lee H N 2018 Evaluating the effect of various speckle reduction filters on ultrasound liver cancer images *International Conference on Electronics, Information and Communication, ICEIC 2018* **2018-Janua** 1–4

[8] Annadate M N and Nagmode M S 2018 Efficient Speckle Reduction Techniques in Ultrasound Video Quality Enhancement-A Comparative Evaluation *International Conference on Current Trends in Computer, Electrical, Electronics and Communication, CTCEEC 2017* 772–8

[9] Shankar P M 2006 Speckle reduction in ultrasonic images through a maximum likelihood based adaptive filter *Physics in Medicine and Biology* **51** 5591–602

[10] Kushwaha S and Singh R K 2017 Performance comparison of different despeckled filters for ultrasound images *Biomedical and Pharmacology Journal* **10** 837–45

[11] Kushwaha S 2017 An efficient filtering approach for speckle reduction in ultrasound images *Biomedical and Pharmacology Journal* **10** 1355–67

[12] Khan A H, Al-Asad J F and Latif G 2018 Speckle suppression in medical ultrasound images through Schur decomposition *IET Image Processing* **12** 307–13

[13] Echocardia, The Echocardiography Resource, https://echocardia.com/en/searchassist.html

[14] Newhauser W, Jones T, Swerdloff S, Newhauser W, Cilia M, Carver R, Halloran A and Zhang R 2014 Anonymization of DICOM electronic medical records for radiation therapy *Computers in Biology and Medicine* **53** 134–40

[15] Shereena V B and Raju G 2019 Role of Filters in Speckle Reduction in Medical Ultrasound Images-A Comparative Study *2019 2nd International Conference on Intelligent Computing, Instrumentation and Control Technologies, ICICICT 2019* 271–8

[16] Hosseini Z and Bibalan M H 2018 Speckle noise reduction of ultrasound images based on neighbor pixels averaging *2018 25th Iranian Conference on Biomedical Engineering and 2018 3rd International Iranian Conference on Biomedical Engineering, ICBME 2018* 1–6

[17] Gungor M A and Karagoz I 2017 The effects of the median filter with different window sizes for

ultrasound image *2016 2nd IEEE International Conference on Computer and Communications, ICCC 2016 - Proceedings* 549–52

[18] Zhu Y and Huang C 2012 An Improved Median Filtering Algorithm for Image Noise Reduction *Physics Procedia* **25** 609–16

[19] Mustafa Z A, Abrahim B A, Omara A, Mohammed A A, Hassan I A and Mustafa E A 2020 Reduction of Speckle Noise and Image Enhancement in Ultrasound Image Using Filtering Technique and Edge Detection *Journal of Clinical Engineering* **45** 51–65

[20] Vasavi G and Jyothi S 2019 A comparative study of speckle reduction filters for ultrasound images of poly cystic ovary *International Journal of Recent Technology and Engineering* **7** 2084–9

[21] Shreyamsha Kumar B K 2013 Image denoising based on non-local means filter and its method noise thresholding *Signal, Image and Video Processing* **7** 1211–27

[22] Gupta P K, Lal S, Kiran M S and Husain F 2018 Two dimensional cuckoo search optimization algorithm based despeckling filter for the real ultrasound images *Journal of Ambient Intelligence and Humanized Computing* **0** 1–22

[23] Singh P, Mukundan R and De Ryke R 2020 Feature Enhancement in Medical Ultrasound Videos Using Contrast-Limited Adaptive Histogram Equalization *Journal of Digital Imaging* **33** 273–85

[24] Zheng R, Guo Q, Gao C and Yu M A 2019 A hybrid contrast limited adaptive histogram equalization (clahe) for parathyroid ultrasonic image enhancement *Chinese Control Conference, CCC* **2019-July** 3577–82

[25] Kharel N, Alsadoon A, Prasad P W C and Elchouemi A 2017 Early diagnosis of breast cancer using contrast limited adaptive histogram equalization (CLAHE) and Morphology methods *2017 8th International Conference on Information and Communication Systems, ICICS 2017* 120–4

IOP Conf. Series: Materials Science and Engineering 1085 (2021) 012034 doi:10.1088/1757-899X/1085/1/012034

Energy Efficient Cluster Based Algorithm Technique for Wireless Sensor Networks

G.Ramya[1], R.Nagarajan[2], S.Kannadhasan[3]
[1]Research Scholar, Department of Electrical and Electronics Engineering, Gnanmani College of Technology, Tamilnadu, India, ramya.kgs@gmail.com,
[2]Professor, Department of Electrical and Electronics Engineering, Gnanamani College of Technology, Tamilnadu, India, krnaga71@yahoo.com
[3]Assistant Professor, Department of Electronics and Communication Engineering, Cheran College of Engineering, India, kannadhasan.ece@gmail.com

Abstract: - Wireless sensor network [WSN] energy efficiency is the most critical task of researchers. Power regulation is a very effective technique for interference minimizing in WSN and energy consumption. WSN with several nodes linked to the network, including transmitting power to measure network output. The energy usage is connected to node size and weight. Aggregation data secure in WSN is a highly challenging activity. In this research paper suggested stable aggregation of data using fuzzy logic, based on clustering techniques. The clustering method is done in a network. The distance power consumed and faith value are measured for each cluster. The stable data aggregation using fuzzy logic techniques (FLT) is based on these parameters. The proposed research work would minimize energy usage to increase the lifespan of WSN.

1. Introduction

Wireless sensor networks are one of twenty-first century's important developments. MEMS and wireless networking technology have made recent advancements in light weight, inexpensive, smart sensors with wireless connection for a variety of military, civilian purpose, including environmental tracking, war surveillance, and industrial process management [1]-[5]. Temperature, light, sound and humidity were all included. Network attacks in the categories of spoofed, modified, (or) replayed, limited forwarded sinkhole, HELLO flood and wormholes attack, Spoofing targets of routing protocol aimed at sharing routing information between nodes. Malicious nodes can fail to transmit such messages to ensure the message for multi-hop networks does not spread. Wireless sensors are one of the most prevalent networks that senses our atmosphere by means of different parameters such as heat, temperature, strain, etc.,. Due to the dense deployment of disposable sensor nodes and minimum-cost sensor systems, destruction of such nodes with hostile activity does not affect an operation military and destruction of a conventional sensor conducting an operation [6]-[8].

The transmission between the two nodes minimizes other nodes to boost their efficiency and neglect power controls, which are better than spatial reuse than wireless networks. Adaptive power transfer technologies in wireless sensor networks to maximize network life span [9], [10]. The data from the cluster heads would be sent directly to the sink node. The data from sink nodes for the distance calculation while the cluster head is transmitted through the shortest multi-hop route and the shortest route to the sink node between each cluster.

In order to locate the prevailing node [maximum route number], power transmission strategies boost network efficiency in many respects. In wireless networks, the propagation range can adjust the ranges of each connection [11]-[15]. Capacity reduces as additional nodes are introduced to maximize intervention. Graph theory from a data point to a neighbour's node is redirected through several pathways.

2. Data Gathering For WSN

This paper detects shortest route to relay power from neighbouring network nodes and improve network life through Fuzzy Logic. It consists of three steps for enhancing network capacity using adaptive transmission technologies. Data collection is one of the key sensor node priorities. It includes the processing and transfer of sensed data from various sensors to BS. This is the process by which data from multiple sensors may be aggregated to eliminate redundant data from the BS. The different problems in the following categories such as:

- The energy savings in some nodes for the immediate transmission of data to BS.
- Data accumulation to establish clustering strategies to maintain the capability of the sensor.
- Performance is also increased.

The vertices of the network nodes and direct communication between the edges of the nodes are depicted. Sensor nodes (SN) are the maximum flow of the distance measured from one node to the other node. As a network system the combinatorial structure is named the number of vertices in the network is related to the SN. More number of boundaries network that connects the similar vertical pair was considered several sides, and is called a multi-path network is shown in Figure 1.

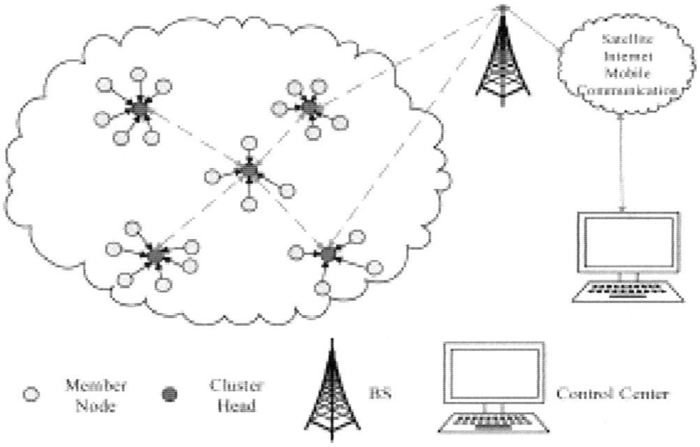

Figure 1. Architecture of WSN

The cluster approaches to find shortest way in WSN using graph theory. Three stages of modules are as follows:

i. Distance Evaluation
ii. Energy Evaluation
iii. Shortest Route Evaluation

2.1 Distance Evaluation

The gap of all other nodes in each cluster, using algorithm is evaluated by the A and B of each node in cluster network.

$$L = \{\{ A_2 - A_1\}^2 + \{ B_2 - B_1\}^2\}\}^{1/2} \quad (i)$$

Where L is distance from (A_2, A_1) and (B_2, B_1) is denoted by Euclidean distance

2.2 Energy Evaluation

The equation (ii) is used to evaluate the energy,

$$E = E_{int} - ((S_t * R_{tx}) + (P_r * T_{rx})) \quad (ii)$$

Where,

E_{int} - Energy level at Initial condition for each cluster
S_t - Node transmitted no. of packets
P_r- Node received no. of packets
R_{tx}- Required energy for transmission of Packet
T_{rx}- Required total energy for each packet

2.3 Shortest Route Evaluation

The energy evaluation and distance measurement centred on their shortest route to relay the data from the CH to the BS by expanding the network life based on their shortest route of transmitting data from CH to the BS by extending the existence of the network. The shortest route approximation using graph representations x and y for two vertices represented by L[x, y]
1. L [x, y] ≤ 0
2. L [x, y] ≥ 0
3. L [x, y] = 0

3. Fuzzy Logic for WSN

The SN is clustered into many clusters with cluster head (CH) and ID. The CH distance calculation and the topology of the packet discovery. The cluster head collects the data of each participant and sets the energy level for the algorithm.

$$Energy = P * T$$

The Fuzzy Logic (FL) is eventually used to pick the right aggregation nodes and parameters such as faith value, distance to each and every node and power via CH as input data and FL rules are created. The FL rules regarded as the good node (GN) and bad node (BN) depending on their performance. The good nodes are added to the CH-ID, which is sent to BS. The CH-ID is related to since the nodes are the largest node of the network. Degree of critical input steps and the state of the fluid logic is high. It is elegant on the basis of the norm. The findings are derived from several fuzzy laws with single cumulative results. The blurry sets of A with X rules membership are calculated. Antecedents 1 and 2 are tiny, which implies that they are strong. Fuzzy set membership function (FSMF) for GN and BN in Cluster-ID for distance, power and faith value evaluation are given below,

Distance (L) = { [GN, h], [BN, i] }
Where,
h- FSMF-GN Cluster-ID for distance evaluation
i- FSMF-BN Cluster-ID for distance evaluation

Power (p) = { [GN, j], [BN, k] }
Where,
j- FSMF-GN Cluster-ID for power evaluation
k- FSMF-BN Cluster-ID for power evaluation

Faith Value (FV) = { [GN, l], [BN, m] }
Where,
l- FSMF-GN Cluster-ID for faith value evaluation
m- FSMF-BN Cluster-ID for faith value evaluation

3.1 Neighbour Identification

Table 1: FLT Decision Making

Distance D	Power Consumed P	Faith Value F	Consequences
Low	High	Low	Worst
High	High	High	Worst
Low	Less	Low	Worst
High	Less	Low	Best
Low	Less	High	Best
High	Less	Low	Best
Low	Less	High	Best
High	High	High	Worst

The vertices of the network nodes and direct communication between the edges of the nodes are depicted. Sensor nodes are the maximum flow of the distance measured from one node to the other node. The composite system is referred to as a network structure. The number of vertices in the network is related to the source node and height of edges. Two or three of the boundaries of network that connects the same vertical pair are considered several sides, and is called a multi-path network. The majority of the weakest nodes are replaced in the network is discussed in Table-1.

4. Results and Discussion

Network Simulator Version 2, generally known as NS2, is a platform that can be used to model incidents and to analyze the complex existence of communications networks. Wired and wireless network functions and protocols can be simulated using NS2, i.e. routing algorithms, TCP and UDP. In general, NS2 allows users a way to define and simulate network protocols is discussed in Table 2. A node performs two main functions in NS2. It provides packages as a host to the port-attached transport layer agent Specified in the header. NS2 packet configuration Link to its Ns Artifacts downstream only.

A Node does not have to be linked to it Upstream NS Entity (e.g., transportation agent sending A connection upstream). Its upstream NS Entity instead creates a node relation. NS2 default the node is based on static and flat addressing. With flat addressing, any new node has an address Increased by one from the previous generated node. The routing table is determined at once Simulation process start and does not start Shift after that. NS2 hires the shortest Dijikstra Road algorithm for estimating maximal routes for all pairs of Nodes. It's a host Delivers packets connected to the transport layer agent and the NS2 environment Tcl, script and C++ implementation. Nodes were transmitted one after the other at a given interval. The clustering strategy of relay data to the (BS) is seen in Figure 2.

Table 2: Simulation Results of the Network

S.No	Parameter	Value
1.	Transmitting Power	**80mW**
2.	Receiving Power	**60mW**
3.	Idle Power	**8.5mW**
4.	Inactive Power	**16.2mW**
5.	Energy per Individual node	**50J**
6.	**Weighting Parameters W_1 and W_2**	**0.6, 0.8**

Figure 3 shows energy consumption of network and red colour indicates the existing work of energy consumption and also green indicates the proposed work of energy consumption. Throughput secured data aggregation using FLT is shown in Figure 4. Figure 5 shows that the Packet Drop Ratio in secured data aggregation using FLT.

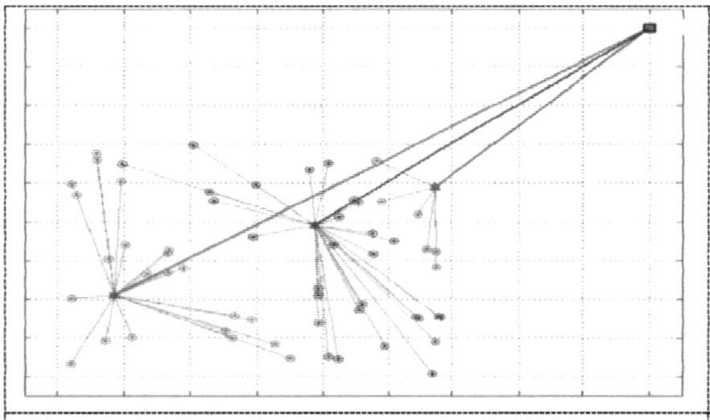

Figure 2. Clustering Techniques of WSN

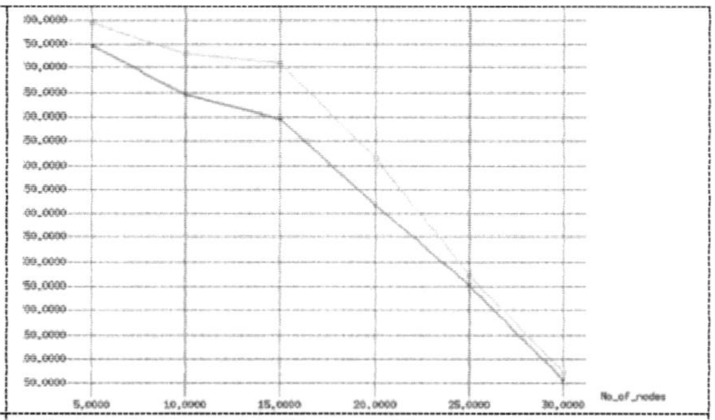

Figure 3. Average Energy Consumption.

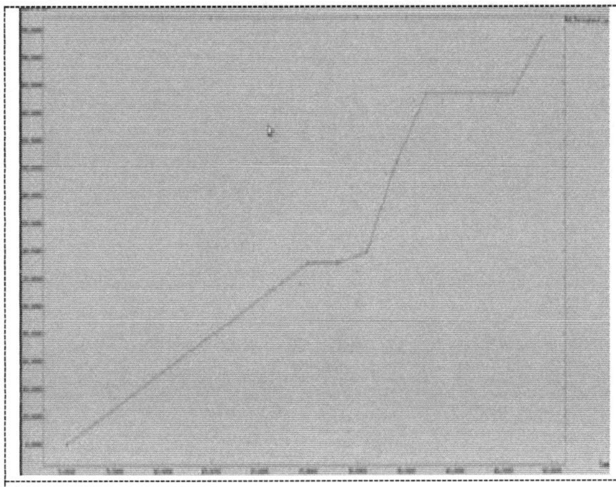

Figure 4. Throughput of the Network

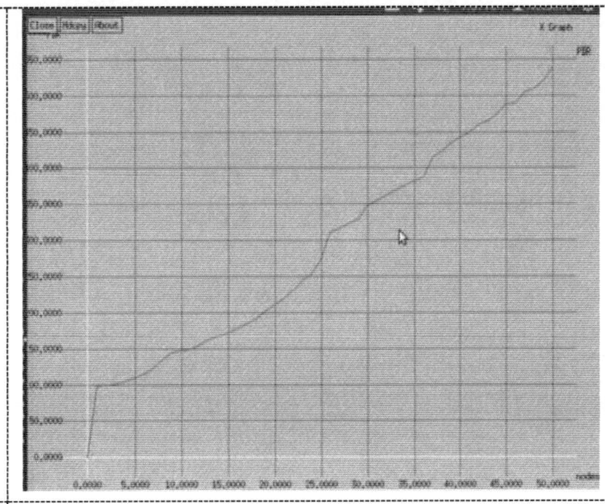

Figure 5. Packet Drop Ratio of the Network

5. Conclusion

The energy consumption rates for sensors in a wireless sensor network differ significantly in compliance with the protocols that the sensors use. Results on energy usage, transmitted and obtained minimum voltage supply required for service, energy consumption impact of power transmission and numerous life time calculation methods of a sensor node are discussed. The activity of sensor nodes is identified and evaluated near to their end of existence.

References

1. G. Arumugam, T. Ponnuchamy, 2015, "EE-LEACH: development of energy-efficient LEACH Protocol for data gathering in WSN", *EURASIP Journal on Wireless Communications and Networking* ,76, DOI 10.1186/s13638-015-0306-5.
2. Hassan Oudani, Salahddine Krit, Lahoucine Elmaimouni and Jalal Laassiri, 2016, "Comparative Study and Simulation of Flat and Hierarchical Routing Protocols for Wireless Sensor Network", *International Conference In Engineering & MIS (ICEMIS), IEEE*, pp. 1-9.
3. J. Jia, J. Chen, G. Chang, Y. Wen and J. Song, 2009, "Multi-objective optimization for coverage control in wireless sensor network with adjustable sensing radius", *Elsevier, Computers and Mathematics with Applications*, Vol. 57, pp. 1767-1775.
4. T. Qiu, F. Xia, L. Feng, G. Wu and B. Jin, 2011, "Queueing theory based path delay analysis of wireless sensor networks", *Advances in Electrical and Computer Engineering (AECE)*, vol. 11, no. 2,.
5. X. Zhang, X. Ding, S. Lu and G. Chen, 2009, "Principles for Energy-Efficient Topology Control in Wireless Sensor Networks", *WiCom '09. 5th International Conference on Wireless Communications, Networking and Mobile Computing, IEEE*, pp. 24-26.
6. Murat Dener, 2014, "Security Analysis in Wireless Sensor Networks", *International Journal of Distributed Sensor Networks,* Vol.10, No.10, pp.303501,.
7. David Martins and Herve Guyennet, 2010, "Wireless Sensor Network Attacks and Security Mechanisms: A Short Survey", *13th International Conference on Network-Based Information Systems*, pp. 313-320,.
8. C. Dhivya Devi and B.Santhi, 2013, "Study on Security Protocols in Wireless Sensor Networks", *International Journal of Engineering and Technology*, Vol.5, No.5, pp.200-207,.
9. M. Revathi and Dr. B. Amutha, 2017, "A Survey on Security Protocols in Wireless Sensor Network", *International Journal of Pure and Applied Mathematics*, Vol.5, No.5, pp.250-257
10. Fasee Ullah; Masood Ahmad; Masood Habib; Jawad Muhammad, 2011, "Analysis of Security Protocols in Wireless Sensor Network", *3rd International Conference on Computer Research and Development,* DOI**:** 10.1109/ICCRD.2011.5764156

11. Yifan Hu, Yongsheng Ding, Lihong Ren, Kuangrong Hao and Hua Han, 2015, "An endocrine cooperative particle swarm optimization algorithm for routing recovery problem of wireless sensor networks with multiple mobile sinks", *Information Sciences, Elsevier*, Volume 300, pp. 100-113
12. Shusen Yang, Usman Adeel, Yad Tahir, and Julie A. McCann, 2017, "Practical Opportunistic Data Collection in Wireless Sensor Networks with Mobile Sinks", *IEEE Transactions on Mobile Computing*, Volume 16, Issue 5, pp.1420 – 1433
13. Mai Abdelhakim, Yuan Liang and Tongtong Li, 2017, "Mobile Access Coordinated Wireless Sensor Networks – Design and Analysis", *IEEE Transactions on Signal and Information Processing over Networks,* Volume 3, Issue, pp.172 – 186
14. S. Kannadhasan and R.Nagarajan , 2018, "Performance Improvement of Slot Antenna Using Various Parameters and Band Pass Filter," 2018 *International Conference on Circuits and Systems in Digital Enterprise Technology (ICCSDET), IEEE Press*, pp-128-131.
15. Yongmin Zhang, Shibo He and Jiming Chen, 2017, "Near Optimal Data Gathering in Rechargeable Sensor Networks with a Mobile Sink", *IEEE Transactions on Mobile Computing*, Volume 16, Issue 6, pp.1718-1729

Real time supervision of faults in converter of PV system

P R Jadhav[1], S B Chavan*[2]
[1]Student, Department of Technology, Shivaji University, Kolhapur, 416004
[2]Department of Technology, Shivaji University, Kolhapur, India, 416004
*Corresponding author's Email ID- sbc_tech@unishivaji.ac.in

Abstract- PV systems are remotely installed at unmanned places. They have sub-systems like converters, inverters etc. Electronic components in these circuits work under electrical stress conditions. PV systems occupy huge land area; therefore system supervision and fault diagnosis are critical in these applications. In this work internet connectivity based supervising system is presented for boost converter to monitor its working status. The system implemented at site monitors the converter output, voltage across the devices, compares it with the known standard values and records it. Whenever any remote node accesses it, the formatted information is displayed which shows the converter status. Due to this remote site supervision and electrical parameter observation is easy.

1. Introduction

Solar PV arrays are in great demand for energy generation. Converters and inverters are important power processing part of PV systems. Electronic devices in power processing circuits work under electrical stress and changing atmospheric conditions. Due to this these devices are failure prone. Researchers carried industrial surveys and signified the reliability aspects in converters of PV systems, as described in references [1] to [7]. Few researchers developed fault tolerant converter topologies in which redundant components were activated on arrival of fault for fault recovery [8],[9]. Various techniques are presented for fault detection, faulty component identification and studying fault signatures and precursors [10]-[13]. A trend is also seen in this area to regularly supervise the system performance and the real time faults using communication technologies like internet, GSM, GPRS, ZigBee etc., such systems are described in [14]-[19]. Considering the need of remote supervision, fault prone behavior and reliability issues in remotely located PV systems, this work focus on design of internet based supervising system for boost converter in PV system for remote monitoring the component abnormalities. In this the voltage at device nodes are measured to monitor the converter status remotely using smart phone or computer.

2. System implementation

Boost converter is designed for PV panel; output voltage, voltage across IGBT, diode, and output capacitor is monitored and processed. The electrical parameters across the devices under observation are monitored and recorded after certain fixed interval. The existing parameters are also compared with the known standard values. The information is formatted and recorded systematically. The boost converter design specifications are given in table 1.

Table 1- Boost converter design specifications

Electrical Parameter	Value
PV input voltage (typical)	19 V
Boost converter Voutput	53 V
Converter efficiency	80 %
Duty cycle	65 %

The system is designed around Arduino Uno board, TL 494 PWM controller IC is used for IGBT gate control, ACS 712 current sensor is used for PV current sensing. Figure 1 shows system block diagram. System simulation model is shown in figure 2. The model is used to find the voltages across devices in normal mode and component faulty modes. Component level OC and SC faults and their fault signatures are studied in simulation model, these fault signatures are implemented in actual system to find the faulty component. Deviation in the voltage gain of converter indicates the occurrence of any component fault. When such voltage gain deviation is detected, the voltage across components of the converter like IGBT, diode, capacitor is measured. If the measured voltage across component under test is not in the desired range, it indicates the possibility of that component fault. System flow chart is shown in figure 3.

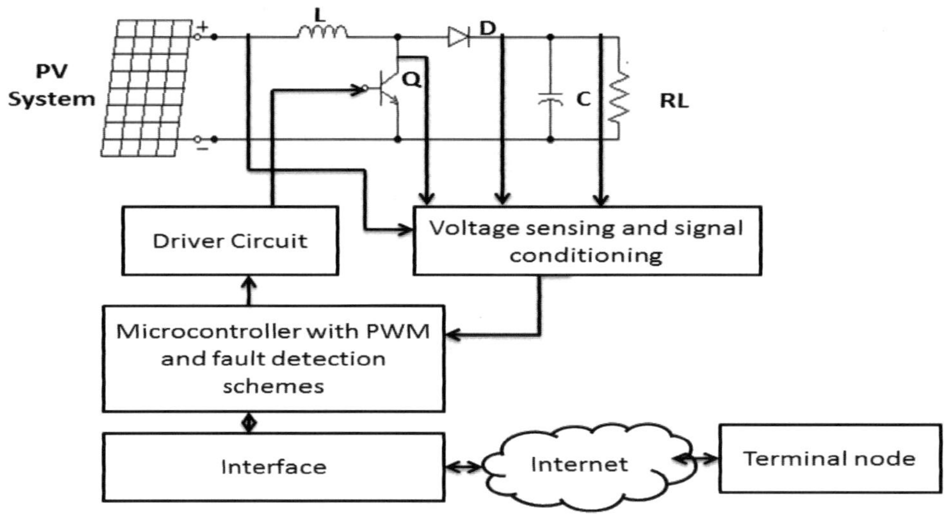

Figure 1- System block diagram

IOP Conf. Series: Materials Science and Engineering 1085 (2021) 012035 doi:10.1088/1757-899X/1085/1/012035

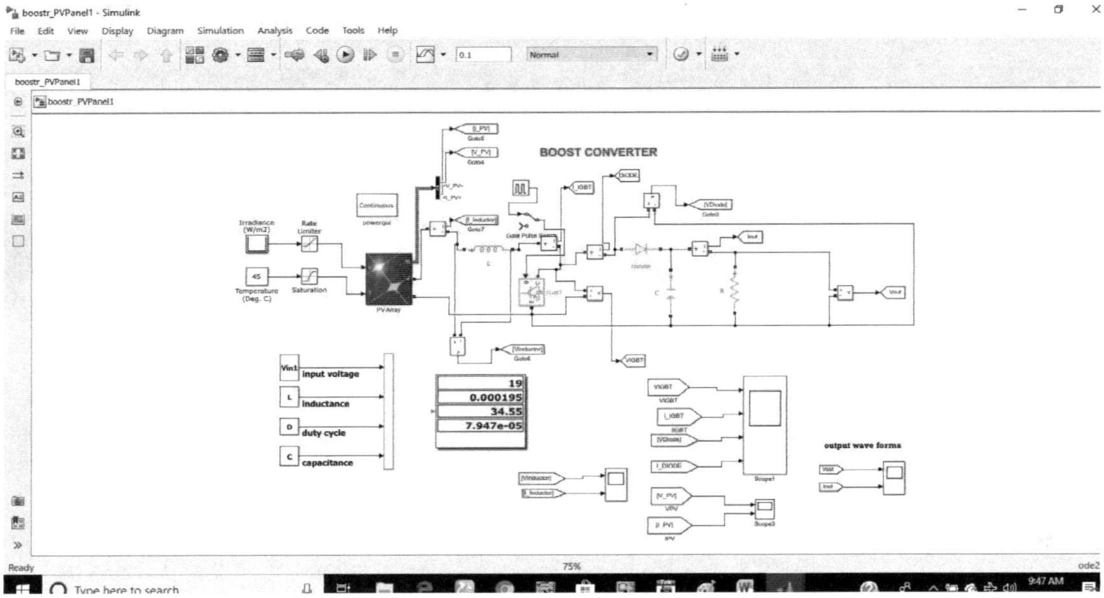

Figure 2- System simulation model

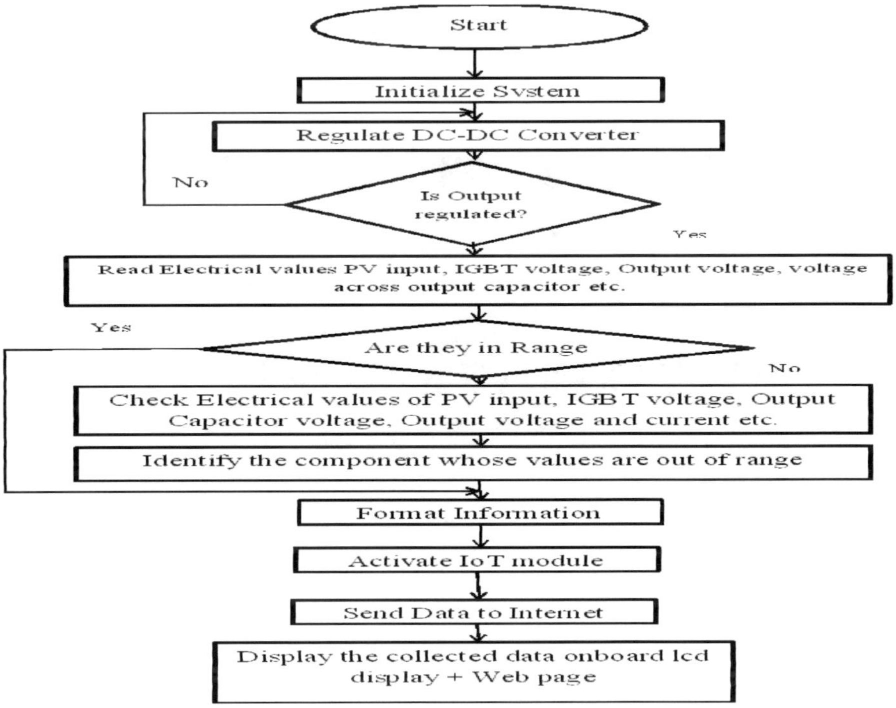

Figure 3- System flow chart

IOP Conf. Series: Materials Science and Engineering 1085 (2021) 012035 doi:10.1088/1757-899X/1085/1/012035

Prototype developed in laboratory is shown in figure 4.

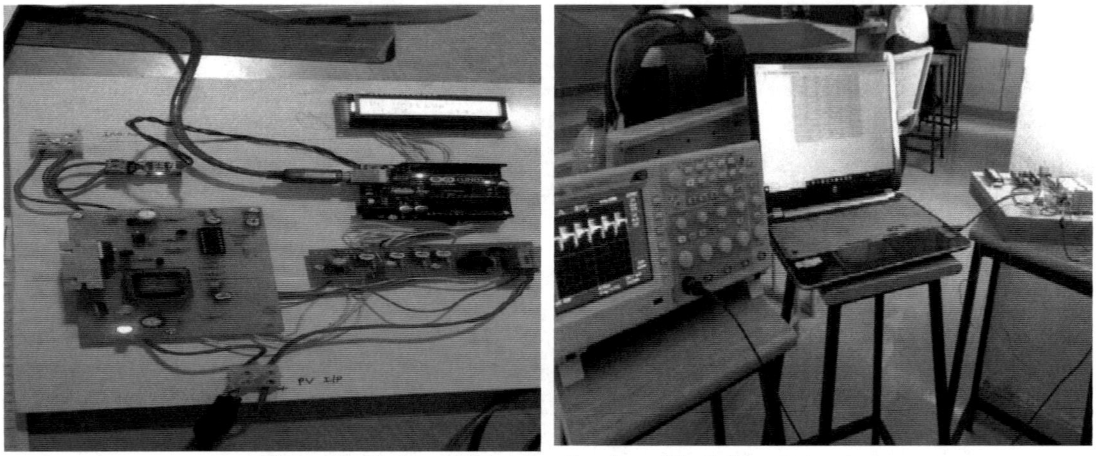

Figure 4- System prototype developed in laboratory

The electrical parameters of the system can be remotely monitored via computer or smart phone. The sample screen shots of observations taken for faulty condition appear as shown in figures 5 and 6.

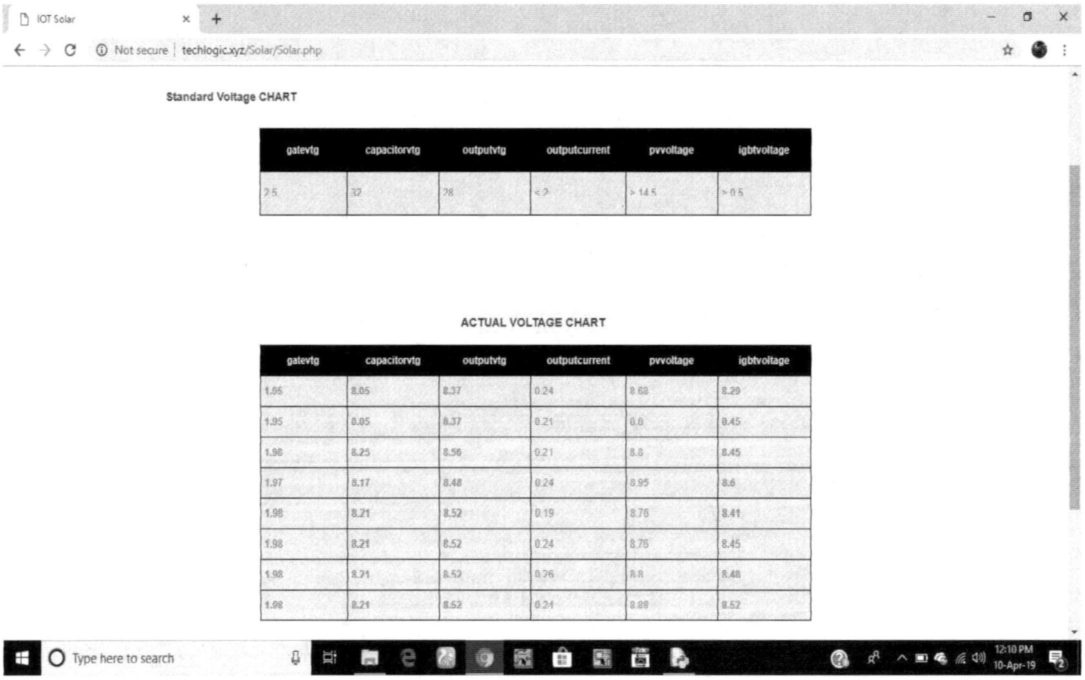

Standard Voltage CHART

gatevtg	capacitorvtg	outputvtg	outputcurrent	pvvoltage	igbtvoltage
2.5	32	28	< 2	> 14.5	> 0.5

ACTUAL VOLTAGE CHART

gatevtg	capacitorvtg	outputvtg	outputcurrent	pvvoltage	igbtvoltage
1.95	8.05	8.37	0.24	8.68	8.29
1.95	8.05	8.37	0.21	0.0	0.45
1.98	8.25	8.56	0.21	8.8	8.45
1.97	8.17	8.48	0.24	8.95	8.6
1.98	8.21	8.52	0.19	8.76	8.41
1.98	8.21	8.52	0.24	8.76	8.45
1.98	8.21	8.52	0.26	8.8	8.48
1.98	8.21	8.52	0.24	8.88	8.52

Figure 5- System parameters observed on computer

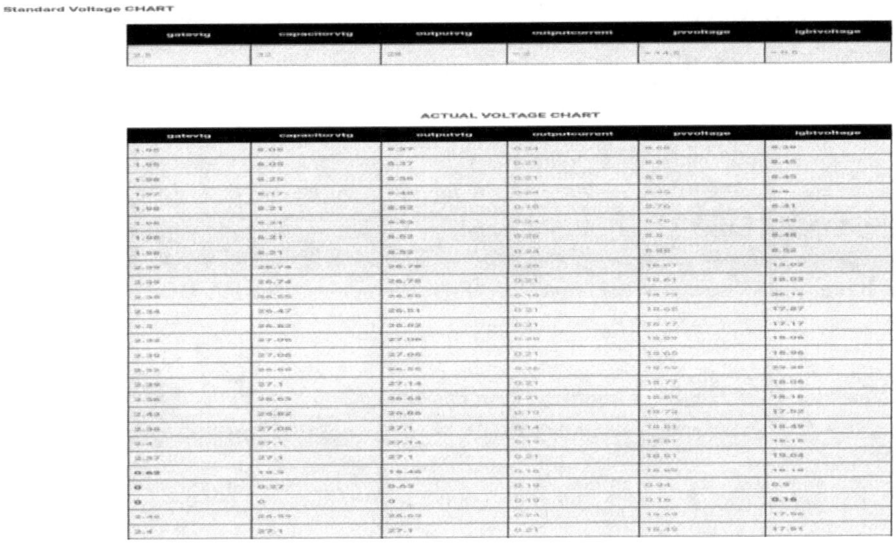

Figure 6- System parameters observed on smart phone

The abnormalities in the electrical parameters can be highlighted with red colors as shown in sample screen shot in figure 6. Thus remotely the system supervision can be done, prediction about any faulty component can be done from observed parameters and possible remedial action can be planned.

3. Conclusion

The system developed in this work provides onsite real time information. For exact location of faulty component, correct information of fault signatures is required. Typically deviation in output voltage against input voltage provides idea about system malfunction. The voltage across device and current through it provides exact status of component. The internet connectivity based systems facilitates remote parameter monitoring and fault detection in PV systems. The PV systems are deployed on huge land area and they are remotely located. Hence implementation of internet connectivity based systems for fault detection will inform the component abnormalities at earliest; due to this quick repairing or component replacement is possible which will avoid loss of energy generation.

4. References

[1] Petrone G Spagnuolo G Teodorescu R Veerachary M Vitelli M 2009 *IEEE transactions on Industrial Electronics* **55** 2569-80

[2] Wang H Ma K Blaabjerg F 2018 *Proc. 38th Annual conference of the IEEE Industrial Electronics Society* 33-44

[3] Wang H Blaabjerg F Ma K Wu R 2013 *4th International Conference on Power Engineering Energy and Electrical Drives* 1846-51

[4] Yang S Bryant A Mawby P Xiang D Ran L Tavner P 2011 *IEEE transactions on Industry Applications* **47** 1441-51
[5] Dhopale S V Davoudi A Dominguez A D Chapman P L 2012 *IEEE transactions on power electronics* **27** 739-51
[6] Callega H Chan F Uribe I 2007 *IEEE power electronics specialists conference* 1522-27
[7] Chavan S B Chavan M S 2014 *International journal of advanced research in Electrical Electronics and Instrumentation Engineering* **3** 11729-37
[8] Pei X Nie S Chen Y Kang Y 2012 *IEEE transactions on Power Electronics* **27** 2550-65
[9] Pei X Nie S Kang Y 2015 *IEEE transactions on power electronics* **30** 996-04
[10] Chavan S B Chavan M S 2014 *IEEE global conference on Wireless Computing and Networking* 112-15
[11] Firth S K Lomas K J Rees S J 2010 *Solar Energy* **84** 624-35
[12] Zhao Y Lehman B Ball R Mosesian J Palma J 2013 *28th IEEE applied Power electronics conference and exposition* 2913-20
[13] Houssein A Heraud N Souleiman I Pellet G 2010 *IEEE International Energy Conference and Exhibition* 389-94
[14] Tejwani R Kumar G Solanki C 2014 *ISES Solar world congress Energy Procedia* **57** 1526-35
[15] Chavan S B Kadam P A Sawant S R 2009 *Instruments and experimental techniques* **52** 784-86
[16] Zahran M Atia Y Al-Hussain El-Sayed I 2010 *12th WSEAS International Conference on automatic control, modelling & simulation* 65-70
[17] Meliones A Apostolacos S Nouvaki A 2014 *Applied computing and informatics* **20** 14-37
[18] Chavan S B Chavan M S *Information and communication technology for sustainable development Lecture notes in Networks and Systems Springer* **10** 133-41
[19] Shariff F Rahim N A Ping H W 2015 *Expert systems with applications* **42** 1730-42

Acknowledgement

The authors are thankful to Embedded systems and VLSI Design laboratory of Electronics and Telecommunication Engineering program of Department of Technology, Shivaji University, Kolhapur for proving necessary resources for completion of this project work.

IOP Conf. Series: Materials Science and Engineering 1085 (2021) 012036 doi:10.1088/1757-899X/1085/1/012036

Analysis of Security Parameters and Network Infrastructure on Cloud

S.Praveen Kumar[1], T.Jaya[2]

[1]Department of Electronics and Communication, VELS Institute of Science, Technology and Advanced Studies, India

[2] Department of Electronics and Communication, VELS Institute of Science, Technology and Advanced Studies, India

Email: praveens1@srmist.edu.in , jaya.se@velsuniv.ac.in

Abstract—This paper pertains to an analytical approach towards establishing standards for network performance on the cloud. Cloud being the most promising domain, is yet hindered by the lack of standardization. To resolve this issue we propose a methodology suitable for all essential protocols in contrasting network architectures[1]. In extension to this, security breaches also severe the threat to cloud data and its privacy. To bridge the two imperative aspects of a cloud service, we intend to deploy a deep learning classification model on Google collab, acting as an interface between the cloud database and client-server. Essential parameters including throughput(X), average delay(ζ), loss function(l) and accuracy. We extend to present our results as the optimized cloud standards.

1 Introduction

Cloud computing is the new phase of technology that has revolutionized the way computing is done rudimentarily. Cloud Computing is the on-demand delivery of computational resources via the Internet. The computational resources generally provided are data storage and computation capability. Cloud Computing, in general, solves many of the obstacles to efficient computing such as the requirement of powerful hardware, computation of excess cycles, redundant cycles of software updates, or incompetent handling of high loads. This new technology has also created certain new challenges and obstacles in terms of network infrastructure and security parameters which pose to be extremely hazardous in terms of security and privacy[2]. The lack of standardization in the industry of cloud computing has also led to additional problems which have been the motivation for the work on this project.

Cloud computing is the new phase of technology that has revolutionized the way computing is done rudimentarily. Cloud Computing is the on-demand delivery of computational resources via the Internet. The computational resources generally provided are data storage and computation capability. Cloud Computing, in general, solves many of the obstacles to efficient computing such as the requirement of powerful hardware, computation of excess cycles, redundant cycles of software updates, or incompetent handling of high loads. This new technology has also created certain new challenges and obstacles in terms of network infrastructure and security parameters which pose to be extremely hazardous in terms of security and privacy.

Cloud architecture refers to the various constituents in terms of databases, capabilities, and applications which leverage the power of cloud resources to direct towards a specific complication that defines the relationship between the components. Based on the occurring relationship between the components, the types of architecture are Software as Service (SaaS), Infrastructure as Service (IaaS), Platform as a Service (Paas). The cloud deployment model designates how cloud services are made available to users. The deployment models corresponding to cloud computation services are Public Cloud, Private Cloud, Hybrid Cloud, Community Cloud[3].
In the rush to improve the cloud services and their interfaces, many companies often end up ignoring the most critical aspect of their service, that being the network infrastructure. For many companies, it is imperative that they massively improve their network infrastructure as it is evident that with the expansion of cloud services the data traffic is only bound to increase in the networks which will make data access difficult be it from a remote location in the world or an institutions headquarters. Some of the principal aspects of a network are Bandwidth, Throughput, Latency. The mere action of improvement of infrastructure does not necessarily yield better network performance. Certain strategies could be implemented beforehand to further improve network performance. Some of the optimization techniques include Caching, Compression, Traffic Management, Reduplication[4]. Various security mechanisms in the cloud can be implemented to ensure data privacy and security. Encryption is the most common method used for security mechanisms as various algorithms can be executed with them.

When encryption is applied on the plaintext data, the data is paired with a string of characters which is termed as an encryption key, a secret key that is established and shared among the relevant parties involved. Then the encryption key is used to decrypt the ciphertext back into its plaintext which was its original format. If any external third-party attempts to eavesdrop, it would be unsuccessful as it would not have the encryption key[5]. The two most preferred forms of encryption are Symmetric Encryption, Asymmetric Encryption. Based on this the various algorithms which are implemented for cloud security are RSA Algorithm, DES Algorithm, AES Algorithm, Blowfish Algorithm.

2 METHODS

2.1 Process Flow

The process implemented is as the pilot study was conducted to ensure the smooth ongoing of the project. The project was started with a prior study for project specification, Cost Estimation, Resources implemented, and the timeline. The project specifications include the analysis of the various protocols in two different topologies that have to be carried out in terms of the Throughput, Average Delay, Accuracy, Loss function, Complexity, and Security. We are proud to mention that our project is economic as we are using the easily available resources in our department labs. For the cloud simulations, we are using the Tensorflow library in Google collab which is free of cost available open-source platform for cloud processing purposes[6]. For the security parameters also, we are using the Jupyter notebooks on Google collab. So our estimated cost is really low and it only includes the paperwork and other hardware.

2.2 Experimental Data Obtained

Primary data is the data being obtained from the first-hand experiments on the cloud environment. For network performance, we have taken the throughput, Average Delay, and comparative graphs. To obtain the performance on the cloud we have checked the loss function and accuracy for the Deep Learning Model deployed on the cloud platform of Google collab. The inference obtained from the primary data and the conclusion drawn is secondary data.

2.3 Interpreted Comparative Study

For the project, our secondary data is the comparative data we obtained by comparing the various results and analyzing them. The result will be a comparative study of the parameters._Data analysis is an imperative part of our study which includes the measurement of data, recording of data into concise tabular formats.

Then for comparative analysis, the obtained results are plotted into graphical representations. For the execution of all these tasks, we are implementing Google Sheets as the primary data analysis tool. The application was used to graphically represent the obtained results for a clearer understanding and better inference to derive a conclusion for the results[7]. We have analyzed the network performance and the security parameters separately. The network parameters were analyzed on an available LAN server-client system. The security parameters were analyzed by implementing various algorithms on cloud compilers and measuring the executing times with their complexity. As a common platform to bridge them both, we have used Google collab as a cloud platform for implementing a real-time deep learning classification model with Tensorflow library and obtaining the loss function, accuracy, and training time per sample.

.

3 SYSTEM DESIGN

The study had 2 different parts involving the network infrastructure and security parameters. Here's a brief description of the system design of the network infrastructure.

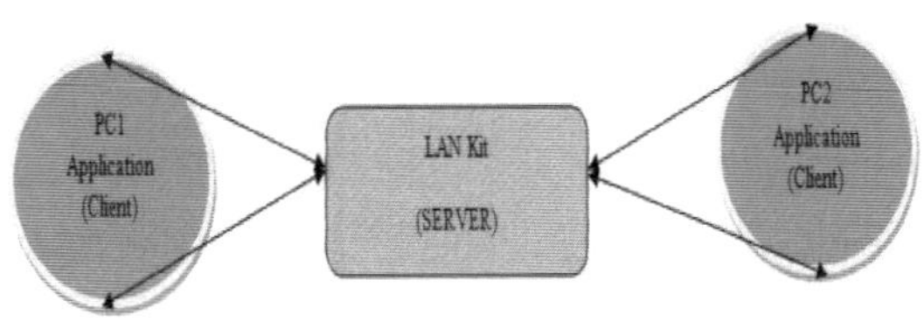

Fig 1: System design of the network infrastructure

Fig 2: The above image is a pictographic representation of the LAN Kit[9]. The analysis of the network infrastructure was done using a LAN client-server kit. The client PC and the server/sender PC were connected to the LAN Kit with connecting wires.

These kits provide a menu-driven interface and a C library which can give access to the programming interface to the NIU. This kit has 6 nodes per NEU, meaning 3 parties can be connected but we have connected for a concise and clear result. The parameters recorded in this setup are offered load, throughput, and average delay[8].
Next, we evaluate the cloud system design which involves the use of Tensor flow library and the cloud platform, Google collab.

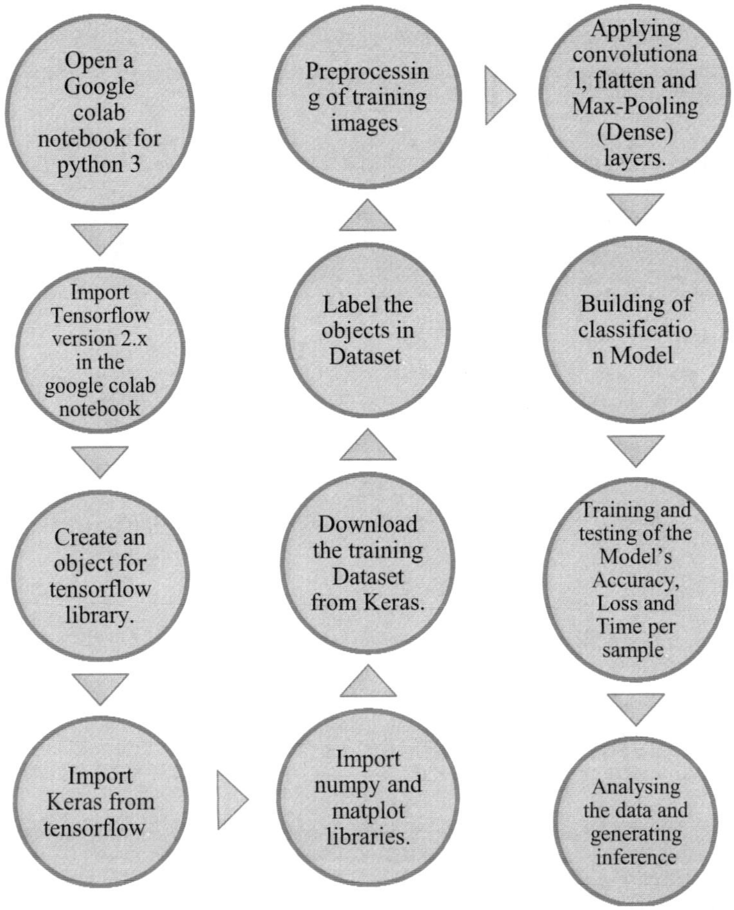

Fig 3: The Real time Cloud Implementation Model

To begin, we first import the Tensorflow library in the Google Colab and create an object for the Tensorflow library[10]. Then using Keras we download a training dataset where we import the metaplot and other relevant libraries. After importing the Keras, we label the objects in the data set succeeding in which the pre-processing of training images takes place. Then we apply convolution flatten and max-pooling layers to analyze the data and to generate an inference. Then we obtain the model's accuracy, loss, and time per sample to build a classification model and obtain the relevant results. Here we implement the model with 10 - 1000 Epochs. Epoch in this algorithm is defined as the number of training instances per training attempt for a model with the same training data set. The concept of Epoch is applied here to demonstrate the optimization technique of Caching which refers to caching of frequently used images in a user's local disk[11]. To analyze the security parameters, we proposed four different security encryption methods for the security of data.

4 Results

The data from the system was recorded concisely with a tabular representation as follows.

IPD	G1	G2	Sum G	Ack 3	Ack 4	Sum Tx	X	G
8000	13	13	26	11	9	20	0.02	0.0003 25
7000	14	14	28	11	11	22	0.022	0.4
6000	18	19	37	15	14	29	0.029	0.6166 66666 7
4000	30	30	60	24	25	49	0.049	1.5

Table 1: Bus - CSMA/CA (8 Kbps)

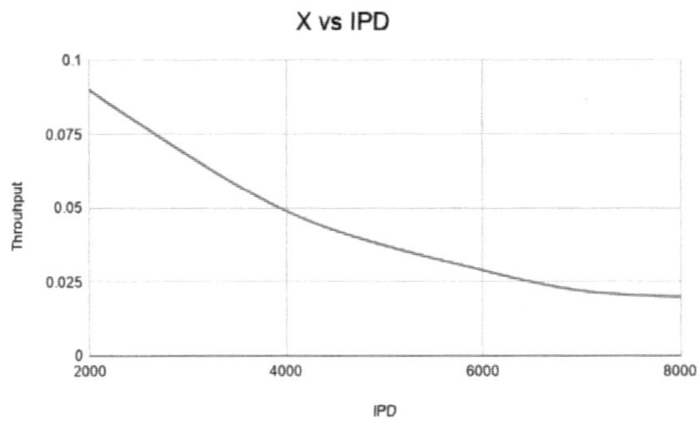

Fig 4: Interpacket Delay Vs Throughput

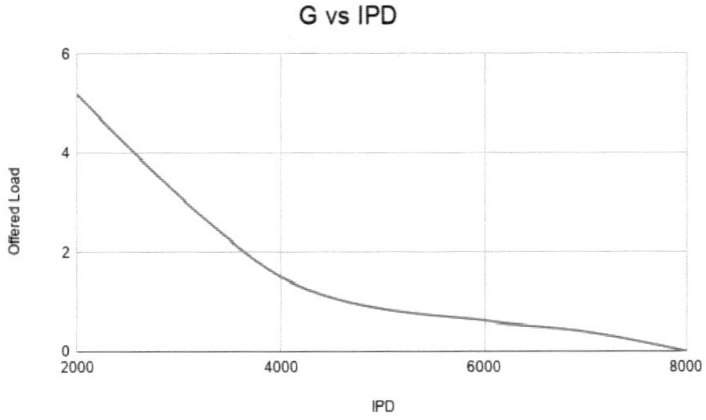

Fig 5: Interpacket Delay Vs Offered Load

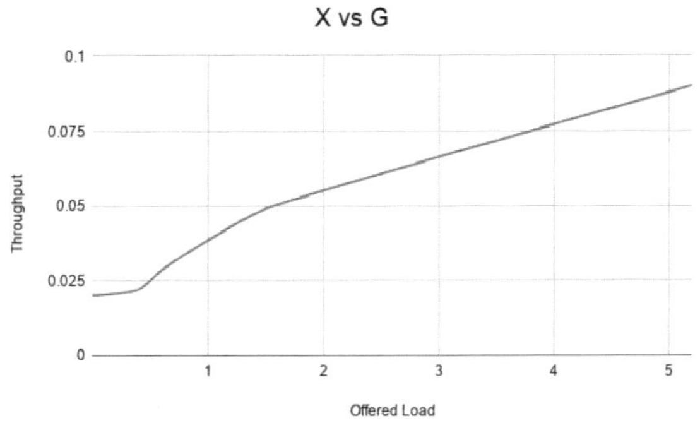

Fig 6: Offered Load Vs Throughput

The model was deployed and the following results were obtained

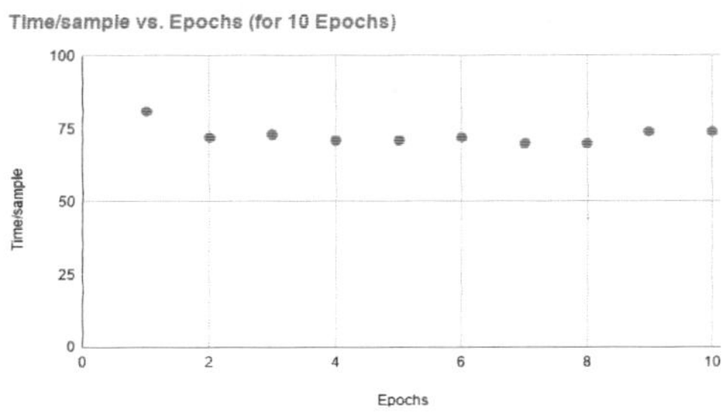

Fig 7: Comparative Study of Throughput with Epochs

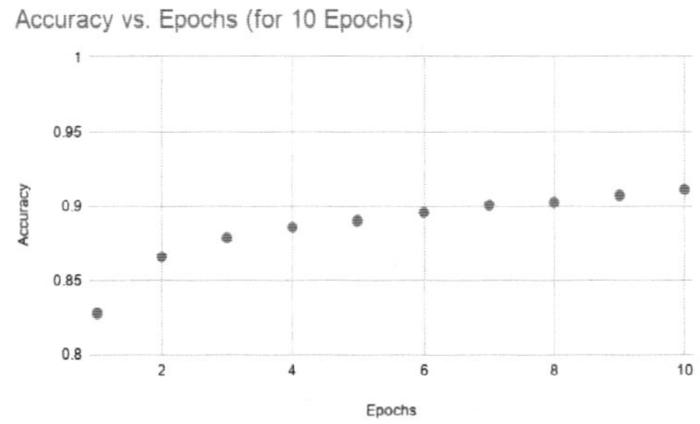

Fig 8: Comparative Study of Accuracy with Epochs

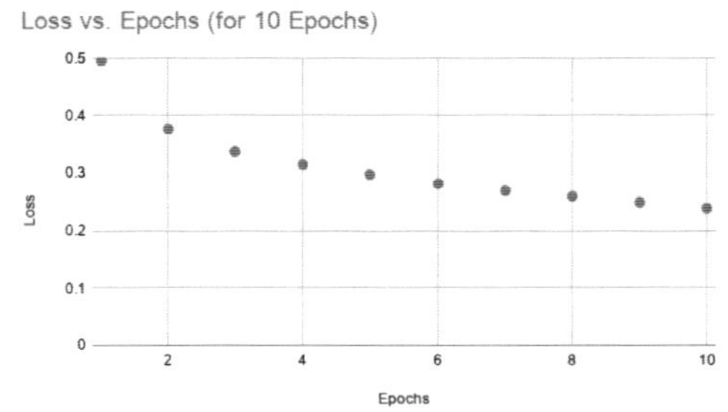

Fig 9: Comparative Study of Loss with Epochs

5 INFERRED DATA

For the network infrastructure analysis, the following inference was drawn-

<u>Throughput v/s Offered load</u>

- CSMA/CA: - It has a stiff slope & saturates after the threshold. As Data rate increases - Stiffness increases in CSMA/CA.
- CSMA/CD: - Slope increases gradually & later decreases after the threshold. As the Data rate increases - The slope gradually tends towards stiffness.

<u>Throughput v/s IPD</u>

- CSMA/CA: As IPD increases Throughput decreases but changes are slower. As the Data rate increases, no significant change in the graph.

- CSMA/CD: As IPD increases Throughput decreases but changes are faster in CSMA/CA compared to CSMA/CD. As the Data rate increases, no significant change in the graph.[12]

IPD v/s Offered Load

- CSMA/CA: - As IPD increases offered load decreases & vice versa. Data rate increases no significant change in the graph
- CSMA/CD: -As IPD increases offered load decreases & vice versa. Data rate increases the rate of change of slope also increases

For security when given the same binary input, most of the algorithms took over 4 seconds to execute. AES Algorithm took only 1.5 seconds to encrypt the data which was the fastest. AES Algorithm also requires the lowest memory to be run, making it the most viable option

6 Conclusion

This research focuses on preliminary study for analyze network performance of cloud computing infrastructure that designed for rendering requirement . When given the same binary input, most of the algorithms took over 4 seconds to execute. AES Algorithm took only 1.5 seconds to encrypt the data which was the fastest AES Algorithm also requires the lowest memory to be run, making it the most viable option. The analytical study of the security parameters and network performance established standards about the cloud. CSMA/CA was found to be the most suitable and optimized protocol for networking in the cloud. In terms of security, the AES algorithm performed most efficiently in terms of execution time.

Acknowledgment

We would extend our gratitude towards our supervisor, Dr.T.Jaya, Assistant Professor, Department of Electronics and Communication Engineering, VISTAS, Chennai. Also, we thank the ECE Department, SRMIST for helping us in getting all the resources.

References

[1] R. Gargista Gustamas and Guruh Fajar Shidik - Analysis of Network Performace on Cloud Computing, *IEEE 2017.*
[2] Hamed Tabrizchi · Marjan Kuchaki Rafsanjani- A Survey On Security Challenges In Cloud Computing: Issues, Threats, And Solutions, *The Journal Of Supercomputing (2020)*
[3] Cheikh Brahim Ould Mohamed El Mocta and Karim Konaté – Survey of Security Challenges in Cloud Computing, *IEEE 2016*
[4] G. Sousa, W. Rudametkin, and L. Duchien, "Automated setup of multicloud environments for microservices applications," *in Proc. IEEE 9th Int. Conf. Cloud Comput. (CLOUD), Jun./Jul. 2016, pp. 327–334*
[5] Z. Hao, E. Novak, S. Yi, and Q. Li, "Challenges and software architecture for fog computing," *IEEE Internet Comput.,* ***vol. 21,*** *no. 2, pp. 44–53, Mar./Apr. 2017*
[6] W. Cai, V. C. Leung, and L. Hu, "A cloudlet-assisted multiplayer cloud gaming system,"

*Mobile Netw. Appl., **vol. 19**, no. 2, pp. 144–152, 2014.*

[7] H. Hong, D. Chen, C. Huang, K. Chen, and C. Hsu, "Placing virtual machines to optimize cloud gaming experience," *IEEE Trans. Cloud Comput., **vol. 3,** no. 1, pp. 42–53, Jan.-Mar. 2014.*

[8] T. Verbelen, P. Simoens, F. De Turck, and B. Dhoedt, "Leveraging cloudlets for immersive collaborative applications," *IEEE Pervasive Comput., **vol. 12,** no. 4, pp. 30–38, Oct.-Dec. 2013.*

[9] I. 802, "Ieee standard for local and metropolitan area networks: Overview and architecture," *IEEE Std 802–2014, **p. 8,** 2014.*

[10] S. Das, M. Khatua, S. Misra, and M. Obaidat, "Quality-assured secured load sharing in mo- bile cloud networking environment," *IEEE Trans. Cloud Comput., 2015, DOI: 10.1109/TCC.2015.2457416.*

[11] M. Aazam and E.-N. Huh, "Dynamic resource provisioning through fog micro datacen- ter," *in Proc. IEEE Int. Conf. Pervasive Comput. Commun. Workshops, Mar. 2015, pp. 105–110.*

[12] E. Baccarelli, N. Cordeschi, A. Mei, M. Panella, M. Shojafar, and J. Stefa, "Energy- efficient dynamic traffic offloading and reconfiguration of networked data centers for big datastream mobile computing: review, challenges, and a case study," *IEEE Netw., **vol. 30,** no. 2, pp. 54–61, Mar./Apr. 2016.*

IOP Conf. Series: Materials Science and Engineering 1085 (2021) 012037 doi:10.1088/1757-899X/1085/1/012037

An Intuitive Remote Monitoring Framework for Water Quality in Fish Pond using Cloud Computing

S. Sivakumar[1], **Dr.V. Ramya**[1]

[1]Department of Computer Science and Engineering, FEAT, Annamalai University, Annamalainagar, Tamilnadu, India

Abstract. Accuracy in prediction of water quality is a recent research domain gaining popularity in smart aquaculture. In aquaculture, the modifications in water quality parameters possess nonlinearity, dynamicity, unstableness and complexity due to the open environment with its surroundings. Conventional prediction methods have several disadvantages like poor generalization, lower accuracy and high time complexity. By considering these issues, a Novel water quality prediction method interfaced with (IoT)Internet of Things termed as Low-Cost Real-Time Monitoring System (LCRTMS) is proposed to predict water temperature, pH, DO, and Ammonia. Multi-sensors and Blynk private cloud integrated framework are used for collecting data in real-time and enhancing the remote monitoring capabilities. Experimental results indicate that the accuracy in predicting temperature, DO, pH and ammonia can attain 98.56% in 0.257 seconds and 98.97% in 0.301seconds in the short-term prediction.

1. Introduction

Aquaculture has a significant part in thriving sectors of the food industry. In 2019, worldwide fish production reached its highest point at around 200 million tons while aquaculture contributing 47 percent of the overall production. Due to the stability in outbreak of fishery production through capturing in the previous 1980s, aquaculture is being accountable for the continuous magnificent development in the surplus circulation of fish for human utilization. As per the UNFAO (United Nations Food and Agriculture Organization) report, currently a person intake an average of around 19 to 20 kg of fish products per year. This mean consumption is estimated to grow by 16.7 kg per year in 2030. Aquaculture needs to be upgraded more to provide aquatic eatable products in abundance to fulfil the demands of society. Availability of multiple varieties in fish made aquaculture to provide a profitable yield since fish products are always delicious and healthier to human. An approximation intimates that, fish demands in the country would be around 16 million tons by 2025. Natural habitat of fishing is getting attenuated because of overfishing. It resulted in the arrival of commercial aquaculture that faces several problems because of the uncertainty in climatic conditions causing modifications in water quality parameters. In comparison with animal production and land-based crops, fish are completely affected by the presence of pollutants in water; Hence aquaculture production is prone to water-quality degradation. In addition to this, uncooperative fish farming leads to the outbreak of several viral infections in fish and results in huge loss to aquaculture. The water quality of the fishpond is one of the parameters that should be given significant importance because the mobility of fish is affected when the water quality becomes poor in an environment where the fish is grown inside a pond or in a closed water surface [1]. Water quality has a direct impact in rate of growth, efficiency in feeding and overall progress of the fish. There are certain factors like pH, water level, dissolved oxygen and temperature are required to be monitored to improve the activeness of fish that sequentially enhances profitability and sustainability of fish farming. The present scenario of aqua farmers is that they are entirely relied on manual testing, which results in ineffective, laborious and irrelevant predictions along with manual field measurement. The major defect in this type is the identification of any abnormality in the water quality untimely causes heavy costs and well framed methodologies to return and preserve the quality of water. With the upgradation in present day's technology, enhancing the monitoring process, representing a wide variety of problems regarding reachability, insufficient measurements, or even temporal and spatial scaling. Therefore, it increases productiveness and decreases the losses by keen observation of water quality constraints instantaneously. New sensor automations provide remote monitoring, data telemetry, and wireless communication technology in real-time to forward the aquaculture practices to next higher level. LCRTMS Architecture depicted in Figure 1.

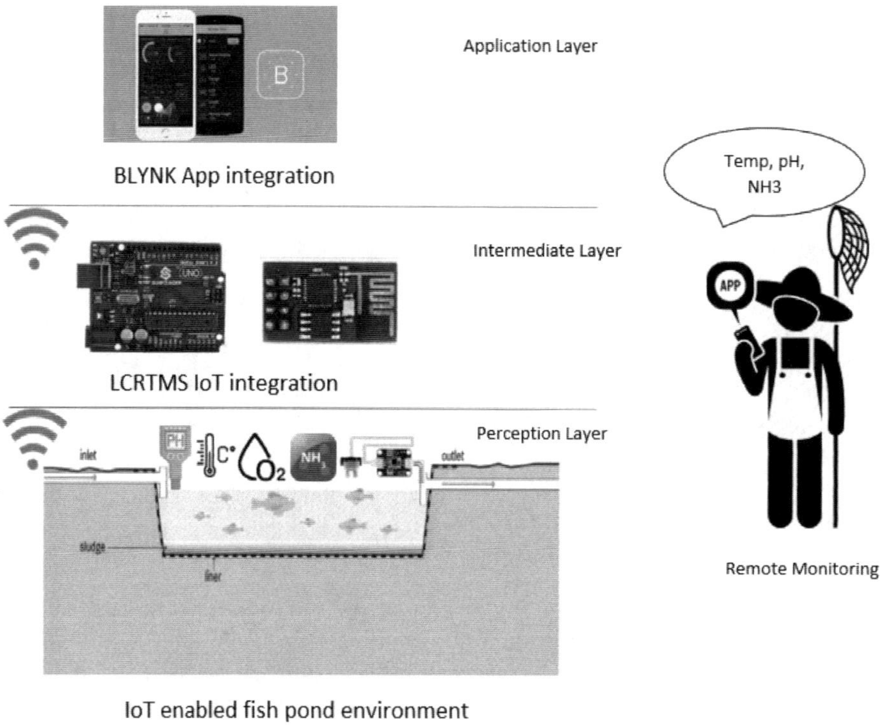

Figure 1. LCRTMS Architecture scenario

Different types of fishpond monitoring systems are in existence nowadays. A remote online monitoring system using artificial neural networks (ANNs) to forecast the constraints for water quality is developed by [1]. An Internet of Things (IoT)-based technological model can track and regulate the essential constraints of a fish- pond. The data gathered with the help of IoT system is utilized in long standing computation and to derive certain conclusions for future reference. Therefore, it enhances the usage of resources which in turn increases the profit. In this research article, The LCRTMS: Low-Cost Real-Time Monitoring, which acquires the quality, proper level and temperature of water using the factors such as dissolved oxygen, pH, electrical conductivity, water body turbidity.

2. Related Works

Water quality and fish habitual behaviour are the deciding factors in the improvement of efficiency in aquaculture.[2] developed an embedded system- (Raspberry Pi) for monitoring and regulating aquaculture that regulates quality in water continuously and gathers, sends the corresponding parameter values to the server present in the cloud for further computation. [4] employed wireless sensor network (WSN) in the aquaculture monitoring system that sensed the water-quality constraints like DO content, water level and pH in real-time. Ullah et al. designed an intelligent fish farm system where an enhancement methodology is used to stabilize the time required for pumping and magnitude of water flow by selecting the proper level in water [3]. It is a hypothesis that continual disclosure to high unionized ammonia concentrations produce disastrous effects in the wellness of fish (Becke et al., 2019). Therefore, proper monitoring of these factors should be carried out to have an effective fisheries management

2.1 Enhancement of reliability in GPRS communication

Mingfei Zhang et al. presented WSN-based water monitoring system capable of calculating pH, dissolved oxygen, temperature and level of water. The measured data was sent to a database that provides the necessary information to the software for real time monitoring. [6] developed a pond controller using specific sensors to track the quality of pond water which is controlled remotely through closed-circuit television (CCTV). Daudi S. [9] measured water quality parameters like pH, temperature, dissolved oxygen and level of water by deploying ZigBee to pass on the data and for displaying the acquired information, the LabView software (National Instruments Corporation, Austin, TX, USA) was used.).[7] implemented an effective submerged acoustic network appropriate for persistent monitoring of fisheries environment. [5] developed an aquaculture monitoring system with ZigBee and the general packet radio service (GPRS) protocol in real-time for communicating from the sensors to the main server for enhancing dependability in communication.

2.2 IoT Based Aquaculture Monitoring

[10] presented an IoT-interfaced monitoring system for water quality. An analog temperature sensor named Atlas Ph Probe digital sensor (Atlas Scientific, Long Island City, NY, USA) was employed by them to send the acquired information to a MySQL database along with an Arduino node using a ZigBee module. Furthermore, the system used 200 mA/h rechargeable batteries and can operate continuously for 8 hours. A mobile application or desktop can be used to visualize the acquired information. The work in Reference by [8] proposed a method for monitoring quality using aquatic drone enhanced with a Raspberry Pi fixed with an arrangement of water and air quality sensors. The air quality is measured by employing temperature, humidity and gas sensing channels. Water, temperature, and conductivity sensors are employed to assess the water quality. The gathered data is kept inside the drone's computing platform and connected with a remote server database. Each indicator's varying trend is forecasted by this model. The functioning of several fishpond farming devices is controlled in relation with the predicted results for maintaining the stability in fish pond's water parameters. To conserve energy, this model provides an improved control over the equipment [11,12]. Therefore, it is essential to gather, compute, and process the various parameters like water level, dissolved oxygen, pH, electrical conductivity, turbidity, water temperature. [13] focuses at developing a monitoring system for the water quality of aquaculture ponds based on NB-IoT methodology. It is evident that NB-IoT possess quality factors like stability, reliability, comfortability in acquiring the data, speedy responses for commands. These can support the methodology technically in several fields, specifically in agricultural monitoring. It can be inferred from the results that remote control device's reacting time falls within 100ms. Also the aerator's intelligence control is realized. In this proposed work, an IoT-based smart fishpond Monitoring system that analyses and predicts several parameters to maintain the water quality effectively.

3. LCRTMS Low-Cost Real-Time Monitoring System

Monitoring the quality of water has a key role in fisheries management. The long-lasting nature of fish diseases has a direct impact on the yield. Poor quality of water tends to delay the fish growth and harvest. At present, water quality monitoring systems are not affordable and lacks in sensitivity. To succeed in aquaculture, maintaining a good quality in water is the most essential factor to be considered. Although the biggest challenge is to refine the water quality specifically in small fish sectors in developing countries. We have developed a LCRTMS equipped with intelligent devices and integration of networks. The LCRTMS enables real-time water monitoring capabilities by reducing the enormous investment in aquaculture. Variations in water quality may occur rapidly irrespective of any time. Therefore, monitoring of water with pre-defined alerts intimates the fisher to do the needful immediately. With this intention, the LCRTMS allows the fisher to refine the environmental management and product quality thereby reducing huge loss and expenditures in production. The most crucial parameters like temperature, dissolved oxygen, pH and ammonia need to be intensively monitored and managed in aquaculture. As per the traditional techniques of analysing water quality, water specimen data are extracted and sent to a chemical lab to diagnose the presence of harmful substances. Contrarily, personal experience of an individual impacts in maintaining the measurements and control process. The proposed system can overcome these limitations. Appropriate and specific sensors for temperature, pH, Ammonia, Dissolved oxygen and turbidity are employed to measure the parameters affecting the water quality.

3.1 LCRTMS parameters

Arduino is a non-proprietary microcontroller package that works with adaptable hardware and coding. Arduino reads input data from various sensors and activates an action concerning the input then publishing something online. So, it is employed as a core controller in proposes model.

3.1.1 Temperature

By using the DS18B20 temperature sensor, the system finds whether the water is hot or cold. The value received by DS18B20 is forwarded to a controller for further action. However, the major constraint to take into consideration is that the various phases in fish growth like incubation of eggs, development of larval fish and production of a food-sized fish involve distinctive temperatures. LCRTMS gives better performance to collect temperature whenever a change happens in water.

3.1.2 Dissolved Oxygen (DO)

Oxygen plays the foremost role in limiting the development and wellbeing of fish. Oxygen is the principal gas used for respiration and denoted in milligram(mg) of oxygen absorbed per kilogram(kg) of fish per hour (mg O2/kg/h) framed by the physiologists. Inflowing water distributes the oxygen in tanks and raceways having the necessity of saturation value closer to altitude and temperature. When the range exceeds the limits, the LCRTMS notify the user to replace the water.

3.1.3 pH

pH value is collected by considering the below guideline: The proportion of pH is indicated by the accumulation of hydrogen ion (H+) in water. The pH scale becomes logarithmic when it ranges between 0 to 14 with base 10. It is highly important to consider since a drop in 1 pH unit leads to 10 times doubling up of hydrogen ions (H+) in water. A pH value tends to drop at any value on a scale from 0 indicating the strongly acidic nature to 14 indicating strongly basic or alkaline nature, and the value 7 indicating neutral nature (= 10-7 moles/litre of H+ ions).

The Fertile natural waters free from pollution has the pH value in the range of 6.5 to 8.5 at sunrise, usually closer to 7 than 8. Photosynthesis plays a role in deciding the diurnal variations.

$$\text{PH is calculated as pH} = -\log [H3O^{+}] \tag{1}$$

3.1.4 Ammonia

The concentration of ammonia has a significant role in assessing the water quality. An extremely stocked culture pond or the discharge of a raceway possess fish excretions through the gills in most cases forms the principal source of ammonia. Generally, the animals produce ammonia as a by-product of protein metabolic process. Nessler method is adopted for chemical analysis and measurement of ammonia and the resulting component is termed as total ammonia nitrogen (TAN) since it possesses two ionic concentrations of Ammonia: the unionized form - Ammonia (NH3), and the ionized form- ammonium ion (NH4 +). The UIA - unionized Ammonia has poisonous reactivity towards fish. As per the following expression given the ammonia level is collected for analysis.

$$NH_3 + H_2O = NH_4{+} + OH \quad (2)$$

$NH_3 + H_2O$ at low temperature and pH

$NH_4^+ + OH^-$ at high temperature and pH

The parameters collected from the water source is analyzed with the normal range, then the same is notified to the fishers to take the needful action. The LCRTMS takes a significant role in monitoring the ponds at round the clock with minimum cost.

4. Experimental results

4.1. The data and environment for experimentation

Nochikadu (Cuddalore Dist, Tamilnadu, India longitude11.6515N latitude 79.7540E) is the site chosen for data collection in the proposed study. The fish pond occupies an area of 70ft long,35ft width, and 4-5 depth along with inlet and outlet for water. The various sensor nodes such as pH sensor (Shanghai Leici, Model E-201-C), temperature sensor (DS18B20), water turbidity sensor (Water WT-RCOT) and dissolved oxygen sensor (YHT-8402) were installed in the fish- pond for evaluation. The sensors are attached to Arduino which is then interconnected with the Blynk app for monitoring through Wi-Fi. The Blynk app is used to inform the fish farming environmental parameters, etc.) the prototype of LCRTMS shown in Figure 2.

Figure 2. LCRTMS Prototype Model

4.2 Data cleaning for reliability

The experiment was conducted from March 04, 2020, to April 22, 2020. The data of pH, temperature, ammonia, and dissolved oxygen of fishpond are collected periodically on the time points of 6:00, 9:00, 15:00 and 21:00. These time points are chosen to find the lower and higher values of temperature and oxygen within a day. Initially, the time was fixed at 6:00 for first data collection during which the measurements of temperature and dissolved oxygen were in lowest values on a daily basis. In order to find the start of feeding secondly, the data collection time begun at 9:00. Then, the third data collection time was fixed at 15:00 during which the values of pH, temperature and dissolved oxygen were in highest form. The final data collection time was fixed at 21:00 for determining whether the aerators should be turned on. Nochikadu is an area having acceptable water quality, medium temperature, high level of dissolved oxygen, and surplus flow of water. In accordance with these features of nochikadu, the fishpond water-quality constraints and parameters must be regulated instantly. Few changes and irrelevant data are encountered in the testing phase due to the disruptions in sensor. It might lead to considerable variations in the range and consistency of the data which in turn affects data analysis results. The outliers are eliminated using an appropriate method to ensure data reliability. If the water level drops below than the sensor's position, the lower level of water is sensed and intimated to the fisher for starting the process of water pumping, which refills the water up to the extent of pre-set sensor position automatically. Hence the level of water is always monitored and made to remain stable. Therefore, the water level sensor data are not considered for further computation and analysis.

4.3 LCRTMS visualization and indication

The proposed low-cost real-time monitoring of fish and pond system facilitates the fish farmers by giving several sensors collected data that act as the indicators for maintaining the water quality parameters. Data visualization is also performed to intimate the functional report of water quality parameters. It delivers valuable suggestions by using information and communication system with the help of functional report. The LCRTMS analyses the water quality as per the instructions embedded in the board. Further, it refers to the average ranges from Table 1 for abnormal notification. All the correspondence between the sensor and Blynk is automated.

4.4 Optimal requirements of water quality in a fish pond

In accordance with (ICAR) Indian Council of Aquaculture Research and Central Institute of Fisheries Technology guidelines, the LCRTMS refers to the water quality parameters presented in Table 1.

Table 1. Water Quality Parameter Range

S.No	Evaluation of Water Quality Parameters	Range(units)
1	p^H	(6.5-8.5) ppm
2	Ammonia	(0-0.1) ppm
3	Dissolved Oxygen (DO)	(4-10) ppm
4	Salt	(0-2) ppt
5	Temperature	21^0C-33^0C
6	Carbonates (CO_3^{2-})	(20-40) ppm
7	Bicarbonates (HCO_3^-)	(150-500) ppm
8	Sour gas (H_2S)	(0-0.4) ppm
9	Nitrates (NO_2)	(0-0.3) ppm

4.5 Blynk Application Scenario

Blynk app is designed for the Internet of Things to control and monitor the hardware in cloud and visualize it. The major components working the cloud platform is App, Server and libraries to enable cloud communications. Here the figure 3 shows the application platform, figure 4 shows the water level indication, figure 5 shows the temperature indication, figure 6 shows the oxygen level indication, figure 7 & figure shows the ammonia and pH level. Every data is been stored in the Blynk cloud.

IOP Conf. Series: Materials Science and Engineering 1085 (2021) 012037 doi:10.1088/1757-899X/1085/1/012037

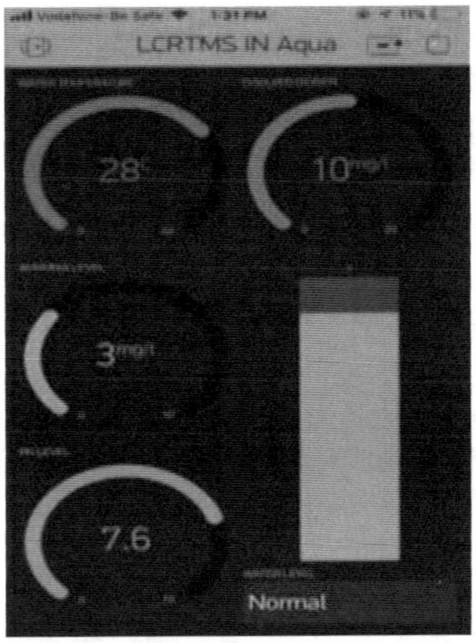

Figure 3. Blynk mobile application

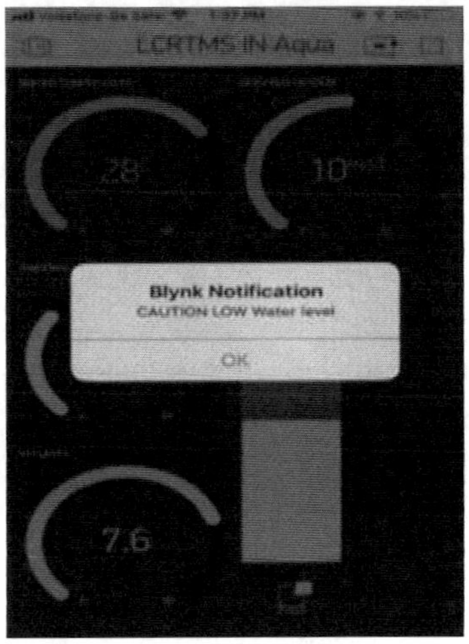

Figure 4. Blynk Low level water Notification

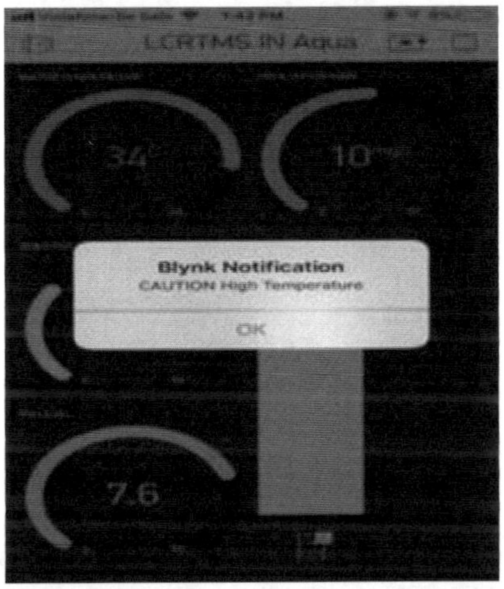

Figure 5. Blynk High Temperature Notification

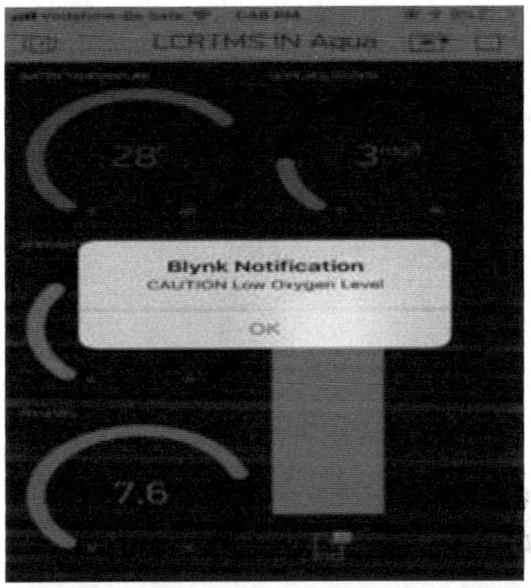

Figure 6. Blynk Low Oxygen level Notification

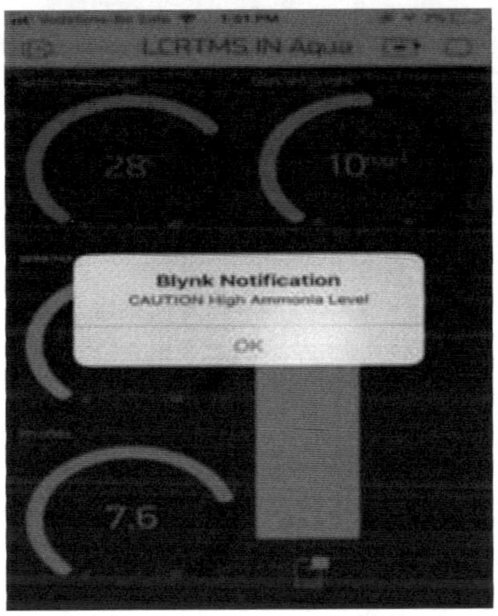

Figure 7. Blynk High Ammonia Notification

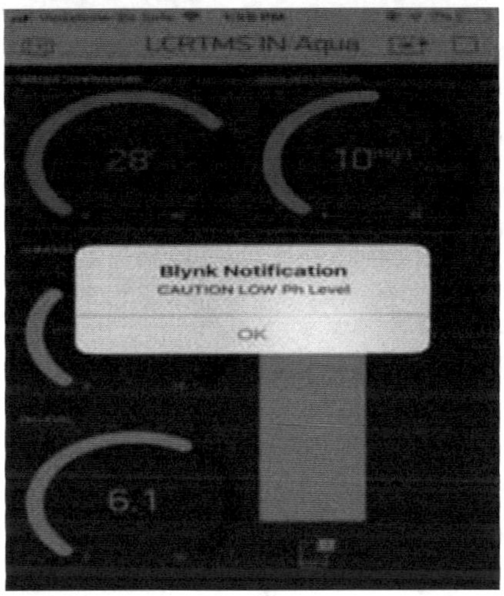

Figure 8. Blynk Low pH level Notification

4.6 Results of the LCRTMS

The sample data was gathered from the aquaculture ponds by using sensor devices and sent to the private cloud in a wireless manner. The sampling frequency of the data was taken once in every 5 mins for instant predictions. The LCRTMS attain 98% of accuracy within earlier of 0.032 secs, which means the system initialization time is very minimal while comparing to conventional method however, the LCRTMS begins the sensing with the above mentioned time span(start-up time). The data associated with water quality of about 610 groups (about 45d), including pH, temperature, DO, and ammonia parameters were used for the prediction. figure 9., shows the graph of Average Temperature in between 6h to 00h at all the days. figure 10., shows the average graph of Dissolved Oxygen Level in between 6h to 00h at all the days. figure 11., shows the average graph of pH Level in between 6h to 00h at all the days. Figure 12, shows the graph of average Ammonia Level in between 6h to 00h at all the days.

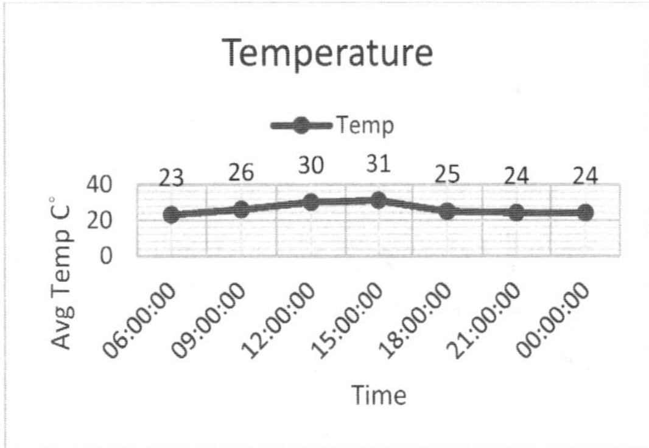

Figure 9. Fish pond temperature

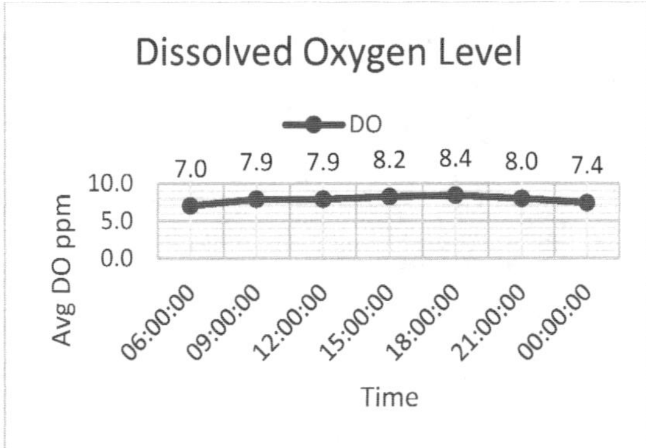

Figure 10. Fish pond DO Parameter

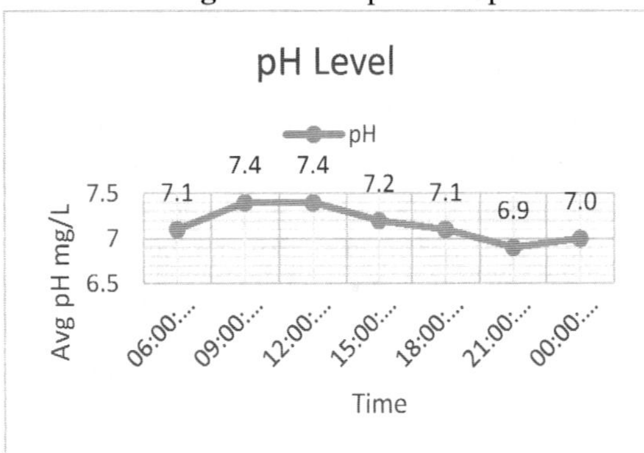

Figure 11. Fish pond pH parameter

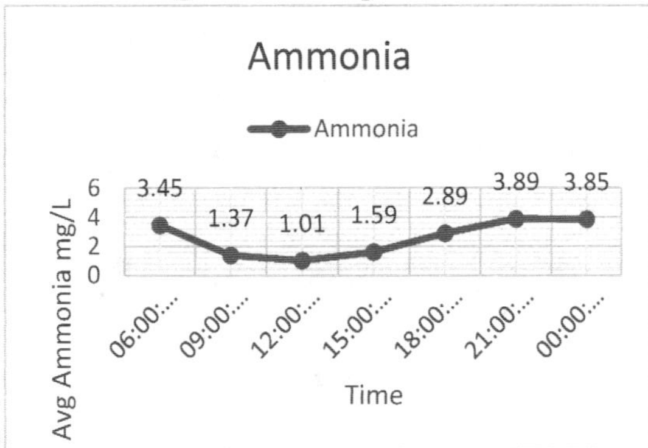

Figure 12. Fish pond Ammonia parameter

5. Conclusion

The quality of water has a significant importance in aquaculture as it is one of the deciding factors in the growth and well-being of fish. Therefore, it is necessary to measure and monitor the water quality using a finest and exclusive water testing kit. In this paper, a low-cost real-time monitoring system (LCRTMS) is developed to monitor the quality of water in aquaculture. The proposed model can monitor the variations in water parameters like pH, temperature, ammonia and dissolved oxygen by the deployment of affordable sensors. The proposed system gives highly reliable and accurate data with minimal human intervention for immediate action.it is evident that the accuracy is nearing to 98% in short term prediction of LCRTMS. Proactive steps are taken further to avoid any harmful occurrence. Although the initial set up cost is higher, there will not be any added expenses for maintenance after it is installed. Thus, the proposed implementation could reach the farmers by alleviating the harm related to climatic changes and assures growth and wellbeing of aquatic life. As a future direction, this proposal can be used at a higher level in various sites.

6. References

[1] Beckea, C., Schumanna, M., Steinhagenb, D., Rojas-Tiradoc, P., Geiste, J., Brinkera, A., 2019. Effects of unionized ammonia and suspended solids on rainbow trout (Oncorhynchus mykiss) in recirculating aquaculture systems. *Aquaculture* 499, 348–357

[2] Raju K.R.S.R.Varma, G.H.K., 2017. Knowledge based real time monitoring system for aquaculture using IoT. *In Proceedings of the 2017 IEEE 7th International Advance Computing Conference (IACC), Hyderabad, India*, 5–7 , pp. 318–321

[3] Ullah, Israr, Kim, DoHyeun, 2018. An optimization scheme for water pump control in smart fish farm with efficient energy consumption. *Processes* 6 (6), 65.

[4] Simbeye, D.S., Zhao, J., Yang, S., 2014. Design and deployment of wireless sensor networks for aquaculture monitoring and control based on virtual instruments. *Comput. Electron. Agric.* 102, 31–40

[5] Luo, Hongpin, Guanglin, L., Weifeng, P., Jie, S., Qiuwei, B., 2015. Real-time remote monitoring system for aquaculture water quality. *Int. J. Agric. Biol. Eng.* 8 (6), 136–143.

[6] Idachaba, Francis E., Olowoleni, Joseph O., Ibhaze, Augustus E., Oni, Oluyinka O., 2017. IoT enabled real-time fishpond management system. *Proceedings of the World Congress on Engineering and Computer Science 2017 Vol I*, San Francisco, USA.

[7] Cario, G., Casavola, A., Gjanci, P., Lupia, M., Petrioli, C., Spaccini, D., 2017. Long lasting underwater wireless sensors network for water quality monitoring in fish farms. Oceans Aberdeen Conference, Aberdeen, England, Jun 19–22, 2017.

[8] Zhang M. Li D.Wang, L. Ma, D.Ding, Q., 22–25 October 2010, Design and Development of Water Quality Monitoring System Based on Wireless Sensor Network in Aquaculture. *In Proceedings of the International Conference on Computer and Computing Technologies in Agriculture*, Nanchang, China,pp. 629–641.

[9] Simbeye, D.S., Yang S.F. ,2014, Water Quality Monitoring and Control for Aquaculture Based on Wireless Sensor Networks. *J. Netw.* 9, 840–849.

[10] Encinas, C.; Ruiz, E.; Cortez, J.; Espinoza, A. ,26–28 April 2017, Design and implementation of a distributed IoT system for the monitoring of water quality in aquaculture. *In Proceedings of the Wireless Telecommunications Symposium, Chicago, IL, USA,* pp. 1–7.

[11] Matos J,Postolache O,20–22 October 2016,IoT enabled aquatic drone for environmental monitoring. *In Proceedings of the 2016 International Conference and Exposition on Electrical and Power Engineering (EPE), Iasi, Romania,* pp. 598–603.

[12] Adhikari S, 2006. Soil and water quality management in aquaculture, p. 1-30. *In. Hand Book of Fisheries and Aquaculture. Indian Council of Agricultural Research*, New Delhi.

[13] Juan Huan, Hui Li, Fan Wu, Weijian Cao, Design of water quality monitoring system for aquaculture ponds based on NB-IoT,*Aquacultural Engineering journal*, P. 90, April 2020.

Soft morphological filtering using hypergraphs

Nuja M Unnikrishnan[1], Mini Tom[2], V Bino Sebastian[3] and K V Thomas[4]

[1]Research Scholar, Mathematics Department, Bharata Mata College, Thrikkakara, Kerala, India, nuja.mu@gmail.com
[2]Professor, Basic Science and Humanities Department, SCMS School of Engineering and Technology, Karukutty, Kerala, India, minitom@scmsgroup.org
[3]Assistant Professor, Mathematics Department, Mar Athanasius College, Kothamangalam, Kerala, India, binosebastianv@gmail.com
[4]Assosiate Professor, Mathematics Department, Nirmala College, Muvattupuzha, Kerala, India, kvtbmc@gmail.com

Abstract. A new framework of soft mathematical morphology on hypergraph spaces is studied. Application in image processing for filtering objects defined in hypergraph spaces are illustrated using several soft morphological operators- openings, closings, granulometries and ASF acting (a) on the subset of vertex and hyperedge set of a hypergraph and (b) on the subhypergraphs of a hypergraph. Experimental results dealing with the extension of soft morphological operators to gray scale images are presented in this paper. The results obtained are promising and is a better substitute for the prevailing methods.

1. Introduction and Related Works

Mathematical morphology is an emerging area of image processing. In the area where image processing can be applied, mathematical morphology have provided solutions to different tasks. Soft mathematical morphology is another approach to mathematical morphology that was introduced by Koskinen et al. [1]. In this method, weighted order statistics are used instead of minima and maxima [3]. Soft morphological operators show better performance than primitive morphological operators as they are less sensitive to additive noise and to small variations in object shape [4]. Soft mathematical morphology have also been extended to fuzzy sets [3].

Morphological operators can be defined on graphs and hypergraphs. In this paper a new context is introduced that lengthens the perceptions of soft morphological filters into hypergraphs. This work is an extension of our preliminary work [2].

2. Preliminaries

In this section, we see some important definitions and properties that will be needed in the sequel.

2.1. Soft mathematical morphology using hypergraphs [2]

Hypergraph [5], [6] is defined as a pair H= ($H^\bullet$, H^X), where $H^\bullet$ is a set of points called vertices and H^X is a family of subsets of $H^\bullet$ called hyperedges ie; $H^X = (e_i)$, $i \in I$ where I is a finite set of indices. The sets $H^\bullet$,

IOP Conf. Series: Materials Science and Engineering 1085 (2021) 012038 doi:10.1088/1757-899X/1085/1/012038

$H^{\Gamma X}$ and H^{Γ} are the subsets of $H^{\bullet}$, subsets of H^X and subhypergraphs of H respectively. Here $H^{\Gamma X}$ and $H^{\Gamma\bullet}$ are Boolean lattices. The set H^{Γ} of all sub hypergraphs of H forms a complete lattice [5], [6]. Morphological operators are defined on these lattices. In soft mathematical morphology, the structural element is divided into two subsets- the core and the soft boundary. In the development of the final output, weightage of core is more than the soft boundary. Hence we subdivide the hyperedges (along with the vertices belonging to it) into core B_1 and soft boundary B_2.

2.1.1. Definition 1 [2]
The operators $\delta^{\bullet}$, $\in^{\bullet}$ are defined from $H^{\Gamma X}$ into $H^{\Gamma\bullet}$ and the operators δ^X, $\in^X$ are defined from $H^{\Gamma\bullet}$ into $H^{\Gamma X}$ as follows: For any $X^{\bullet} \subseteq H^{\bullet}$ and and $X^X \subseteq H^X$, where $X^X = e_i \cup e_i^*$; $e_i \in B_1$ and $e_i^* \in B_2$, $i \in J$ such that $J \subseteq I$,

$\delta^{\bullet}: H^X \to H^{\bullet}$ is defined as $\delta^{\bullet}(X^X) = k^{th}\ largest\ of\ [\cup_{j\in J}\{k \lozenge v(e_j), v(e_j^*)\ ; e_j \in B_1, e_j^* \in B_2\}]$

$\in^X$: $H^{\bullet} \to H^X$ is defined as $\in^X (X^{\bullet}) = k^{th}$smallest of $\{k \lozenge e_i, e_i^*, i \in I \mid v(e_i), v(e_i^*) \subseteq X^{\bullet}, e_i \in B_1, e_i^* \in B_2\}$

$\in^{\bullet}: H^X \to H^{\bullet}$ is defined as $\in^{\bullet} (X^X) = k^{th} smallest\ of [\cap_{j\notin J}\{k \lozenge \overline{v(e_j)}\ ,\ \overline{v(e_j^*)}; e_j \in B_1, e_j^* \in B_2\}]$

$\delta^X: H^{\bullet} \to H^X$is defined as $\delta^X(X^{\bullet}) = k^{th}\ largest\ of\ \{k \lozenge e_i, e_i^*, i \in I \mid v(e_i) \cap X^{\bullet} \neq \emptyset\ and\ v(e_i^*) \cap X^{\bullet} \neq \emptyset; e_i \in B_1, e_i^* \in B_2\}$.

Here, k◊ x is read as x is repeated k times. Vertex soft dilation (δ) and vertex soft erosion($\mathcal{E}$) that act on $H^{\Gamma\bullet}$, also hyper edge soft erosion ($\boldsymbol{\varepsilon}$), hyper edge soft dilation(Δ) that act on $H^{\Gamma X}$ are defined as follows: The opertors δ and $\in$ that act on $H^{\Gamma\bullet}$ are defined by $\delta = \delta^{\bullet} \circ \delta^X$ and $\mathcal{E} = \in^{\bullet} \circ \in^X$ and the operators Δ and $\boldsymbol{\varepsilon}$ that act on $H^{\Gamma X}$ are defined by $\Delta = \delta^X \circ \delta^{\bullet}$ and $\boldsymbol{\varepsilon} = \in^X \circ \in^{\bullet}$.Furthermore the operators $[\delta, \Delta]$ and $[\mathcal{E}, \boldsymbol{\varepsilon}]$ are defined as

$[\delta, \Delta](X) = (\delta(X^{\bullet}), \Delta(X^X))$ and $[\mathcal{E}, \boldsymbol{\varepsilon}](X) = (\mathcal{E}(X^{\bullet}), \boldsymbol{\varepsilon}(X^X)$ for any $X \in H^{\Gamma}$.

These operators are called the soft dilation and soft erosion acting on the lattice (H^{Γ}, ⊆)

3. Property

(a) Operators $\in^X$ and δ^X are dual of each other. Similar duality concept hold for $\in^{\bullet}$ and $\delta^{\bullet}$.
(b) $(\in^X, \delta^{\bullet})$ and $(\in^{\bullet}, \delta^X)$ are adjunctions.
(c) $\delta^{\bullet}$ and $\delta^X$are soft dilations.
(d) $\in^{\bullet}$ and $\in^X$are soft erosions.

Proof (a): This property is proved in [2].

(b): Let $X^X \subseteq \in^X (Y^{\bullet})$. Then,

$x \in \delta^{\bullet}(X^X) \Rightarrow x \in k^{th}\ largest\ of\ [\cup_{j\in J}\{k \lozenge v(e_j), v(e_j^*)\}]$
$\Rightarrow x \in v(e_j), x \in v(e_j^*)$ for some $j \in J$
$\Rightarrow \exists\ e \in X^X$ such that $x \in v(e)$
$\Rightarrow e \in \in^X (Y^{\bullet})$ {Since $X^X \subseteq \in^X (Y^{\bullet})$}
$\Rightarrow e \in k^{th}$ smallest of $\{k \lozenge e_i, e_i^*, i \in I \mid v(e_i), v(e_i^*) \subseteq Y^{\bullet}\}$
$\Rightarrow v(e) \subseteq Y^{\bullet}$
$\Rightarrow x \in Y^{\bullet}$ {Since $x \in v(e)$}

Thus, $\delta^{\bullet}(X^X) \subseteq Y^{\bullet}$.
Conversely, if $\delta^{\bullet}(X^X) \subseteq Y^{\bullet}$.

Then, e$\in X^X \Rightarrow v(e) \subseteq \delta^\bullet(X^X)$
$\Rightarrow v(e) \subseteq Y^\bullet$ {Since $\delta^\bullet(X^X) \subseteq Y^\bullet$}
$\Rightarrow e \in \in^X (Y^\bullet)$
Therefore, $X^X \subseteq \in^X (Y^\bullet)$.
Hence, $(\in^X, \delta^\bullet)$ is an adjunction.
In a similar manner, $(\in^\bullet, \delta^X)$ is an adjunction can be proved using the information that operators $\in^X$ and δ^X are dual of each other, also $\in^\bullet$ and $\delta^\bullet$ are dual of each other and $(\in^X, \delta^\bullet)$ is an adjunction. Properties (c) and (d) follows directly using property (b).

4. Soft morphological filters

A morphological filter [5], [8] is an operator α acting on a lattice $\mathcal{L}$, which is increasing and idempotent. If (α, β) is an adjunction [5], [8] then α is an erosion, β is a dilation. Also, $\beta \circ \alpha$ is called an opening and $\alpha \circ \beta$ is called a closing on $\mathcal{L}$. Opening and closing are two commonly used filters.

4.1. Definition 2

- Soft opening $\gamma_1 = \delta \circ \varepsilon$ and closing $\Phi_1 = \varepsilon \circ \delta$.
- Soft half opening $\gamma_{1/2} = \delta^\bullet \circ \in^X$ and half closing $\Phi_{1/2} = \in^\bullet \circ \delta^X$.

5. Soft flat morphological Operators on weighted hypergraphs

In real phase, gray- scale soft morphological operations are difficult to work with. Threshold decomposition and stacking principle can be successfully applied on gray-scale soft morphological operations. This property allows the gray scale signals to be decomposed into binary signals and the results so obtained by processing are combined to obtain the desired gray-scale output [7]. By threshold decomposition, [8] the lattice of all subhypergraphs of H induces a lattice Fun($H^\bullet$)$\otimes$Fun($H^\times$) of pairs of functions weighting respectively the vertices and the hyperedges of H such that the simultaneous threshold of these two functions at any given level produces a subhypergraph of H and the properties for hypergraph operators on the lattices $\mathcal{H}^\bullet$, $\mathcal{H}^X$, or , $\mathcal{H}$ also hold for operators on the lattices Fun($H^\bullet$), Fun($H^\times$) and Fun($H^\bullet$) $\otimes$ Fun($H^\times$). Thus we can define a set of soft operators which are stack analogues to the soft operators $\in^X, \delta^\bullet, \in^\bullet$ *and* δ^X defined in [2].

5.1. Definition 3.

Let $F^\bullet \in$ Fun($H^\bullet$) and let $F^X \in$ Fun($H^\times$), we define

$$\delta^\bullet(F^X)(x) = k^{th}\ largest\ of_{\ x\in v(e_i), x\in v(e_i^*)} \left\{k \lozenge F^X(e_i), F^X(e_i^*) \middle| e_i \in B_1, e_i^* \in B_2, B_1 \cup B_2 = H^X\right\} \forall x \in H^\bullet$$
$$\in^X (F^\bullet)(e) = k^{th} \text{ smallest of } \{k \lozenge F^\bullet(x), F^\bullet(x) | x \in v(e), e \in B_1 \cup B_2\}$$
$$\in^\bullet (F^X)(x) = k^{th} smallest\ of_{\ x\in v(e_i), x\in v(e_i^*)} \left\{k \lozenge F^X(e_i), F^X(e_i^*) \middle| e_i \in B_1, e_i^* \in B_2, B_1 \cup B_2 = H^X\right\} \forall\ x \in H^\bullet$$
$$\delta^X(F^\bullet)(e) = k^{th} \text{ smallest of } \{k \lozenge F^\bullet(x), F^\bullet(x) | x \in v(e), e \in B_1 \cup B_2\}.$$

By using soft flat morphology, it is possible to work with gray scale soft dilation, erosion, opening, closing, granulometries and Alternating Sequential Filters.

6. Experimental Result

To represent the hyperedges we use 4-uniform hypergraph illustrated in Figure 1. A hyperedge together with the 4 vertices belonging to it is taken as core, remaining vertices and hypereges are taken as soft boundary. Using this simple hypergraph structure, experimental result for k=1 and k=2 are tabulated. Dilated and eroded gray scale image results are obtained to get the soft flat morphological operators defined in the previous section. Composition of these operators generates alternating sequential filters(ASF) which are capable of removing noise effectively from binary and gray scale images. Figure 2 represent the gray scale image taken for the experimental purpose. The noisy image obtained by adding 5% salt and pepper noise is shown in figure 3. PSNR of noised image with original image is 17.24. Figure 4 and figure 5 shows the noise removed images obtained by applying alternating sequential filters(ASF) $\gamma_1 \circ \Phi_1$and $\gamma_{1/2} \circ \Phi_{1/2}$ for k=1. The resultant PSNR of the noised removed images with the original image are respectively 40.53 and 40.23. Similarly figure 6 and figure 7 shows the results obtained after applying $\gamma_1 \circ \Phi_1$and $\gamma_{1/2} \circ \Phi_{1/2}$ for k=2. In this case, the resultant PSNR of the noised removed images with the original image are 39.79 and 41.82 respectively. Experiment results depicts that $\gamma_{1/2} \circ \Phi_{1/2}$ for k = 2 is giving better approximations than $\gamma_1 \circ \Phi_1$ and $\gamma_{1/2} \circ \Phi_{1/2}$ for k =1.

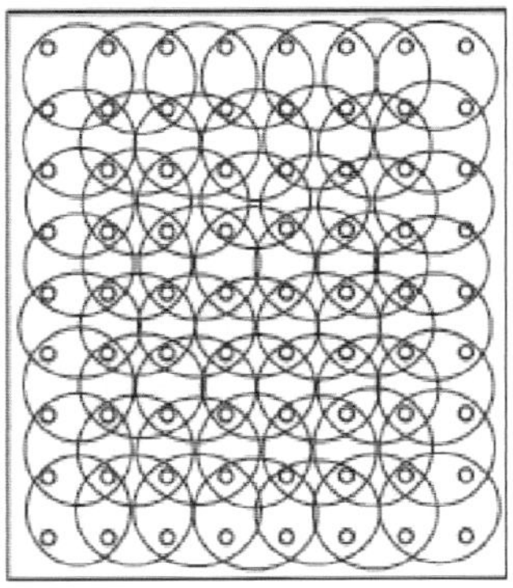

Figure 1 Four Uniform Hypergraph structure used to represent an image

PSNR of noised image with original image : 17.24

Figure 2. Original Image

Figure 3. Noised Image (Added Noise 5%)

IOP Conf. Series: Materials Science and Engineering 1085 (2021) 012038 doi:10.1088/1757-899X/1085/1/012038

Noise Removed Images using Alternating Sequential Filters

For K =1

Figure 4. $\gamma_1 \circ \Phi_1$

PSNR = 40.53

Figure 5. $\gamma_{1/2} \circ \Phi_{1/2}$

PSNR = 40.23

Noise Removed Images using Alternating Sequential Filters

For K = 2

Figure 6. $\gamma_1 \circ \Phi_1$

PSNR = 39.79

Figure 7. $\gamma_{1/2} \circ \Phi_{1/2}$

PSNR = 41.82

7. Conclusion

The aim of this work is to recognize the prospects of using soft morphological operators under the framework of hypergraph. In our study a four uniform hypergraph structure was taken. It was then separated into core and soft boundary. But, it's a well-known fact that 3-unifrom hypergraph structure gives good result for binary, gray scale and colour image filtering applications [8]. So by the choice of different hypergraph structures along with the suitable choice of core and soft boundary, results of the above method can be improved for different and bigger values of k. The initial results are promising and future work is to be done in this regard using different hypergraph structures.

References

[1] Koskinen L, Astola J and Neuvo Y 1991 Soft morphological filters *Proc. SPIE- Int. Soc. Opt. Eng.* 262-70.

[2] Unnikrishnan M Nuja, Sebastian V Bino and Thomas K V 2019 Soft Morphological Operators on Fuzzy Hypergraphs *Mathematical Science International Research Journal (MSIRJ)* **8** Iss.2 72-78.

[3] Gasteratos A, Andreadis I and Ph. Tsalides 1998 Fuzzy soft mathematical morphology *IEE Proc.-Vis. Image Signal Processing*. **145** 41-49.

[4] Kuosmanen P and Astola J 1995 Soft morphological filtering *J.Math. Imag. Vis.***5** 231-62.

[5] Sebastian V Bino, Unnikrishnan A, Balakriishnan K and Ramkumar P B 2017 Morphological filtering on hypergraphs *Discrete Applied Mathematics Elsevier* 307-20.

[6] Sebastian V Bino, Unnikrishnan A, Balakrishnan K, and Ramkumar P B 2014 Mathematical morphology on hypergraphs using vertex-hyperedge correspondence *ISRN Discrete Mathematics.*

[7] Shih F Y and Mitchell O K 1989 Threshold decomposition of gray-scale morphology into binary morphology *IEEE Trans. Pattern Anal. Mach. Intell.* **11** Iss.1 31-42.

[8] Sebastian V Bino, Unnikrishnan M Nuja, Sebastian Neenu and Paul Rosebell 2020 Morphological operators on hypergraphs for colour image processing *Advanced Computing and Communication Technologies for High Performance Applications(ACCTHPA) IEEE Xplore* 217-20.

IOP Conf. Series: Materials Science and Engineering 1085 (2021) 012039 doi:10.1088/1757-899X/1085/1/012039

Lucky and Proper Lucky Labeling of Quadrilateral Snake Graphs

T. V. Sateesh Kumar[1] and S. Meenakshi[2]
[1]Ressearch scholar, Department of Mathematics, Vels Institute of Science, Technology and Advanced studies, Pallavaram, Chennai, India.
[2]Associate Professor, Department of Mathematics, Vels Institute of Science, Technology and Advanced studies, Pallavaram, Chennai, India.
Email: [1]sateeshkumarsks@gmail.com and [2]meenakshikarthikeyan@yahoo.in

Abstract. The labeling is said to be lucky labeling of the graph if the vertices of the graph are labeled by natural number with satisfying the condition that sum of labels over the adjacent of the vertices in the graph are not equal and if vertices are isolated vertex then the sum of the vertex is zero. The least natural number which labelled the graph is the lucky number. The Lucky Number of graph G is denoted by $\eta(G)$. The labeling defined as proper labeling if the vertices of the graph are labeled by natural number with fulfilling the condition that label of adjacent vertices is not the same. The labeling is defined as proper lucky labeling if labeling is proper and also lucky. The proper lucky number of graph G is denoted by $\eta_p(G)$. Here we obtain a lucky number and proper lucky number for family of quadrilateral snake graph such as quadrilateral snake graph, double quadrilateral snake graph, alternate quadrilateral snake graph and double alternate quadrilateral snake graph.

1. Introduction

In graph theory, graph labeling has a wide area. It has a well-built connection among numbers and the structure of the graph. It was introduced by Rosa in 1967 [10]. Labeling is a function to mark the number to a graph vertex or edge or both with the specific condition. Gallian [4] gave a dynamic survey of graph labeling. The proper labeling was initiated by Karonsik (2004) [7] which mean naming adjacent vertices by a different natural number. Lucky labeling for 3 colourable graphs was studied by Ahadi (2012) [1]. Lucky labeling is compared with proper vertex coloring by Akbari (2013) [2]. The graph which fulfill the lucky labeling conditions are called lucky graph. The graph which fulfill the proper lucky labeling conditions are called proper lucky graph. Proper lucky labeling has a very close association with graph colouring.

Vidya computed that alternate quadrilateral snake graphs are product cordial in 2014 [13]. Lavanya had detected that quadrilateral snake graphs are graceful in 2017 [11]. Prajapati computed that quadrilateral snake graph is SD-Prime Cordial in 2019 [9]. Jadav computed snake graphs are strongly-graph in 2016 [6]. Labeling application is computer networking, clustering, image segmentation etc., it also applicable in astronomy, circuit designing, image process, cryptography and information security. The objective of this paper is to study the lucky number and proper lucky number for quadrilateral snake graph, double quadrilateral snake graph, alternate quadrilateral snake graph and double alternate quadrilateral snake graph.

2. Preliminaries

2.1 Quadrilateral snake graph

A quadrilateral snake graph QS_n is obtained from $3n-2$ vertices where $n>1, u_i$ have n vertices, v_i have $2n-2$ vertices and the edge set $E(QS_n)=\{u_iu_{i+1}: 1\leq i\leq n\}\cup\{v_{2i-1}v_{2i}: 1\leq i\leq n\}\cup\{u_iv_{2i-1}: 1\leq i\leq n\}\cup\{u_{i+1}v_{2i}: 1\leq i\leq n\}$ refer figure 1 [5,12].

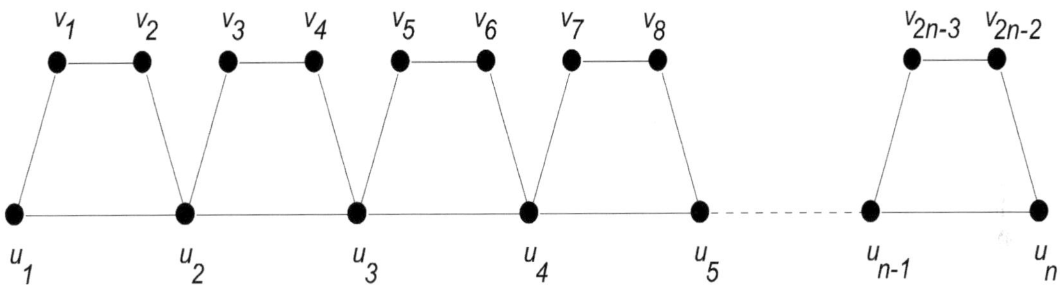

Figure 1: Quadrilateral snake graph QS_n

2.2 Alternate quadrilateral snake graph

An alternate quadrilateral snake graph AQS_n is consist of a $2n$ vertices where $n>1, u_i$ have n vertices, v_i have n vertices and the edge set $E(AQS_n)=\{u_iu_{i+1}: 1\leq i\leq n\}\cup\{v_{2i-1}v_{2i}: 1\leq i\leq n\}\cup\{u_{2i}v_{2i}: 1\leq i\leq n\}\cup\{u_{2i-1}v_{2i-1}: 1\leq i\leq n\}$ refer figure 2 [5,12].

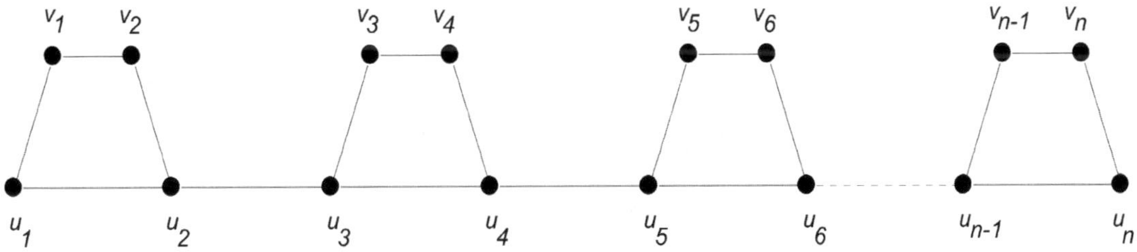

Figure 2: Alternate quadrilateral snake graph AQS_n

2.3 Double quadrilateral snake graph

A double quadrilateral snake graph DQS_n is consist of a $5n-2$ vertices where $n>1, u_i$ have n vertices, v_i have $2n-2$ vertices, w_i have $2n-2$ vertices and the edge set $E(DQS_n)=E(QS_n)\cup\{w_{2i-1}w_{2i}: 1\leq i\leq n\}\cup\{u_iw_{2i-1}: 1\leq i\leq n\}\cup\{u_{i+1}w_{2i}: 1\leq i\leq n\}$ refer figure 3 [5,12].

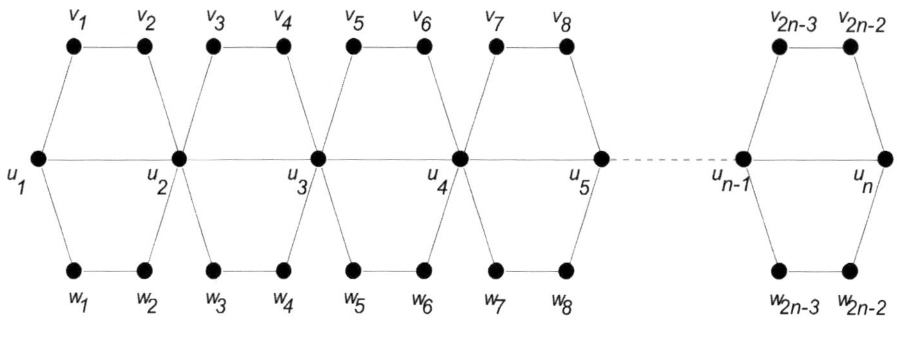

Figure 3: Double quadrilateral snake graph DQS_n

2.4 Double alternate quadrilateral snake graph

An alternate double quadrilateral snake graph $DAQS_n$ is consist of a $3n$ vertices where $n > 1, u_i$ have n vertices, v_i have n vertices, w_i have n vertices and the edge set $E(DAQS_n) = E(AQS_n) \cup \{w_{2i-1}w_{2i}: 1 \leq i \leq n\} \cup \{u_{2i}w_{2i}: 1 \leq i \leq n\} \cup \{u_{2i-1}w_{2i-1}: 1 \leq i \leq n\}$ refer figure 4 [5,12].

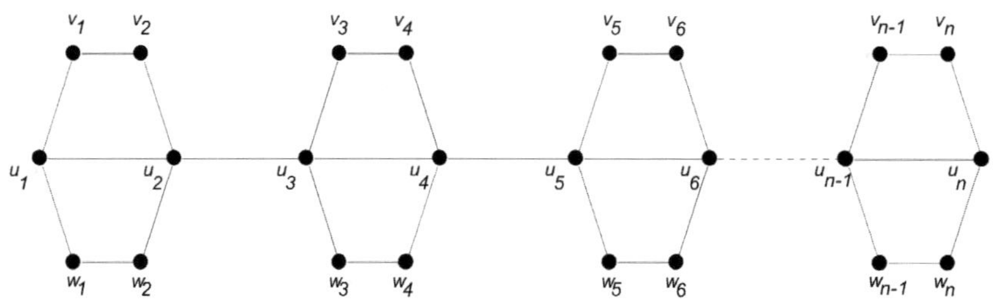

Figure 4: Double alternate quadrilateral snake graph $DAQS_n$

2.5 Lucky labeling

Let $f: V(G) \to \mathbb{N}$ be a labeling of the vertices of a graph G by a natural number. Let $S(v)$ indicate the sum of labels over the neighbours of the vertex v in G. If v is an isolated vertex of G we put $S(v) = 0$. A labeling f is lucky if $S(u) \neq S(v)$ for an adjacent vertex u and v [3,8].

2.6 Proper lucky labeling

A labeling f is proper lucky labeling, if u and v are adjacent in G then $f(u) \neq f(v)$ & $S(u) \neq S(v)$. The proper Lucky number of G is denoted by $\eta_p(G)$, is the least natural number k such that G has proper lucky labeling with $\{1,2, \dots, k\}$ as the set of labels [3,8].

3. Main results

3.1 Theorem

The quadrilateral snake QS_n with $n > 1$ admits lucky labeling with lucky number $\eta(QS_n) = 2$.

Proof:

Let $f: V(QS_n) \to \{1,2\}$ quadrilateral snake graph with $3n-2$ vertices where $n > 1, u_i$ have n vertices, v_i have $2n-2$ vertices and the edge set $E(QS_n) = \{u_i u_{i+1}: 1 \le i \le n\} \cup \{v_{2i-1}v_{2i}: 1 \le i \le n\} \cup \{u_i v_{2i-1}: 1 \le i \le n\} \cup \{u_{i+1}v_{2i}: 1 \le i \le n\}$, be defined by

$$f(u_i) = \begin{cases} 1 & even\ i \\ 2 & odd\ i \end{cases}$$
$$f(v_i) = 1$$

Here we obtain the sum of neighbour vertices as,

$$s(u_i) = \begin{cases} 2 & i = 1, i = odd\ n \\ 3 & i = even\ n \\ 4 & odd\ i > 1, i < n \\ 6 & even\ i < n \end{cases}$$
$$s(v_i) = \begin{cases} 2 & i = 4k-1, i = 4k-2 \\ 3 & i = 4k, i = 4k-3 \end{cases}, k \in \mathbb{N}$$

Therefore sum of adjacent vertices are not same. So, quadrilateral snake QS_n with $n > 1$ admits lucky labeling with lucky number $\eta(QS_n) = 2$.

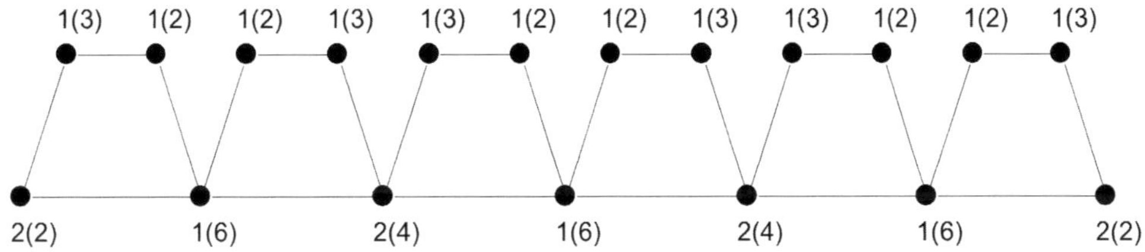

Figure 5: Lucky quadrilateral snake graph QS_7

3.2 Illustration: The lucky number 2 of quadrilateral snake graph $\eta(QS_7) = 2$. is shown in figure 5.

3.3 Theorem

The double quadrilateral snake DQS_n with $n > 1$ admits lucky labeling with lucky number $\eta(DQS_n) = 2$.

Proof:

Given double quadrilateral snake graph with $5n-2$ vertices where $n > 1, u_i$ have n vertices, v_i have $2n-2$ vertices, w_i have $2n-2$ vertices and the edge set $E(DQS_n) = E(QS_n) \cup \{w_{2i-1}w_{2i}: 1 \le i \le n\} \cup \{u_i w_{2i-1}: 1 \le i \le n\} \cup \{u_{i+1}w_{2i}: 1 \le i \le n\}$

Case (i): n = even

Let $f: V(DQS_n) \to \{1,2\}$ for double quadrilateral snake graph be defined by

$$f(u_i) = \begin{cases} 1 & even\ i \\ 2 & odd\ i \end{cases}$$
$$f(v_i) = \begin{cases} 1 & i \ne 2 \\ 2 & i = 2 \end{cases}$$
$$f(w_i) = \begin{cases} 1 & i \ne 2 \\ 2 & i = 2 \end{cases}$$

Here we obtain the sum of neighbour vertices as,

$$s(u_1) = 3$$
$$s(u_n) = 4$$
$$s(u_i) = \begin{cases} 6 & odd\ i > 1 \\ 8 & even\ i < n, i > 2 \\ 10 & i = 2 \end{cases}$$

$$s(v_i)=\begin{cases}2 & i=4k-1, i=4k-2\\ 3 & i=4k, i=4k-3, i\neq 1\\ 4 & i=1\end{cases}\quad ,k\in\mathbb{N}$$

$$s(w_i)=\begin{cases}2 & i=4k-1, i=4k-2\\ 3 & i=4k, i=4k-3, i\neq 1\\ 4 & i=1\end{cases}\quad ,k\in\mathbb{N}$$

Therefore sum of adjacent vertices are not same. So, double quadrilateral snake DQS_n with $n>1$ admits lucky labeling for even n with lucky number $\eta(DQS_n)=2$.

Case (ii): $n=odd$

Let $f: V(DQS_n)\rightarrow\{1,2\}$ for double quadrilateral snake graph be defined by

$$f(u_i)=\begin{cases}1 & even\, i\\ 2 & odd\, i\end{cases}$$

$$f(v_i)=\begin{cases}1 & i\neq n-1, i\neq 2\\ 2 & i=n-1, i=2\end{cases}$$

$$f(w_i)=\begin{cases}1 & i\neq n-1, i\neq 2\\ 2 & i=n-1, i=2\end{cases}$$

Here we obtain the sum of neighbour vertices as,

$$s(u_1)=3$$

$$s(u_n)=3$$

$$s(u_i)=\begin{cases}6 & odd\, i>1, \mathrm{i}<\mathrm{n}\\ 8 & even\, i<n-1, i>2\\ 10 & i=n-1, i=2\end{cases}$$

$$s(v_i)=\begin{cases}2 & i=4k-1, i=4k-2, i\neq n\\ 3 & i=4k, i=4k-3, i\neq n, i\neq 1\\ 4 & i=n, i=1\end{cases}\quad ,k\in\mathbb{N}$$

$$s(w_i)=\begin{cases}2 & i=4k-1, i=4k-2, i\neq n\\ 3 & i=4k, i=4k-3, i\neq n, i\neq 1\\ 4 & i=n, i=1\end{cases}\quad ,k\in\mathbb{N}$$

Therefore sum of adjacent vertices are not same. So, double quadrilateral snake DQS_n with $n>1$ admits lucky labeling for odd n with lucky number $\eta(DQS_n)=2$. Hence, the double quadrilateral Snake DQS_n with $n>2$ admits lucky labeling with lucky number $\eta(DQS_n)=2$.

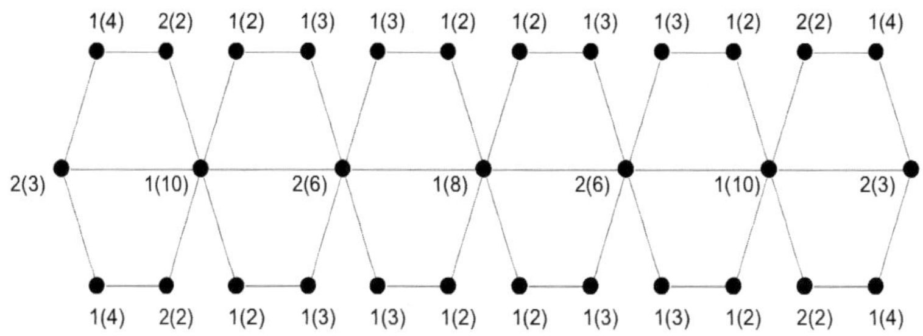

Figure 6: Lucky double quadrilateral snake graph DQS_7

3.4 Illustration: The lucky number 2 of double quadrilateral snake graph $\eta(DQS_7)=2$. is shown in figure 6.

3.5 Theorem

The alternate quadrilateral snake AQS_n with $n>1$ admits lucky labeling with lucky number $\eta(AQS_n)=2$.

Proof:
Let $f: V(AQS_n) \to \{1,2\}$ for alternate quadrilateral snake graph with $2n$ vertices where $n > 1, u_i$ have n vertices, v_i have n vertices and the edge set $E(AQS_n) = \{u_i u_{i+1}: 1 \le i \le n\} \cup \{v_{2i-1}v_{2i}: 1 \le i \le n\} \cup \{u_{2i}v_{2i}: 1 \le i \le n\} \cup \{u_{2i-1}v_{2i-1}: 1 \le i \le n\}$, be defined by

$$f(u_i) = \begin{cases} 1 & odd\ i \\ 2 & even\ i \end{cases}$$

$$f(v_i) = \begin{cases} 1 & even\ i \\ 2 & odd\ i \end{cases}$$

Here we obtain the sum of neighbour vertices as,

$$s(u_1) = 4$$

$$s(u_n) = 2$$

$$s(u_i) = \begin{cases} 3 & even\ i, i < n \\ 6 & odd\ i, i > 1 \end{cases}$$

$$s(v_i) = \begin{cases} 2 & odd\ i \\ 4 & even\ i \end{cases}$$

Therefore sum of adjacent vertices are not same. So, alternate quadrilateral snake AQS_n with $n > 1$ admits lucky labeling with lucky number $\eta(AQS_n) = 2$.

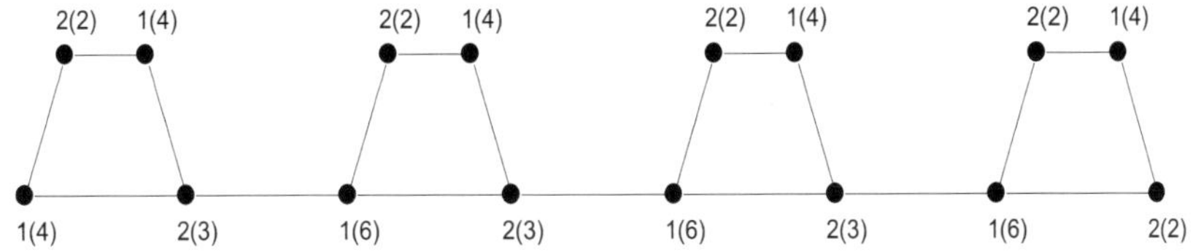

Figure 7: Lucky alternate quadrilateral snake graph AQS_8

3.6 Illustration: The lucky number 2 of alternate quadrilateral snake graph $\eta(AQS_8) = 2$. is shown in figure 7.

3.7 Theorem
The double alternate quadrilateral snake $DAQS_n$ with $n > 1$ admits lucky labeling with lucky number $\eta(DAQS_n) = 2$.

Proof:
Let $f: V(DAQS_n) \to \{1,2\}$ for alternate quadrilateral snake graph with $3n$ vertices where $n > 1, u_i$ have n vertices, v_i have n vertices, w_i have n vertices and the edge set $E(DAQS_n) = E(AQS_n) \cup \{w_{2i-1}w_{2i}: 1 \le i \le n\} \cup \{u_{2i}w_{2i}: 1 \le i \le n\} \cup \{u_{2i-1}w_{2i-1}: 1 \le i \le n\}$, be defined by

$$f(u_i) = \begin{cases} 1 & odd\ i \\ 2 & even\ i \end{cases}$$

$$f(v_i) = \begin{cases} 1 & even\ i \\ 2 & odd\ i \end{cases}$$

$$f(w_i) = \begin{cases} 1 & even\ i \\ 2 & odd\ i \end{cases}$$

Here we obtain the sum of neighbour vertices as,

$$s(u_1) = 6$$

$$s(u_n) = 3$$

$$s(u_i) = \begin{cases} 5 & even\ i, i < n \\ 8 & odd\ i, i > 1 \end{cases}$$

$$s(v_i) = \begin{cases} 2 & odd\ i \\ 4 & even\ i \end{cases}$$

$$s(w_i) = \begin{cases} 2 & i = n-1 \\ 3 & odd\ i, i \neq n-1 \\ 4 & even\ i \end{cases}$$

Therefore sum of adjacent vertices are not same. So, double alternate quadrilateral snake $DAQS_n$ with $n > 1$ admits lucky labeling with lucky number $\eta(DAQS_n) = 2$.

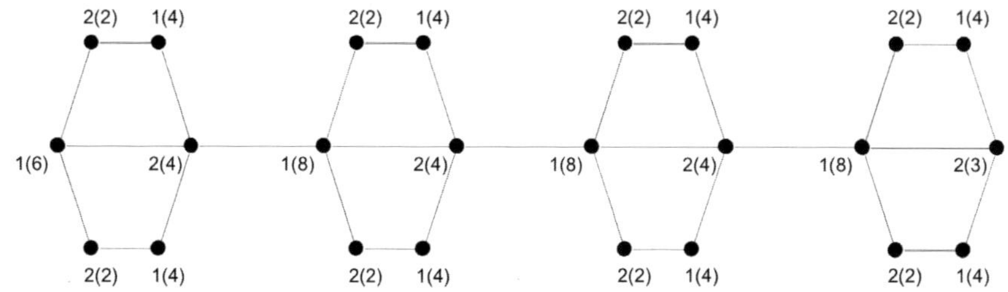

Figure 8: Lucky double alternate quadrilateral snake graph $DAQS_8$

3.8 Illustration: The lucky number 2 of double alternate quadrilateral snake graph $\eta(DAQS_8) = 2$. is shown in figure 8.

3.9 Theorem

The quadrilateral snake QS_n with $n > 1$ admits proper lucky labeling with proper lucky number $\eta_p(QS_n) = 3$.

Proof:

Let $f: V(QS_n) \to \{1,2,3\}$ for quadrilateral snake graph with $3n-2$ vertices where $n > 1, u_i$ have n vertices, v_i have $2n-2$ vertices and the edge set $E(QS_n) = \{u_i u_{i+1}: 1 \leq i \leq n\} \cup \{v_{2i-1}v_{2i}: 1 \leq i \leq n\} \cup \{u_i v_{2i-1}: 1 \leq i \leq n\} \cup \{u_{i+1}v_{2i}: 1 \leq i \leq n\}$, be defined by

$$f(u_i) = \begin{cases} 1 & odd\ i \\ 3 & even\ i \end{cases}$$

$$f(v_i) = \begin{cases} 1 & i = 4k-1, i = 4k-2 \\ 2 & i = 4k, i = 4k-3 \end{cases}, \quad k \in \mathbb{N}$$

Here we obtain the sum of neighbour vertices as,

$$s(u_1) = 5$$

$$s(u_n) = \begin{cases} 2 & even\ n \\ 5 & odd\ n \end{cases}$$

$$s(u_i) = \begin{cases} 4 & even\ i < n \\ 10 & odd\ i > 1, i < n \end{cases}$$

$$s(v_i) = \begin{cases} 2 & i = 4k, i = 4k-3 \\ 5 & i = 4k-1, i = 4k-2 \end{cases}, k \in \mathbb{N}$$

Therefore sum of adjacent vertices and adjacent labeling are not same. So, quadrilateral snake QS_n with $n > 1$ admits proper lucky labeling with proper lucky number $\eta_p(QS_n) = 3$.

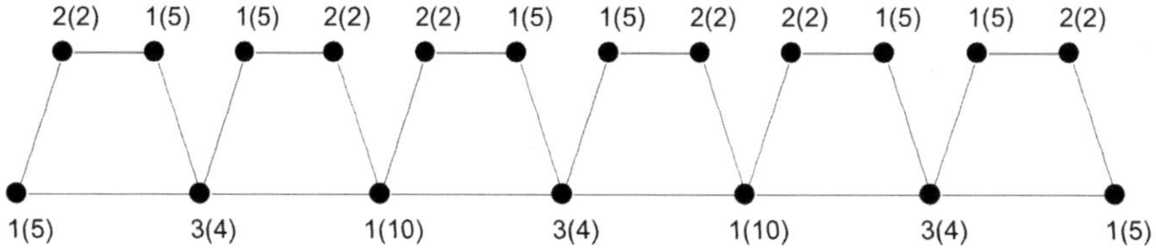

Figure 9: Proper lucky quadrilateral snake graph QS_7

3.10 Illustration: The proper lucky number 3 of quadrilateral snake graph $\eta_p(QS_7) = 3$. is shown in figure 9.

3.11 Theorem
The double quadrilateral snake DQS_n with $n > 1$, for odd n it admits proper lucky labeling with proper lucky number $\eta_p(DQS_n) = 2$.
Proof:
Let $f: V(DQS_n) \to \{1,2\}$ for odd n double quadrilateral snake graph with $5n-2$ vertices where $n > 1, u_i$ have n vertices, v_i have $2n-2$ vertices, w_i have $2n-2$ vertices and the edge set $E(DQS_n) = E(QS_n) \cup \{w_{2i-1}w_{2i}: 1 \le i \le n\} \cup \{u_i w_{2i-1}: 1 \le i \le n\} \cup \{u_{i+1}w_{2i}: 1 \le i \le n\}$, be defined by

$$f(u_i) = \begin{cases} 1 & even\, i \\ 2 & odd\, i \end{cases}$$

$$f(v_i) = \begin{cases} 1 & i = 4k, i = 4k-3 \\ 2 & i = 4k-1, i = 4k-2 \end{cases}, \quad k \in \mathbb{N}$$

$$f(w_i) = \begin{cases} 1 & i = 4k, i = 4k-3 \\ 2 & i = 4k-1, i = 4k-2 \end{cases}, \quad k \in \mathbb{N}$$

Here we obtain the sum of neighbour vertices as,

$$s(u_1) = 3$$

$$s(u_n) = 3$$

$$s(u_i) = \begin{cases} 6 & odd\, i > 1, i < n \\ 12 & even\, i \end{cases}$$

$$s(v_i) = \begin{cases} 2 & i = 4k, i = 4k-3 \\ 4 & i = 4k-1, i = 4k-2 \end{cases}, k \in \mathbb{N}$$

$$s(w_i) = \begin{cases} 2 & i = 4k, i = 4k-3 \\ 4 & i = 4k-1, i = 4k-2 \end{cases}, k \in \mathbb{N}$$

Therefore sum of adjacent vertices and adjacent labeling are not same. So, double quadrilateral snake DQS_n with $n > 1$ admits proper lucky labeling for odd n with proper lucky number $\eta_p(DQS_n) = 2$.

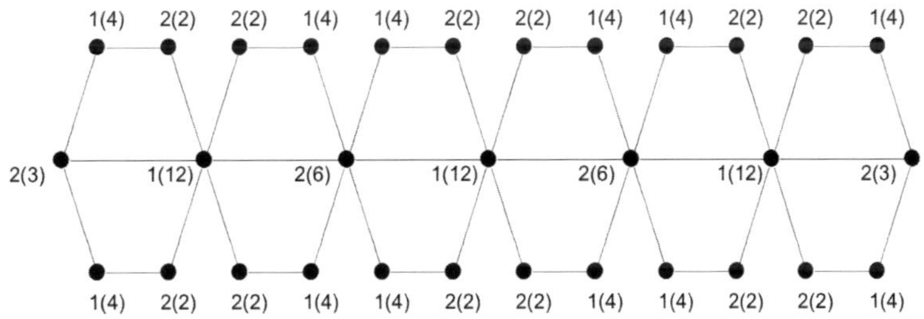

Figure 10: Proper lucky double quadrilateral snake graph DQS_7

3.12 Illustration: The proper lucky number 2 of double quadrilateral snake graph $\eta_p(DQS_7) = 2$. is shown in figure 10.

3.13 Theorem
The double quadrilateral snake DQS_n with $n > 1$, for even n it admits proper lucky labeling with proper lucky number $\eta_p(DQS_n) = 3$.
Proof:

Let $f: V(DQS_n) \rightarrow \{1,2,3\}$ for even n double quadrilateral snake graph with $5n-2$ vertices where $n > 1, u_i$ have n vertices, v_i have $2n-2$ vertices, w_i have $2n-2$ vertices and the edge set $E(DQS_n) = E(QS_n) \cup \{w_{2i-1}w_{2i}: 1 \le i \le n\} \cup \{u_i w_{2i-1}: 1 \le i \le n\} \cup \{u_{i+1}w_{2i}: 1 \le i \le n\}$, be defined by

$$f(u_i) = \begin{cases} 1 & even\, i \\ 2 & odd\, i \end{cases}$$

$$f(v_i) = \begin{cases} 1 & i = 4k, i = 4k-3, i \ne 2n-2 \\ 2 & i = 4k-1, i = 4k-2, i \ne 2n-2, \quad k \in \mathbb{N} \\ 3 & i = 2n-2 \end{cases}$$

$$f(v_i) = \begin{cases} 1 & i = 4k, i = 4k-3, i \ne 2n-2 \\ 2 & i = 4k-1, i = 4k-2, i \ne 2n-2, \quad k \in \mathbb{N} \\ 3 & i = 2n-2 \end{cases}$$

Here we obtain the sum of neighbour vertices as,

$$s(u_1) = 3$$

$$s(u_n) = 8$$

$$s(u_i) = \begin{cases} 6 & odd\, i > 1 \\ 12 & even\, i < n \end{cases}$$

$$s(v_i) = \begin{cases} 2 & i = 4k-1, i = 4k-2, i \ne 2n-3 \\ 4 & i = 4k, i = 4k-3, i \ne 2n-3 \quad ,k \in \mathbb{N} \\ 5 & i = 2n-3 \end{cases}$$

$$s(w_i) = \begin{cases} 2 & i = 4k-1, i = 4k-2, i \ne 2n-3 \\ 4 & i = 4k, i = 4k-3, i \ne 2n-3 \quad ,k \in \mathbb{N} \\ 5 & i = 2n-3 \end{cases}$$

Therefore sum of adjacent vertices and labeling are not same. So, double quadrilateral snake DQS_n with $n > 1$ admits lucky labeling for even n with proper lucky number $\eta(DQS_n) = 3$.

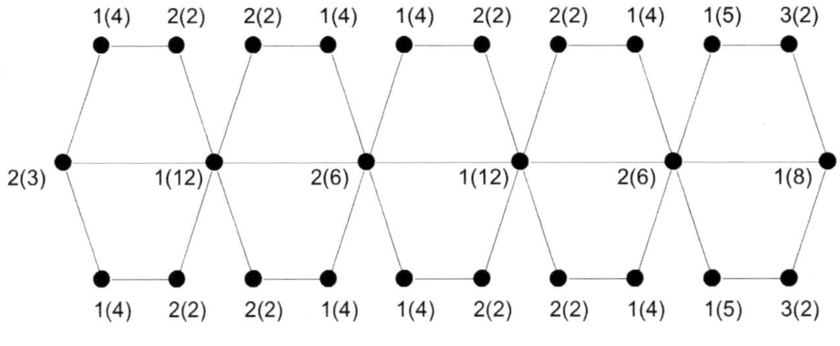

Figure 11: Proper double quadrilateral snake graph DQS_6

3.14 Illustration: The proper lucky number 3 of double quadrilateral snake graph $\eta_p(DQS_6) = 3$. is shown in figure 11.

3.15 Theorem

The alternate quadrilateral snake AQS_n with $n > 1$ admits proper lucky labeling with proper lucky number $\eta_p(AQS_n) = 2$.

Proof:

Refer theorem 3.5, since it satisfies the condition the sum of adjacent vertices and adjacent labeling are not same. So, alternate quadrilateral Snake AQS_n with $n > 1$ admits proper lucky labeling with proper lucky number $\eta_p(AQS_n) = 2$.

3.16 Theorem
The double alternate quadrilateral snake $DAQS_n$ with $n > 1$ admits proper lucky labeling with proper lucky number $\eta_p(DAQS_n) = 2$.
Proof:
Refer theorem 3.7, since it satisfies the condition the sum of adjacent vertices and adjacent labeling are not same. So, Double alternate quadrilateral snake graph $DAQS_n$ with $n > 1$ admits proper lucky labeling with lucky number $\eta_p(DAQS_n) = 2$.

3.17 Corollary
Lucky number and minimum degree of quadrilateral snake graph, alternate quadrilateral snake graph, double quadrilateral snake graph and double alternate quadrilateral snake graph are equal i.e., $\eta(G) = \delta(G), G \in QS_n, AQS_n, DQS_n, DAQS_n$.

3.19 Corollary
Lucky labeling is half of maximum degree of quadrilateral snake graph and double alternate quadrilateral snake graph i.e., $\eta(G) = \frac{1}{2}\Delta(G), G \in QS_n, DAQS_n$.

3.20 Corollary
Lucky number is one third of maximum degree of double quadrilateral snake graph i.e., $\eta(DQS_n) = \frac{1}{3}\Delta(DQS_n)$.

3.21 Corollary
Lucky number is one less than maximum degree of alternate quadrilateral snake graph i.e., $\eta(AQS_n) = \Delta(AQS_n) - 1$.

3.22 Corollary
Proper lucky number of quadrilateral snake graph is one more than minimum degree i.e., $\eta_p(QS_n) = \delta(QS_n) + 1$.

3.23 Corollary
Proper lucky number and minimum degree of double quadrilateral snake graph are equal, if n is odd i.e., $\eta_p(DQS_n) = \delta(DQS_n), for\ odd\ n$.

3.23 Corollary
Proper lucky number is one more than minimum degree of double quadrilateral snake graph, if n is even i.e., $\eta_p(DQS_n) = \delta(DQS_n) + 1, for\ even\ n$.

3.24 Corollary
Proper lucky number and minimum degree of alternate quadrilateral snake graph and double alternate quadrilateral snake graph are equal i.e., $\eta_p(G) = \delta(G), G \in AQS_n, DAQS_n$.

3.27 Corollary
Proper lucky number of quadrilateral and alternate quadrilateral snake graph are one less than maximum degree i.e., $\eta_p(G) = \Delta(G) - 1, G \in QS_n, AQS_n$.

3.28 Corollary

Proper lucky number of double quadrilateral snake graph is one third of maximum degree if n is odd i.e., $\eta_p(DQS_n) = \frac{1}{3}\Delta(DQS_n), for\ odd\ n.$

3.29 Corollary
Proper lucky number of double quadrilateral snake graph is half of maximum degree if n is even i.e., $\eta_p(DQS_n) = \frac{1}{2}\Delta(DQS_n), for\ even\ n.$

3.30 Corollary
Proper lucky number of double alternate quadrilateral snake graph is half of the maximum degree i.e., $\eta_p(DATS_n) = \frac{1}{2}\Delta(DATS_n).$

4. CONCLUSION

In this paper we compute the lucky number and proper lucky number of the quadrilateral snake graph, double quadrilateral snake graph, alternate quadrilateral snake graph and double alternate quadrilateral snake graph. Also we connected it with minimum and maximum degree. Further we take this work to standard graph and networks.

References

[1] Ahadi A, Dehghan A, Kazemi M, and Mollaahmadi E 2012 Computation of lucky number of planar graphs is NP-hard *Information processing letters* **112** 109-112
[2] Akbari S, Ghanbari M, Manaviyat R and Zare S 2013 On the lucky choice number of graphs *Graphs and combinatorics* **29** 157-163
[3] Chiranjilal Kujur, Antony Xavier D and Arul Amirtha Raja S 2017 Proper Lucky labeling of K-Identified Triangular Mesh and K- Identified Sierpinski Gasket graphs *Int. J. of Scientific and Engineering Research* **8** 249-261
[4] Gallian Joseph 2019 A Dynamic Survey of Graph Labeling *Electron J. Combin.* **DS6**
[5] Gopi R 2016 Odd Sum Labeling of Alternative Quadrilateral Snake *Int. J. of Engineering Sci. Advanced Computing and Bio-Technology* **7** 73-77
[6] Jadav I I and Ghodasara G V 2016 Snakes related strongly*-graphs *Int. J. of Advanced Engineering Research and Sci.* **3** 240-245
[7] Karoński M, Łuczak T, and Thomason A 2004 Edge weights and vertex colours *J. of Combinatorial Theory* B **91** 151-157
[8] Kujur Chiranjilal 2017 Lucky Labeling and Proper Lucky Labeling for Bloom Graph *IOSR J. of Mathematics* **13** 52-59
[9] Prajapati U M and Vantiya A V 2019 SD-Prime cordial labeling of some snake graphs *J. of Applied Sci. and Computation* **6** 1857-1868
[10] Rosa A 1966 On certain valuations of the vertices of a graph *In Theory of Graphs* (Rome: Int. Symp.) pp 349-355
[11] Sankari G and Lavanya S 2017 Graceful Labeling of Quadrilateral Snakes *Int. J. of Mathematics Trends and Technnology* **51** 377-380
[12] Senthurpriya S and Meenakshi S 2019 Independent Domination Number in Triangular and Quadrilateral Snake graph *Int. J. of Recent Technology and Engineering (IJRTE)* **8** 20-23
[13] Vaidya S K and Vyas N B 2014 Product cordial labeling for alternate snake graphs *Malaya J. of Mathematik* **2** 188-196

AUTHOR INDEX